Tech. Info. Ctr.
Hewlett Packard
Corvallis

DISCARDED

ט . 97

H. P. Corvallis
Research Library

ANALYSIS AND DESIGN OF ELECTRONIC CIRCUITS USING PCs

ANALYSIS AND DESIGN OF ELECTRONIC CIRCUITS USING PCs

John R. Greenbaum
Les Besser / Brian Biehl
Bruce D. Pollard / Robert Osann

VNR VAN NOSTRAND REINHOLD COMPANY
_____ New York

Copyright © 1988 by Van Nostrand Reinhold Company Inc.

Library of Congress Catalog Card Number 87-21598

ISBN 0-442-22773-6

All rights reserved. No part of this work covered by the copyright hereon may be produced or used in any form or by any means—graphic, electronic, or mechanical, including photocopying, recording, taping, or information storage and retrieval systems—without written permission of the publisher.

Printed in the United States of America

Van Nostrand Reinhold Company Inc.
115 Fifth Avenue
New York, New York 10003

Van Nostrand Reinhold Company Limited
Molly Millars Lane
Wokingham, Berkshire RG11 2PY, England

Van Nostrand Reinhold
480 La Trobe Street
Melbourne, Victoria 3000, Australia

Macmillan of Canada
Division of Canada Publishing Corporation
164 Commander Boulevard
Agincourt, Ontario M1S 3C7, Canada

16 15 14 13 12 11 10 9 8 7 6 5 4 3 2 1

Library of Congress Cataloging-in-Publication Data

Analysis and design of electronic circuits using PCs.
  (Van Nostrand Reinhold electrical/computer science and engineering series)
  Includes index.
  1. Electronic circuit design—Data processing.
2. Computer-aided design. 3. Microcomputers.
I. Greenbaum, John. II. Series.
TK7867.A56 1988   621.3815′3′0285416   87-21598
ISBN 0-442-22773-6

Preface

Since the mid 1960s, the digital computer has been used as a design tool by electronic circuit designers. Computer software programs called ECAP[1] and SCEPTRE[2] were among the earliest circuit analysis codes to gain general acceptance by the design community. These programs permitted circuit performance to be simulated for small-signal frequency responses, dc operation points, and transient responses to varying input stimulii. Unfortunately, accessability to programs such as these by the design community of that era was quite limited since they could be used solely on large, expensive mainframe computers. Only a fraction of the circuit designers at that time were employed by companies large enough to afford the acquisition and maintainance costs of these large computers.

The availability of personal computers (PCs) at moderate prices has dramatically changed this picture. The sophistication of the PCs as well as the software that can be run on them has potentially put circuit performance simulation at every designer's desk. Since the early days of ECAP and SCEPTRE, the amount of software for circuit design and analysis has grown enormously. At the same time, the sophistication of the analyses provided by this software has correspondingly increased. In addition, the accuracy of simulation software has improved to where laboratory measurements have become a verification of the analyses, rather than vice versa.

This dramatic turnabout has encouraged the use of PCs for circuit simulation and design by both computer novices and experts, that is, practicing engineers who are mainly proficient computer users rather than programmers. Concurrently, they are demanding an increased understanding of the mathematical and theoretical concepts behind these CAD (computer-aided design) tools. This book has been written to satisfy these needs of both novices and experts. It will provide guidance on the use of the PC as a circuit design and analysis tool as well as describe recently developed techniques for more efficient use of both the PCs and the programs.

The primary objective of this book is to present an application-oriented approach to the use of the PC as a design medium. It will describe available programs, their strengths and limitations, as well as techniques to insure optimum results from simulations. Analog, digital, and microwave programs will be covered. Also, computer models for bipolar, FET, and other types of semiconductors as required by the different programs and by digital and analog ICs, will be described and explained. The text is geared for practicing electrical and electronic engineers as well as senior-level university students who have an understanding of electronic circuit fundamentals. In its primary focus, this book is not intended to be another electrical engineering text since there are numerous excellent ones already available.

Many types of computers are available, and every reasonable effort has been made to provide guidance as to which programs run on what computer. Many examples demonstrating the use of the programs will be included to properly explain CAD techniques as well as attempt to define both computer and program limits. Computer lisitings for both mathematical analysis programs and graphical techniques are provided.

Since an enhanced appreciation of both CAD applications and CAD techniques will result from even a minimal understanding of how computers behave, a review of the workings of computers is included. Also provided are descriptions of hard-copy and graphic devices that are used to interface the designer and the PC. Since many PC applications such as production control, robotics, and the like are industrially oriented, a chapter is dedicated to the industrial uses of the PC and to the design of circuits that must operate in nonideal environments.

Copies of all codes listed in the book are available on disk from the principal author for a small fee.

John R. Greenbaum
Bala Cynwyd, PA

References

1. Jensen R. W., and Lieberman. M. D., *IBM Electronic Circuit Analysis Program*. Englewood Cliffs, NJ: Prentice Hall, 1968.
2. *SCEPTRE Users Manual*, vol. 1, revised April 1968, AFWL-TR-67-124.

Contents

PREFACE / v

1. INTRODUCTION / 1
1.1. The Role of Pcs in Circuit Analysis and Design: An Overview / 1
1.2. Desirable Characteristics of Circuit Analysis Programs / 2
1.3. From Analysis to Design / 3
1.4. Organization of Book / 3

2. PC BASICS / 6
2.1. Introduction / 6
2.2. Number Systems, Codes, Bits, Bytes, and Words / 7
2.3. Hardware / 9
2.4. Software / 17
2.5. Special Topics / 21
2.6. Conclusion / 25

3. AC, DC, AND TRANSIENT CIRCUIT SIMULATION PROGRAMS / 28
3.1. Introduction / 28
3.2. Three Circuit Simulators for PCs / 29
3.3. Architecture of a Circuit Simulator / 37
3.4. Modeling / 53

4. COMPUTER-AIDED DESIGN AT MICROWAVE FREQUENCIES / 85
4.1. Historical Review / 85
4.2. Circuit Analysis / 87
4.3. Circuit Synthesis / 92
4.4. Transmission Line Analysis and Synthesis / 110
4.5. Circuit Optimization / 113
4.6. Statistical Analysis / 117
4.7. Active Device Modeling / 119
4.8. Circuit Contents Layout and Mask Design / 122
4.9. Nonlinear Circuit Design / 124
4.10. Illustrative Examples / 126
4.11. The Future / 137

5. DIGITAL DESIGN AND ANALYSIS / 138
5.1. Introduction / 138
5.2. PLDs Proliferate CAE on PCs / 139
5.3. PALASM, the First PLD Design Language / 144

viii CONTENTS

5.4. CUPL, the First High-level Universal Language for PLDs / 149
5.5. ABEL, first High-level PDL Language with State Machine Capability / 154
5.6. Arrival of High-level Design on PCs / 157

6. **THE USE OF PCS FOR INDUSTRIAL/LABORATORY AUTOMATION** / 158
6.1. Introduction / 158
6.2. Basics / 159
6.3. Connecting a Digital pH Meter to a Computer / 166
6.4. Serial Interfacing (Asynchronous RS-232C) / 171
6.5. Analog Signal Acquisition / 176
6.6. PC-based Robotic Workstations / 181
6.7. Summary and Conclusion / 184

7. **SPECIAL COMPUTER CODES** / 187
7.1. Introduction / 187
7.2. Fitting General-order Polynomials / 192
7.3. Designing T- and Pi-Pads / 196
7.4. Calculating "L"-Networks / 210
7.5. Bandpass Filters / 222
7.6. Cad Cuts Math in RF Amplifier Design / 226
7.7. Simulating a Servo System / 236

APPENDIX A GLOSSARY OF TERMS / 253
APPENDIX B BIAS-D REFERENCE MANUAL / 267
E-1. Introduction / 267
E-2. Input Data / 267
E-3. Miscellaneous / 274

INDEX / 275

1
Introduction

John R. Greenbaum

1.1. THE ROLE OF PCs IN CIRCUIT ANALYSIS AND DESIGN: AN OVERVIEW

This book's ultimate goal is to facilitate and optimize the use of personal computers (PCs) for electronic circuit analysis and design applications. This chapter is an overview of both the material in the book and its organization; the book as a whole will present the practical, application-oriented use of PCs for (1) circuit simulation and subsequent performance evaluation, and (2) control of industrial and laboratory processes; and it will discuss available computer programs and their usage in these areas.

The major advantage of a PC over a mainframe computer is that the PC is all yours. There is no competition for computer memory, no timesharing, no prioritizing or queueing of the job. An extremely large percentage of circuits of reasonable size can be simulated on a PC in less than 15 minutes. Depending on the type of analysis desired, the detail of requested information, and the different limiting conditions imposed by the user, simulation results can be produced in shorter times. Obviously PC CAD programs cannot, and should not, be expected to solve all problems. Large circuit and system simulation can be properly executed only on large computers. Yet in many cases the PC is the best tool for the job. Most PC CAD software is available commercially with costs ranging from $250 to $10,000. However, there are many software programs in the public domain and consequently available at a nominal charge; both commercial and public-domain software will be described.

The PC is continually being adapted as an integral part of the manufacturing, controlling or measurement process. Its use in commercial robotics, process monitoring and control, automated manufacturing, and so on, is a rapidly growing industry. Methods of applying the PC as well as practical techniques and guidelines for including it in industrial processes will be described. Also included are recommendations for adapting the information provided as well as the sources of such information for hardware and for additional technical data.

To the degree it is possible to divorce the design of circuits from the basics of engineering, the book will attempt to de-emphasize EE theory. However, since one of the advantages of circuit performance simulation is to permit in-depth comprehension as to how different circuit components respond to varied stimulii, attention will be drawn to critical component response behavior. Therefore, although a novice can do circuit design using "cookbook" techniques, the material presented will presume at least a third/fourth-year comprehension of engineering theory. In some instances, previous design experience on the part of the reader may also be presumed. For all classes of circuits, and all of the CAD programs described in the book, the material is intended to supplement, not to replace the user's manual provided with each program. The plan of the book is to guide the user through those areas that require additional explanation or interpretation by means of examples and explanatory text.

Since the scope of circuit analysis programs encompasses dc operating points, small-signal frequency responses, transients, logic, RF and microwave simulation, more than one program can be used to simulate the same circuit. A comparison of some competing programs will be provided, along with examples of analytic techniques which insure proper evaluation of the simulation results. In addition, programs and procedures for the adaptation of the PC to monitor and measure industrial processes will be surveyed and means for their expansion to other areas suggested.

1.2. DESIRABLE CHARACTERISTICS OF CIRCUIT ANALYSIS PROGRAMS

As early as 1966, during the infancy of CAD, F. F. Kuo[1] defined six characteristics that he believed should be incorporated into circuit analysis programs:

1. The program should be easy to use, or, in the current vernacular, should be "user friendly." This means that the input should be convenient to use and permit simple descriptions of circuit topology with associated component values.
2. The program should be capable of handling more than a single model for different classes and types of devices.
3. The program should be capable of simulating a range of analyses, i.e., from dc to RF to nonlinear to digital.
4. The programs should be capable of providing output information in any form and format that a designer desires.
5. In addition to obvious analysis results, the program should be able to expand its limits and provide information on sensitivities of performance to component variations, poles, and zeros for ac analysis, Monte Carlo calculations, and optimization. Also, for high-frequency simulation,

INTRODUCTION 3

various impedances should be calculated automatically and "S" parameter and other similar parameter data should be available.
6. The program should be capable of error-checking in order to permit the user to evaluate the accuracy of the simulation results.

Over the past two decades these characteristics have been incorporated in circuit analysis programs and are today available in most CAD programs that have been written especially for, or adapted to run on, the PC.

1.3. FROM ANALYSIS TO DESIGN

All circuit designers are faced with the problem of insuring the adequacy of their designs. This can be done through analysis, or by breadboarding and testing. The primary function of analysis is to determine the response of a circuit or system, real or conceptual, to various operating conditions *before* commitment to the time and expense of obtaining component parts plus building and testing the circuit. The advent of CAD programs for use on PCs offers the designer the luxury of many trials and "adjustments" before a final circuit is defined. The designer may also attempt, both rapidly and economically, circuit response evaluations for component value variations (i.e., Monte Carlo, worst case), input or power line voltage transients (or both), extreme variations in environmental conditions (i.e., high or low temperature), and high noise conditions, before final circuit definition.

Often, it may be desirable or necessary to transform data provided to the designer or obtained from laboratory measurements into a form appropriate for use with CAD programs. For example, a pulse with a short risetime and a long decay may have been monitored as the output of a circuit that is intended to drive a second circuit. Through obtaining the X-Y coordinates of the signal from a photograph of the CRT presentation, its equational equivalent can be computed and used in a subsequent analysis. To permit the PC CAD user to transform tabular or pictorial data to equations, or to perform other pre-analysis evaluations, much computer coding is included in this book. The bulk is written in BASIC, the most common PC language. This software allows the ready solution of many fundamental problems and provides data for later analyses. Additional software is furnished to be used for certain analyses independently of the traditional CAD programs.

1.4. ORGANIZATION OF BOOK

Separate chapters will be dedicated to each major area of interest to circuit designers, i.e., analog, digital, RF, and so forth. Others will deal with such subjects as the modeling of semiconductors for CAD analysis. In as much as different computer models of the same device types are often used in each of the application areas, practical problems associated with the simulation of circuit types and methods for solving the problems are described. The practical limits of the different programs and methods for coping with the limits are suggested; and judgemental information concerning the form, format, quantity, and detail of output data requests to permit the reader to optimize the use of CAD programs on the PC is provided.

Chapter 2 surveys personal computing technology with emphasis on selecting and operating PC-based systems for circuit design, analysis, and simulation, as well as for data acquisition and control. In addition to reviewing personal computing in general, the chapter highlights those details of PC hardware and software which must be considered when the system is used for the described applications. Special topics focus on graphics, intercomputer communications, and hardware/software compatibility—all very important issues in technical computing.

Chapter 3 describes general-purpose analog circuit simulation analysis programs. It includes the architecture of general-purpose circuit simulation programs as well as a description of the simulator element models. The simulation programs will compute dc operating points as well as small signal frequency responses and transient responses. Examples include different analysis types and classes of circuits. The example circuits demonstrate the use of bipolar and MOS models and operational amplifier macromodels. The examples demonstrate the results obtainable from three simulation programs, BIAS-D™, PSpice™, and MICROCAP-II™.

Chapter 4 describes active and passive circuit analysis techniques using the SUPER-COMPACT PC™, TOUCHSTONE™, and TOUCHSTONE RF™ programs. These programs can be used on the IBM and compatible PCs as well as on the HP 200 computers. Design capabilities include component and topology selection, graphical techniques, device modeling, statistical analysis, and physical layout. Both lumped and distributed components are considered, with an examination of the effects of conductive and dielectric losses, dispersion, discontinuities, parasitics, and surface roughness. Conversion from electrical to physical design is illustrated by examples using MICAD™ and AUTOART™ programs. Although these programs offer only linear, steady-state circuit analyses, they are capable of handling certain classes of nonlinear analyses by approximations.

Chapter 5 covers digital design and analysis, focusing on programmable logic devices (PLDs). It first provides background on the technology as well as the evolution of PLD design tools. A simple applications example (a video controller subsystem) is the vehicle for showing design synthesis and analysis using the three most popular PLD languages, PALASM, CUPL, and ABEL. This example demonstrates advanced techniques for State Machine design and address decoding.

Chapter 6 describes the use of personal computers for automating various industrial and laboratory systems. The focus is on establishing a computer/application interface to facilitate data acquisition and control by the computer. The initial sections of the chapter outline the possible approaches to automation as well as the basic considerations required for choosing and implimenting each approach. Subsequent sections elaborate on connecting devices with serial interfaces or voltage outputs to personal computers. The remainder of the chapter presents concepts and devices applicable to the construction, programming, and operation of microcomputer-based robotic workstations.

Chapter 7 describes methods and procedures while giving examples and code listings for the design and analysis of six circuit types. The software listings have all been written to be "user-firendly"; most are menu-driven. They accommodate fitting general-order polynomials to data available in the form of tables or curves, the design and analysis of T and Pi pads, network impedance matching circuits, bandpass and RF amplifier circuit design and analysis, and feedback systems that range from simple amplifiers to complete servo systems. The chapter concludes with a list of statistical programs especially written for or adapted to run on the PC, and availability references.

Appendix A is a glossary of terms used throughout the book. In general, these terms have been accepted as standards in both the computer and engineering disciplines. Appendix B provides the complete BIAS-D listing, with its users' manual.

REFERENCE

1. Kuo, F. F., "Network Analysis by Digital Computer," *Proceedings of the Institute of Electrical and Electronic Engineers* 54(6), 1966.

2

PC Basics

Bruce D. Pollard, Ph.D.
Technology and Development Center
ARCO Chemical Company
Newtown Square, Pennsylvania

2.1. INTRODUCTION

This chapter is not intended to provide a basic course in computing itself but to present an overview of personal computers with emphasis on the issues pertinent to choosing and operating a PC-based computer-aided circuit analysis, design, simulation, or data acquisition system. The PC is a personal computing tool and as with all tools its true capability can be realized only when one knows its features, its operation, and needed care. This is particularly true when choosing and using PCs for technical applications because special features such as graphics and computing speed are required to prevent the PC from becoming a cumbersome liability rather than a time-saving device.

This chapter's division into sections has been determined by the classical boundary between hardware and software. However, there are places where, due to the nature of technical computing, hardware and software must be discussed together. The hardware and software sections are preceeded by a review of number systems, codes, bits, bytes and words. The final section, "Special Topics," addresses some issues important to technical computing such as graphics, communications, and compatibility.

There is a wealth of information available about personal computing in general and specific to the applications featured in this book, as reflected in the references at the end of this chapter. Basic texts[1,2] and magazines[3,4,5] are a good introduction to the terminology and available equipment. Technical journals[6] treat specific subjects, such as circuit design and analysis, in detail, and trade publications[7] usually present the latest developments in microcomputers and personal computers for use in technical applications.

2.2. NUMBER SYSTEMS, CODES, BITS, BYTES, AND WORDS

Computers are made up of electronic switches or gates having two conditions, or states—ON and OFF. As a result, the *binary*, or base-2, number system is the foundation of computing mathematics. Binary numbers, such as those shown in Table 2-1, are made up of digits each representing a power of two starting on the right with 2^0. The digits can be multiplied and added to give their equivalent base-10 value. For example: $1101_2 = 1 * 2^3 + 1 * 2^2 + 0 * 2^1 + 1 * 2^0 = 1 * 8 + 1 * 4 + 0 * 2 + 1 * 1 = 13_{10}$. (The subscripts indicate the base, i.e. 2 or 10. Base-10 is assumed unless otherwise specified.)

Each digit or switch condition in the binary representation is called a *bit* and the total set of digits or bits is called a *word*. Because it is a common word size or at least a common denominator of most word sizes, a group of 8 bits is called a *byte*. An 8-bit (1-byte) word can have $2^8 = 256$ states, starting at $0000\ 0000_2 = 0_{10}$ and ending at $1111\ 1111_2 = 255_{10}$. Word sizes are determined by the microcomputer's electronic design.

Notice the 8-bit word has been subdivided into two 4-bit sections. This has been done for clarity and because each of the 4-bit *nibbles* can be represented

Table 2-1. Number Systems

Base 10	Base 16 (Hex)	Binary	Binary-Coded Decimal
0	0	0	0000
1	1	1	0001
2	2	10	0010
3	3	11	0011
4	4	100	0100
5	5	101	0101
6	6	110	0110
7	7	111	0111
8	8	1000	1000
9	9	1001	1001
10	A	1010	0001 0000
11	B	1011	0001 0001
12	C	1100	0001 0010
13	D	1101	0001 0011
14	E	1110	0001 0100
15	F	1111	0001 0101
16	10	1 0000	0001 0110
20	14	1 0100	0010 0000
83	53	101 0011	1000 0011
172	AC	1010 1100	0001 0111 0010
255	FF	1111 1111	0010 0101 0101

8 ANALYSIS AND DESIGN OF ELECTRONIC CIRCUITS USING PCs

by a base-16 number. The base-16, or *hexidecimal*, representation, often called "hex", is commonly used in place of the binary representation of each nibble to reduce the space required as well as to increase readability (once you know the hex characters shown in Table 2-1). With hex, the range of states in a byte is from 00_{16} to FF_{16}.

ASCII, the American National Standard Code for Information Interchange, has become the nearly universal method for storing alphanumeric characters and transmitting them between microcomputers[8]. Historically, ASCII has been a 7-bit binary code (Table 2-2) which includes numbers, upper- and lower-case

Table 2-2. The ASCII Character Set (7-Bit Code)

LSD \ MSD	0	1	2	3	4	5	6	7
0	NUL	DLE	SP	0		P		p
1	SOH	DC1	!	1	A	Q	a	q
2	STX	DC2	"	2	B	R	b	r
3	ETX	DC3	#	3	C	S	c	s
4	EOT	DC4	$	4	D	T	d	t
5	ENQ	NAK	%	5	E	U	e	u
6	ACK	SYN	&	6	F	V	f	v
7	BEL	ETB	'	7	G	W	g	w
8	BS	CAN	(	8	H	X	h	x
9	HT	EM	)	9	I	Y	i	y
A	LF	SUB	*	:	J	Z	j	z
B	VT	ESC	+	;	K	[	k	{
C	FF	FS	,	<	L	\	l	|
D	CR	GS	-	=	M	]	m	}
E	SO	RS	.	>	N	^	n	
F	SI	US	/	?	O	_	o	DEL

NUL – Null
SOH – Start of heading
STX – Start of text
ETX – End of text
EOT – End of transmission
ENQ – Enquiry
BEL – Bell
BS – Backspace
HT – Horizontal tab
LF – Line feed
VT – Vertical tab
FF – Form feed
CR – Carriage return
SO – Shift out
SI – Shift in

DLE – Data link escape
DC – Device control
NAK – Negative acknowledge
SYN – Synchronize
ETB – End of transmission block
CAN – Cancel
EM – End of medium
SUB – Substitute
ESC – Escape
FS – File separator
GS – Group separator
RS – Record separator
US – Unit separator
SP – Space
Del – Delete

Example: 43 (hex), or 0100 0011 (binary), prints a "C".

letters, punctuation and codes for control commands such as line feed, carriage return and end of file. Allocating a byte (8 bits) for storing each 7-bit ASCII character has become the standard. The extra bit is sometimes used as a *parity bit* for error-checking (see Section 6.4) or, recently, to expand the 128-character limit to 256, providing entries for printing graphics and foreign letters[8].

There are numerous coding methods for representing integers and nonintegers in binary words. Depending on the word size used in the computer, two (double precision) or more words may be combined to allow the representation of very large integers. The *binary-coded decimal* system (*BCD*) uses 4 bits to represent each digit of a base-10 number ($0000_2 = 0_{10}$ through $1001_2 = 9_{10}$. While it is not an efficient storage code (6 of the 16 possible codes, 1010_2 through 1111_2, are wasted for each 4 bits), BCD is convenient for calculations and for digital display (Section 6.2). In *two's complement*, the leftmost bit of the digital word is designated as a *sign bit* to facilitate negative numbers. A special coding procedure is used to make two's complement continuous and thus suitable for mathematical operations[2].

Floating-point coding is the method used to process noninteger numbers, both positive and negative, over a substantial range (i.e., $\pm 1.00000000 \times 10^{99}$ through $\pm 1.00000000 \times 10^{-99}$. Although not standardized, the typical floating-point system reduces a number to a power of 2 called the *exponent* and a remaining portion called the *mantissa*. The number of bits in the exponent determines the ultimate range and the number of bits in the mantissa determines the number of *significant figures* or digits in the original number. For example, the representation of 0.01 through 9.99 (three significant figures) would require a 10-bit mantissa (1024 states). An excellent treatise on floating-point and two's complement binary mathematics can be found in the handbooks distributed by Digital Equipment Corporation to promote PDP series of minicomputers[9].

Just as eggs are sold by the dozen, binary words are counted using special units: 1024 (2^{10}) bytes are called *1 kilobyte*, or 1K for short; 1024×1024 (2^{20}) bytes are called *1 megabyte*, or 1 meg. If the word size is 8 bits (1 byte), 1 Kbyte is 1K words. But if a word is a multiple of bytes, the number of words must be divided by that multiple to obtain the number of bytes. This may sound very basic, but many owners of computers having 16-bit data words have been surprised to discover the 64-Kbyte memory board they bought on sale has one-half as much memory as the 64K words (when each word equals 16 bits) required for a particular application.

2.3. HARDWARE

Generally, personal computers are marketed as an integrated collection of subsystems. The main subunit is a microcomputer containing a microprocessor chip, support circuitry, some memory chips and interface circuitry, usually all mounted along with a group of sockets on a printed circuit board called the *motherboard*. The sockets facilitate connection to a power supply and to mass storage and peripheral devices which may be built-in or purchased separately. Because PCs are a compromise between quality and cost, it is important to verify that the base unit power supply has both a sufficient current rating and cooling capability to support the plug-in devices which draw power directly from the PC rather than from a separate power source. The general features and operation of the various subsystems in PCs are discussed below.

2.3.1. The Microcomputer

Figure 2-1 is a diagram of a generic microcomputer. It consists of assorted support circuitry grouped about the central feature, the microprocessor. Memory chips hold the binary representations of instructions that make up the computer program and the data to be processed. The memory is arranged by word in a numbered sequence called an *array*. Each memory word has a binary address. Information is passed from the microprocessor to the memory and other integrated circuits over a group of conductors called the *bus*, which is usually divided according to function into the data bus, the address bus, a control and timing bus, a power bus and sometimes an input/output (I/O) bus. Timing

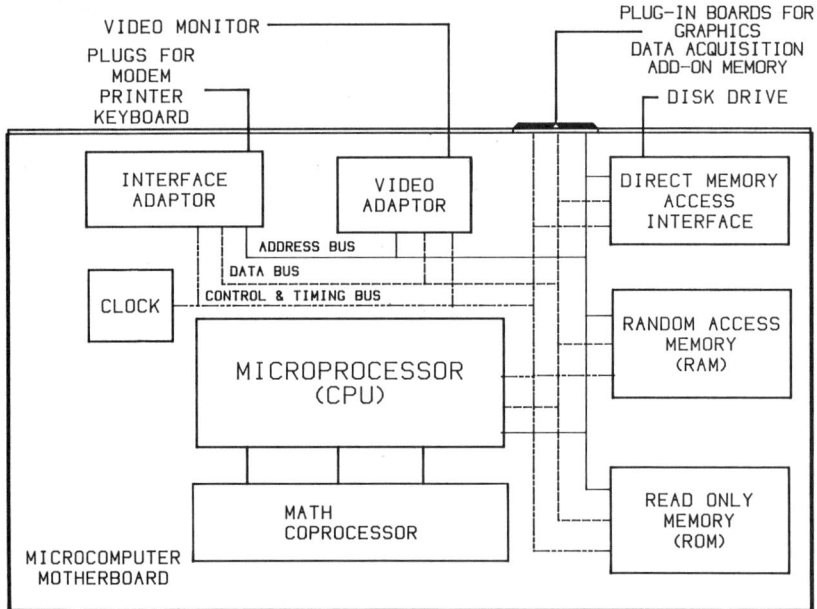

Fig. 2-1. A block diagram of the "generic" microcomputer.

circuits, controlled by a crystal oscillator, provide square-wave signals to coordinate the sequencing of all the components.

The microprocessor, or central processing unit (CPU), is a group of electronic logic gates implemented in large-scale integrated circuitry.[10] The gates are grouped into functional units (Figure 2-2) consisting of an arithmetic and logic unit (ALU), a control unit, and a group of temporary storage locations called *registers*. Some of the registers have specific functions; for example, the program counter (PC) keeps track of the memory address of the next instruction to be executed by the ALU, while the status and flag register reports the conditional results of ALU processes. Most of the registers are used as indexes, pointing to memory locations, or to hold addresses or data which have been read in from memory or I/O. The ultimate power of the microcomputer is determined by the word size of the registers, which may be 8, 16, or even 32 bits. The width of the program counter dictates the amount of addressable memory (a 16-bit PC can access 64K at once). Wider data registers exponentially decrease the time required for mathematical computations. A common design trade-off is to make the registers wide but to shift data in from a narrower data bus thereby requiring narrower, less expensive memory words.

Microprocessors are divided by manufacturer into families, each member having different sized data words (general registers), address words (program counters) and clock speed. Among these families are the Intel 8000/80000 series, which contains the 8088, 20-bit address, 8-bit data bus, 4-MHz clock—foundation of the IBM-PC and many compatible models[11]; and the Motorola 6800/

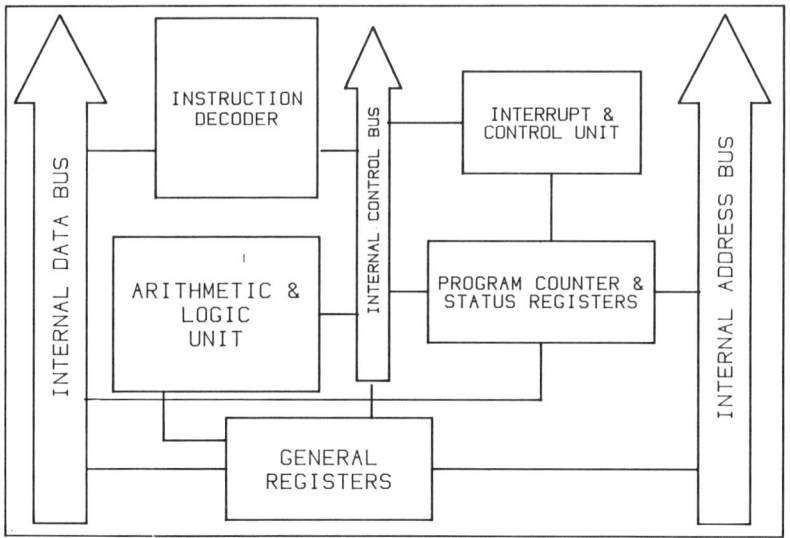

Fig. 2-2. A block diagram of a typical memory-mapped microprocessor.

68000 series, containing the MC68000, 23-bit address, 16-bit data bus, 8-MHz clock—foundation of Apple's Macintosh[12]. Each family has a basic architecture associated with it, each member having some variation in speed or bus width. In the Intel family, the upgraded version of the 8088, the 8086, used in IBM's PC and PS/2 Model 30, has a 16- instead of an 8-bit data bus; while the 80286, used in the IBM PC-AT and PS/2 Models 50 and 60, has a 22-bit address word and much faster clock (10 MHz); and the 80386, used in IBM's PS/2 Model 80 and Compaq's Model 386 has a full 32-bit address, 32-bit data word and 20-MHz clock. The 80286 and 80386 implement a method for expediting execution of instructions called "pipelining"[13]. A more recent version of the MC68000, the MC68020, used in the Macintosh II, expands the data bus to 32 bits and doubles the clock speed to 16 MHz. A comparison of microprocessor architectures is beyond the scope of this chapter but is recommended reading. Such information is available in computer publications and in particular in an excellent column by Dessey[14] for scientists and engineers who use computers.

Usually, a portion of the memory, where program and data codes are stored during use, is located on the motherboard. For mathematic or graphic applications, an extra memory circuit board may be plugged into the motherboard. The two kinds of memory are *read only memory* (*ROM*) and *random access memory* (*RAM*). The data in ROM can not be altered by the CPU and will not be lost if the computer looses power or will stay on the chip even if it is removed. I/O programming, a diagnostic test program, sometimes the operating system, and occasionally a programming language come with the PC in ROM because ideally they will never need to be changed. Having these programs in ROM means thay need not be loaded every time the computer is turned on. There are several variations of ROMs, including PROMs (programmable ROMs) and EPROMS (erasable PROMs), which can be erased and rewritten with special equipment. ROMs should not subjected to excessive heat, electrical fields, ultraviolet light, or static charges, because such exposure may alter or erase the memory contents. RAM is the term used for memory that can be changed as well as read by the microcomputer circuits. Unless a special battery backup is built into the PC, RAM will loose its contents when the computer is turned off or the power fails. Because it is so inexpensive, the microcomputer should be supplied with as much memory as is practical. The maximum amount is determined by the microprocessor type and operating system; for example, the 80286 in an IBM PC-AT can address over 8 Mbytes but MS-DOS supports only 640 Kbtyes without the addition of special software.

Any math function can be implemented by combining sequences of the basic binary addition function built in to all microprocessors, but a PC well suited for technical computing will include provision for a special math processing chip on the motherboard. The 8087, 80287, and 80387 are math coprocessors for the Intel microprocessors used in IBM PCs, PS/2, and related systems; each coprocessor can be connected in parallel with the corresponding CPU (8086,

80286, and 80386 repsectively) and increase the floating-point calculations rate of the host system by an order of magnitude (factor of 10) or more. The analogous chip for the MC68000 micros is the MC68881. There is an obvious benefit to purchasing a PC with a coprocessor, but improved performance can be realized only when the software is written to utilize the coprocessor's capability, including the required floating-point range and precision. When the PC is utilized for high-speed data acquisition, it is important to insure the coprocessor does not alter the timing of the acquisition sequence by prohibiting the CPU from processing interrupts (Section 6.5).

Besides a clock, memory, and processors, input/output circuits to support a keyboard, a video screen, and mass storage are integral parts of every microcomputer. I/O circuits may be located on the motherboard or on boards that plug into it. The CPU, depending upon its architecture, may be connected to its peripherals through a special I/O bus or through a continuation of the control, address and data busses that attach the memory. Special chips called *peripheral interface adapters* (*PIA*), or something similar are supplied to facilitate I/O as part of the microprocessor family. Some I/O circuits have assigned addresses allowing the CPU to read data from and write it to the I/O device as though it were memory. Others which are required to transfer large blocks of data use a technique called *direct memory access* (*DMA*) whereby they take over the address and data busses when the CPU is not using them and transfer the blocks with high speed directly to RAM. DMA I/O processing is particularly useful for mass storage, graphic displays and high-speed data acquisition. Sometimes a video adapter chip for driving a television monitor is included on the motherboard, but most often it is contained on a plug-in. A description of the various display systems is provided in the peripheral hardware section.

2.3.2. Mass Storage

Programs, data, and text not in current use by the microcomputer are permanently saved in subsystems called *mass storage* which have the capacity to store many times the amount of active memory. Although reel-to-reel and cassette tape units have been around a long time, disk systems, which store the data on a circular surface in tracks like a phonograph record, are now the most common. Most mass storage systems are based on magnetic recording technology; however, optical read-only storage systems based on the same technology as that employed for audio compact disk recordings are becoming popular. The mass storage medium may be portable, such as a "floppy" (plastic) disk or a cassette tape, or it may be "hard" (metallic, rigid) disk, usually non removable. PCs used for technical computing usually have one of each type (soft and hard disks) to provide fast access to the data and programs in use as well as to facilitate backup and loading of new programs and data.

Mass storage devices are designed to optimize data storage with respect to hardware and ease of use. While the CPU access *individual* words of memory, mass storage devices transfer *blocks* of words to and from *blocks* of memory. The block size depends upon the medium and the design, but is usually 256 bytes. Disk systems are divided into *tracks* and again into *sectors*, which commonly hold one block each. Where the blocks are located and how many are transferred is handled by the PC operating system. The PC user thinks of data transfer in terms of *files*, which may consist of many blocks. On disk systems, which can be accessed by the computer at any sector, special files called *directories* contain the track and sector locations of all the other files. When a file is called into memory, the computer reads the directory and then retrieves, that is, copies the file itself into active memory. Most tape systems, being only sequential in terms of access, don't have directories, so they must be read from the start until the requested file is found.

Floppy disks are the most common mass storage medium because they have reasonable speed and size, and are portable and inexpensive. Although there are several types, the typical floppy can store 300 kilobytes to 1.4 megabytes on a thin, 3- to 8-inch round plastic sheet coated on two sides with ferric oxide and encased in a hard plastic or paper envelope. Care is required in handling floppy disks. They should not be bent and the surfaces should not be scratched, touched, or exposed to excess heat, magnetic fields, or dust. Rather than stored in a pile, they should be kept upright in a suitable case. To protect the data stored on them from accidental erasure, a write-enable notch, detected by the drive when the disk is inserted, can be covered to indicate read-only status to the microcomputer. Before they can be used, most floppy disks need to have magnetic sector markers stored on them. This process, called *formatting*, is performed using a program supplied with the PC[16]. Floppy disks do wear out eventually. Backup copies should be made of all new programs that come on a floppy before they are used, and of any important data stored on a floppy.

Hard disks until recently were very expensive, but are now standard on most PCs used for technical computing. The main differences between floppy and hard disks are that the oxide coating for the hard disk is on a rigid medium which is not removable from the drive system and one hard disk system may actually contain several disks. This approach allows for much closer alignment of magnetic domains resulting in hard disk storage capacities ranging from 10 to 600 megabytes, as well as much faster data rates from disk to RAM, from disk to disk or tape, and vice versa. Hard disks are available on some portable PCs; however, they are fragile and should be purchased from a reputable source with a guarantee. When they break they are replaced, not repaired. Therefore, if a proper backup has not been made, all the data is lost; backing up important files onto a floppy or tape is also helpful if the PC is shared among many users. Like floppy disks, hard disks must be formatted. During the formatting process, *all previous data are erased*, so formatting should be done only once and the formatting program transferred to a floppy and stored in a secure place. Hard disks have so much more storage that they require better housekeeping, which can be made easy by having subdirectories for each project or kind of file. However, continuous house-cleaning is needed to prevent having to spend say, two hours deleting old files because a group of users have filled up a 30-megabyte drive.

Magnetic tape, one of the oldest mass-storage methods, is still useful because tapes are inexpensive and have large storage capacities. Some of the inexpensive "toy" systems (i.e., Commodore's VIC-20) which provide permanent programs in plug-in ROM cartridges, also come with an interface for storing user programs and data on standard audio cassette tape recorders. Such systems are usually slow, have small block sizes and redundant error-checking, but are very cost-effective. More expensive tape units are available for backing up hard disk systems. These are designed with large block sizes so they go much faster. Error-checking is done by storing the same data on many tracks and comparing the tracks while recording and reading.

Disks read by reflecting infrared light off the storage surface are available for read only applications where a lot of storage is required. A typical $5\frac{1}{4}$-inch optical disk contains 700 megabytes stored using the same technology as audio recording. Magnetic hard disks of comparable size hold much less. Optical disks are suitable for distributing large databases or programs but will be limited to mass-market items until production costs are reduced. Write once/read many times (WORM) optical disks are replacing magnetic tape for backup on larger systems. The future will bring read/write optical disks with integrated text graphics and sound.[17]

2.3.3. Peripherals

Peripherals, attachments that plug into the PC, may come "bundled" with the PC as a system or they can be added. When purchased separately, it is important to insure that peripherals are compatible with the hardware and supported by the operating system and application software. In some cases, this assurance may require extensive testing. The acquisition and installation of boards, which require opening the PC box and sometimes modifying circuitry, should be accomplished with the aid of fully qualified personnel. Although there are many different peripherals, those discussed in this section are the monitor, keyboard, mouse or digitizer, plotters, printers and modems. Graphics boards are described in Section 2.5 and peripherals used for data acquisition and control in Section 6.5.

While the *cathode-ray thermal* (*CRT*) is the standard user interface to mainframe and minicomputers, most PCs integrate the digital functions of the CRT into the microcomputer and require only a keyboard and television monitor as external units. Most PCs come with a keyboard, simply a bunch of switches arranged to operate like a typewriter and interfaced to the microcomputer. There is no standard key configuration; each manufacturer chooses the arrangements that seem the most convenient for the functions supplied. Keyboards designed for similar computers may be interchangable, but should be tested to be sure. Because the choice of monitor depends on the application, it is usually an option. Most PCs come with a special board, or circuit on the motherboard, that generates a signal according to one of several video standards. The basic requirements for the monitor are for it to accept the particular video signal and to provide good enough resolution for the display to be readable. Monochrome (white, amber, or green on black) monitors usually use a 1-volt peak-to-peak composite signal and give the highest resolution for the least cost. Color monitors are attractive but usually the characters they display are less readable than those displayed on a monochrome monitor unless a medium- to high-resolution-color model (see the graphics discussion in Section 2.5) is purchased. Most PCs equipped to support color have a three-signal (red-green-blue) standard output.

The *mouse* is representative of the attachments to PC systems designed to bypass the keyboard to facilitate better interaction between the user and a graphic display. The mouse is a small, handheld electrical gadget that upon being moved across any flat surface sends electrical signals to the PC indicating the amount and direction of motion. Apple's Macintosh employs a mouse to expedite input from display menus: the user points at a menu item rather than entering code. Mice are very useful for freehand drawing and locating figures, lines, and labels with a CAD system. A PC that supports a mouse may come with a special socket for it. Mice designed for other PCs may share the keyboard port, plug into a serial port or require a special interface board. Other devices that serve the same function are a shaft called a "joystick" connected to electronics which converts a X-Y position to a computer signal, and a forerunner of the mouse which uses a flat grid of wires, called a "digitizing pad" to sense position.

A *printer* is the usual hard-copy device attached to PCs. There are numerous printers of several types available but most connect to the PC via a standard parallel connection called a *Centronics Printer Port*. *Dot matrix* printers which print characters or graphics as sets of dots across the page, are the most common. *Ink-jet* printers operate similarly though with a nonimpact technology. Depending on the resolution (number of dots per inch), there are inexpensive matrix printers suitable for printing programs and data for safe-keeping and expensive matrix printers which generate textbook-quality letters and reasonable graphs. *Character* printers sometimes use a "daisy-wheel," print typewriter quality (but text only) and are generally inexpensive. Although they are expensive, *laser* printers (really computer-driven copy machines) have revolutionized computer hard copy. With a resolution of 300-dots per inch, a *laser* printer can print a page of text in any of many character designs, called fonts, or draw a graph with a resolution approaching that of a draftsperson. As with all peripherals, printers can only do what the computer software supports, so testing is recommended. Most printers are one-color devices and are limited in size to $8\frac{1}{2}$ or 14-inch widths. However, ink-jet printers, in particular, are now multicolor devices, as are also some dot matrix printers like the Apple Imagewriter II.

Although laser printers serve the purpose in many cases, *digital plotters* are the most common device used to produce high-resolution drawings and graphs. Most plotters are controlled by digital signals sent from the computer via an RS-232 asynchronous serial port (Sections 6.4). This port may be part of the motherboard or require a plug-in board. The signals cause the plotter to select a pen (for color and thickness) and draw lines from one point on the paper to the next. A point-addressing resolution equivalent to 1000 dots per inch is not uncommon. Among the various plotter manufacturers, some offer a range of models capable of plotting on different-sized media with assorted pens. Within a given manufacturer, they all have the same commands so they are upward-compatible.[18] The user must verify that a plotter is supported by his or her application software.

When one computer is located some distance from another, the easiest way to connect them physically is over telephone lines with a *modem* interface at each end of the line. The modem, (modulator-demodulator) converts digital data signals, which have limited transmission distance, into frequency-phase modulated signals, which can be transmitted virtually any distance over standard telephone lines. Some modems plug directly into the PC motherboard, while others are connected to an RS-232 asynchronous serial port (see Section 6.4). Many modems have built in microprocessors which dial a number, select the transmission speed, and make the connection. Although modems capable of a 2400-baud rate are available, 1200 baud is the most common and reliable speed. Special communications software is usually required to enable the computer to use modems efficiently, that is, to transfer files with proper error-checking and to make the PC emulate a supported terminal type when connected to a minicomputer or mainframe.

2.4. SOFTWARE

The software chosen and how it is utilized gives the PC its personality. Proper selection of a software ensemble requires careful consideration of the specific PC and the peripherals attached to it. Efficient use of any software package can be accomplished only through a complete study of the documentation provided and continued practice, until the rules and commands (called *syntax*) are mastered. This section presents an overview of the spectrum of software available for technical computing with a PC. It suggests how a constellation of programs might be assembled to provide a proper operating environment for design, simulation, and automation activities.

18 ANALYSIS AND DESIGN OF ELECTRONIC CIRCUITS USING PCs

While a knowledge of computer programming is not generally necessary to operate a PC, it is a great advantage when using a PC for technical computing. This is particularly true for simulations and design work in which slight modifications to a program might be required to yield optimal results. If the new PC user views programming as a learnable skill rather than a black art, reference to the recommended texts,[19,20] taking local courses, and consulting friendly gurus should relieve any problem.

Computer *programs* are made up of lines or *statements* which have been written by the programmer according to the syntax (rules) of the programming language. The overall goals of the program are described by the programmer and then divided into smaller subtasks. If possible, the programmer draws a *flow chart* (see Chapter 3 for several examples) for each task and another connecting the tasks. When numbers or alphanumeric strings must be stored or passed from task to task, they are given names according to the syntax. Statements are written using the variable names and operations defined by the language to accomplish the tasks. When they are used repeatedly in the same program, tasks are made into *subroutines*, which can be called numerous times independent of specific variable names. The subroutines and other task modules along with a backbone program called the *main program* are combined into the final complete program. The program is tested through a process called *debugging* and modified until it performs the desired task according to the programmer's specifications. Documenting the program (primarily as a user's manual, commented listings, and flow charts), training the users, and further modifications to suit the wishes of the user take time but assure the utility of the program.

Even with a single-user PC, several pieces of software running simultaneously on various levels are required to accomplish any task. On the highest level is a program or software package called the *application*, chosen by the user for a specific purpose; for example, word processing or numerical integration. The application is supported by the *operating system*, which is a program that manages the flow, storage, and use of information in a particular PC system. The software foundation for any PC, called the *basic input/output system* (BIOS), connects and controls all the software interactions between the PC and its peripherals.

Before any program can run on a microcomputer, it must be reduced to a sequence of steps called *instructions* or *machine code*.[21] The set of instructions available for use is determined by the (hardware) design of the arithmetic and logic unit in the microprocessor. Most PC documentation includes a description of the microprocessor's instruction set and directions on how to decode the binary instructions to find their functions and operands (the memory position or registers upon which the function is performed). Although useful for automation projects, machine language programming is not required for normal PC activities because most application software and operating systems come ready to run. *Compiler* programs are available enabling the user to write application programs in a higher-level language such as FORTRAN or BASIC and convert them to machine code.

When microcomputers were first made available, they were sold with a *monitor* program to facilitate simple functions such as the inspection of memory positions, loading programs into memory from tape, and starting them. The transformation of microcomputers into PCs was accompanied by the development of a program to control system resources and manage the flow of information between the application program, mass storage devices and peripherals. Such programs, called operation systems, are analogous to the master control programs on large multiuser computer systems.[22] They are usually purchased with the PC system; for example, the IBM PC and compatibles always come with Microsoft's PC-DOS (IBM) or MS-DOS (compatibles)[16] although some systems like those by AT&T may also include Unix. From time to time, new versions of the same operating system having additional features and fewer bugs are released. When purchasing an application program, one should check to see if it runs on a particular PC, if it requires a specific operating system, or even a certain version number.

Proper use of any PC requires experience with the functions and syntax of its operating system. A minimal subset will be mentioned here. When a PC is first switched on, a program called the *bootstrap*, stored in ROM, loads the operating system. From then on, the user interacts with the computer through the operating system by typing commands in response to a ready character (for example, in MS-DOS a >) called the *prompt*. The commands facilitate the manipulation of programs, data and text, which are stored and moved throughout the system as *files*. The naming scheme and format of files are determined by the operating system program. Most operating systems come with a complement of *utility* programs which perform basic functions such as formatting or copying disks, editing files, and linking programs. A "full screen" editor program which displays the text on the monitor screen and supports editing and replacing text by way of moving the cursor is preferred; however, the more limited line-oriented editor supplied with most operating systems suffices for simple programs. The linker is required to connect subprograms with the BIOS to form a file called an *executable image* that can be loaded into memory and run on command. Peripherals can be controlled by operating system commands that assign port names and conditions such as the transmission speed to serial and parallel devices. A batch or command file feature is usually included to expedite the execution of groups of commands which are often entered together. A good example of a batch file is the AUTOEXEC.BAT file in MS-DOS which is executed at bootup and sets a customized configuration automatically.[16]

Through the use of special conversion programs the microprocessor's native tongue, machine language, can be replaced with higher level languages having English commands and standard mathematic symbols. Many of the higher level languages have been developed to serve a particular programming need.

BASIC[23] and FORTRAN[24] are the most commonly used for technical computing because many programs as well as graphic and statistical subroutine libraries are available in these languages and most technically oriented workers have been exposed to them during their education. BASIC has become very popular and many PCs come with it. Other languages such as C,[25] Pascal,[26] COBOL and Prolog have been created to fill certain niches and should be considered for use when their special features are needed.

The operational steps in creating a computer program are more or less language independent, the only variation being whether the program used to convert the commands to machine language is an interpreter or a compiler. An *interpreter* program is loaded into memory along with the higher level program. *Every time* the program is run, its statements are translated and executed line by line. A compiled program requires translation into machine language only once; thereafter, regardless of how many times it is used, it remains in machine language. Interpreter-based systems run slower but they detect syntax errors as the program runs and make changing the program easier and much faster. A compiler program converts a special text file, called the *source file*, to an *object file* containing machine language, calls to subroutines and a description of the variables used in a program. Before it can run the first time, the object file must be linked with the subroutines, stored on disk, and rearranged to form an *image file* that the operating system can load and execute. This image file, which remains on disk, is loaded and executed each time the program is run. Compiled programs run much faster and the linking process allows the use of different subroutines depending on the hardware; for example special math subroutines are required to utilize a math coprocessor. The first versions of BASIC were interpreter-based, but now BASIC compilers are available. FORTRAN is almost exclusively a compiled language.

The ability to transport a program between computer systems is an important issue. Certainly identical machines with the same peripherals and operating system should have no problem executing image files from a twin machine. If there are differences, then the probability of a successful exchange is increased when the *source* files are transported and *recompiled* on the second machine. Sometimes changes in the source code are required because the programming language dialect is machine-specific. The 1977 ANSI FORTRAN standard[24] has been accepted by most FORTRAN compiler authors. There are many versions of BASIC but some effort is being made to standardize a subset of it so only minor changes are required to transport a BASIC program.

There are a host of application programs available; some are mass-marketed as PC software and others serve a dedicated purpose in the technical computing area. Once a need for a particular application has been established, the user should "shop around" for the proper package. Although some software (primarily games and graphic simulators) is reasonably transportable and does not require an operating system, great care should be made to assure compatibility. It should be guaranteed or demonstrated to run on the *user's* PC, operating system, and peripherals, and come with complete documentation. Many computer magazines have reviews and comparisons of popular software packages. When a package is used in the workplace, it is a great advantage to use the one recommended by the computer staff if they are expected to give assistance. Most programs that are purchased are copyrighted; and some are copy-proofed to prevent unauthorized copying, which in most cases also makes storing on hard disks difficult or impossible. Although there are methods for defeating copy-proof, or copy-protected, systems, such procedures are not recommended for legal and practical reasons. Fortunately for users, the trend today is to unprotected disks, with reliance by software developers on close monitoring of users and legal action to protect their proprietary interests. The general rule is "one program (copy) per machine" to be purchased, though large users can often purchase "on-site" copying rights.

2.5. SPECIAL TOPICS

Graphics, communications, and compatibility are discussed in this section. These subjects have been selected because they illustrate how hardware and software are interrelated in the PC system. In addition, they are subjects of concern to those using PCs for technical computing applications. The topics are covered through the introduction of terminology to provide a basis for further study.

2.5.1. Graphics

Interactive graphic systems[27] have increased the utility of the PC. However, unless proper choices are made, the display process may severely reduce the system's computing power by utilizing too much CPU or memory access time. The smallest addressable unit in any display is called the *pixel*. The *resolution* of a screen is measured in pixels on the horizontal and vertical axes. The ratio of horizontal to vertical pixels per unit length is called the *aspect ratio*. Unless the aspect ratio is considered when scaling figures, the figures may come out distorted; for example, circles may become ellipses.

The most common approach to displaying a full screen is through *bit-mapped graphics* where each pixel is assigned a bit in the portion of the PC memory allocated for graphics. A special direct-memory-access chip scans the graphic memory and if the bit is ON, the pixel is lit on the display. *Raster scanning*, similar to a television display, is the most popular scanning technique. For color, several bits are required for each pixel. A byte per pixel can address 256 colors, but a medium screen resolution of 640 × 350 with that many colors requires very careful hardware and software design. Separate boards that have a second CPU and lots of memory can be plugged into many PCs giving resolutions approaching 1024 × 1024. Some PCs such as the Amiga by Commodore have been built with graphics in mind. Many mid-priced PCs have 16-color, 310 × 200 resolution graphics, which are acceptable only for business graphics.

Reverting to one color is a sensible trade-off when high resolution is necessary. High-resolution monochrome and color graphics systems require special high-resolution display monitors. Inexpensive display systems save memory and require less sophisticated graphic circuitry by defining an expanded character set consisting of alphanumerics and graphic designs rather than addressing each pixel individually. A compatible graphic input device such as a mouse and an output device having equal or higher resolution, usually a digital plotter, are useful additions to the system. As always, the best choice is made by studying the possibilities and requiring a prepurchase demonstration of the entire system.

There are many software approaches to utilizing graphics hardware.[28] Higher-level language interpreters and compilers usually include a PC-specific set of intrinsic functions or subroutines for addressing pixels in the graphic memory plane, a section of memory set aside for holding the bit map corresponding to the display. Drawing lines and primitives (conic sections and polygons) and shading areas with a continuous color or pattern are examples of intrinsic subroutines. Sets of extrinsic subroutines are available for displaying axes of graphs, scaling data sets, and other routine needs. Some of the programs discussed in the design portions of this book have included graphs as part of the program output. Much of the mass-marketed software, for example, spreadsheet programs, come with graphics support to facilitate interpretation of the stored information. The graphic display of data acquired from laboratory instrumentation, as discussed in Section 6.1, is a mandatory part of those software packages. The game and animation packages available for PCs that are optimized for graphics are stunning.

One particular area of science and technology revolutionized by the PC's graphic display ability has been CAD, specifically, computer-aided drafting, modeling, and printed circuit layout. Good PC software/hardware packages are available[29] which support every phase of the drafting process from drawing lines to labeling. Frequently drawn figures which have been stored as *icons* in a library can be positioned and recalled as desired. Parts of diagrams can be drawn with overlays stored on several *layers* with full dimensions on other layers, all of which can be displayed in any color or sequence. Fine detail can be drawn by *panning* across the drawing and *zooming* in on the desired section. Whole cabinets of drawings can be stored on disk and recalled, rescaled, and plotted as desired. A draftsperson who learns how to use a CAD system can easily double his or her output. Although standard drafting software can be used, a number of PC packages are available for creating and revising printed-circuit-board artwork.

The limitations of two-dimensional drawing are rapidly giving way to three-dimensional *modeling* in which a solid object is constructed by piecing together parts of primitives such as cubes, cones, and toruses (doughnuts). The model is stored in memory as a series of equations describing its three-dimensional boundaries with respect to a reference point. Depending on the program, the display of layers may be used to show several pieces of detailed structure within a solid. Low-cost displays show only the edges and boundaries, (the *wireframe*) of the model, but more sophisticated CAD systems allow the solid to be rotated, cut away, or even shaded, depending on the lighting perspective.

2.5.2. Communications

Although an advantage to using a PC is "you have it all to yourself," sooner or later the need to exchange information with another computer will arise. Communication between computers supports the exchange of documents, messages, and data with coworkers, the search and retrieval of information from literature databases, and access to the facilities of other computer systems. There are two distinct modes of intercomputer communication: (1) with *point-to-point* communications, the user establishes a connection through the PC to a target system, for example, logging onto a mainframe, and after carrying out the desired function, disconnects; and (2) in *networking*, special software and sometimes hardware establish and maintain the connection between any number of systems without the direct intervention of the user. *Connectivity* is the term which indicates system support of communications on all levels.

Configuring a PC for point-to-point communication is not difficult and can be beneficial for many reasons. The simplest approach is the connection of a modem to the serial port of the PC (see Sections 2.3.3 and 6.4). The modem facilitates the hardware connection of the PC to another computer via standard telephone lines. The transmission rate via standard telephony is limited to 2400 baud (~ 240 char/sec) maximum so if the target computer is nearby it is often worth obtaining a direct (hardwired) connection. Sometimes special hardware in the form of a circuit board or external box called a *protocol converter* is required to establish the link with large computers. Until recently, this has been the case when connecting PCs to IBM mainframes, but IBM has developed new mainframe software and hardware that support asynchronous serial, ASCII transmission, and eliminate the need for special external converters.

After the hardware computer-to-computer connection has been established, specialized software is required for the PC and may be required for the target computer as well. The simplest communications package provides a menu to set up the speed and logic parameters for the serial port, sends incoming characters from the serial port to the screen, and also sends characters typed on the keyboard out the port to the remote computer. More advanced packages include autodialing and control for specific kinds of modems.

Terminal emulator programs are necessary to enable the PC to support the full screen features of the software on the target computer. Most computer

24   ANALYSIS AND DESIGN OF ELECTRONIC CIRCUITS USING PCs

centers that encourage PC users to connect to their mainframe distribute a recommended emulator program for which they hold a site license. A "shareware" package,[30] available for the price of copying, emulates the most popular DEC terminal, the VT-100. Most communications packages support the transfer of text and binary files. Special software, for example, KERMIT[31], is required on both ends of the link to check data during transfer for errors. Remote computers containing literature citations and other databases can be accessed with the communication package described above. A feature called *logging* sends everything that appears on the screen to a text file on a disk so it can be edited and printed off-line. To aid on-line literature searching, some data services supply special *gateway* software[32] which assists the user in developing a search strategy before connecting to the target system, thereby saving on connect time and long-distance telephone charges.

In contrast to the two-computer limit of point-to-point connections, a *network* ties together a number of independent computers allowing the exchange of information and the sharing of files.[33,34] The *local area network* (LAN) where PCs are connected to a central server via twisted pairs of wire or coaxial cable, is the most common approach. The central server is really another PC with special hardware and software to support the communication tasks. Some PC networks do not require connected individual systems, called *nodes*, to have their own hard disks because they can share one on the server. However, a separate hard disk is still recommended for PCs used for technical computing; otherwise programs that must access the server disk often will run much slower. Although networks of larger computers work well, connecting PCs in a LAN is an infant technology yet to be proven. Unless the buyer does a rigorous field trial, the purchase of such a network may be financially attractive on paper but in practice a losing proposition due to high maintenance costs and excessive downtime.

2.5.3. Compatibility

Connecting computers to peripherals and other computers as well as transferring information from one technical computing task to the next cannot be accomplished unless each entity's input and output parts are compatible. The compatibility issue translates directly into a financial one because additional hardware is required to convert between hardware protocols and time is required to retype data or write file translation programs. Therefore the question of compatibility must be addressed throughout the development cycle of any technical computing project. The answer in most cases lies in choosing systems that support accepted standards.

Although vendors frequently disagree, the benefits of a larger customer base has induced them to develop and follow hardware standards. Because most serial ports support the RS-232-C definition (Section 6.4), peripherals of almost every function utilize it. If a printer uses a parallel port, it should work when connected to a computer having a "Centronics" parallel interface. Regardless of manufacturer, a modem should support the Bell and CCITT communication standards.[30] Provided they have the same operating system, a disk written on one computer can probably be read on a look-alike, or "clone," model. Graphic hardware technology is developing rapidly, but without standards for resolution, aspect ratios, and the display of character fonts, commerical software developers must support numerous hardware configurations or require their customers to buy a special graphics board.[35] IBM, due to its popular PC hardware, has dictated graphics resolution standards including the low-resolution Color Graphics Adaptor (CGA, 320 × 200) and medium-resolution Enhanced Graphics Adaptor (EGA, 640 × 350). However, the IBM PS/2 models and other manufacturers of graphic boards have confused the issue by offering other resolution ratios in the same range. The ultimate success and cost of optical disk systems depends upon the acceptance of standard data formats.[17] Third-party manufacturers of memory and disk systems have adopted standard bus connection schemes. As always, the promise that a device is standard should guide a purchase, but a complete demonstration continues to be required before acceptance.

Software standards have been developed to facilitate connectivity during programming and the exchange of information between software packages. The ANSI definition of standard syntax and commands has made the transport of FORTRAN[24] and BASIC[23] programs between dissimilar computers possible, but the inclusion of special graphic and control commands by each vendor limits the functionality of transportable programs. When functionality and transportability are of concern, other languages such as C,[25] Pascal,[26] Modula-2,[36] and Ada,[37] should be considered. Standards are required for the transfer of information between software packages which perform similar functions and between packages which process data produced by some other package. Examples of the first type are the WPS format sponsored by IBM for the transfer of word processing documents and the Initial Graphics Exchange Specification (IGES), which describes a drawing produced by any CAD system in a series of standard codes and coordinates.[38] Examples of the second type are the Electronic Design Interchange Format (EDIF), a public-domain format for the interchange of all necessary data for integrated and printed circuit design,[39] and the DIF Format, recommended by LOTUS Development Corporation for exchanging data between data acquisition and database packages and spreadsheet programs.

2.6. Conclusion

This introduction to the hardware, software, and concepts of personal computing should serve as a beginning for those interested in setting up a PC to accomplish the activities discussed in this book as well as for technical computing in general. Because developments in the PC field are so rapid, the reader should continually consult recent publications to obtain the latest information. Although assistance from an experienced veteran is invaluable, PC users who take an active part in choosing the hardware and installing the software will experience increased productivity sooner than those who order a "black box." The instruction manual for each new piece of hardware or software package should be studied, thoroughly tested, and integrated into the system's personality. With the investment of time and patience in mastering the basic skills, the use of PCs for technical computing will become a rewarding and enjoyable activity.

REFERENCES AND NOTES

1. *Inside the Personal Computer*. New York: Abbeville, 1984.
2. Flores, I. *The Professional Microcomputing Handbook*. New York: Van Nostrand Reinhold, 1986.
3. *Byte*. Peterborough, NH: McGraw-Hill Inc.
4. *Compute!*. New York: ABC Pub.
5. *Creative Computing*. Los Angeles: Ahl Computing, Inc.
6. *IEEE Transactions on Circuits and Systems*. New York: IEEE.
7. *Scientific Computing and Automation*. Dover, NJ: Gordon Pub.
8. *VT-240 Graphic Terminal Operator's Guide*. Maynard, MA: Digital Equipment Corp. 1983.
9. *Microcomputer Handbook*. Maynard, MA: Digital Equipment Corp., 1980.
10. Zaks, R. *From Chips to Systems: An Introduction to Microprocessors*. Berkeley, CA: Sybex, 1981.
11. Special Issue: "Inside the IBM PCs." *Byte*, Nov. 15, 1985; *Byte*, Nov. 1986.
12. Special Issue: "The 68000 Family." *Byte*, Sept. 1986.
13. Lemmons, P. "The New Generation: High-Tech Horsepower." *Byte*, Jul. 1987.
14. Dessy, R. E., "Choosing a PC." *Anal. Chem.* 58(1), 78-91A, 1986.
15. Palmer, J. F., and Morse, S. P. *The 8087 Primer*. New York: John Wiley & Sons, 1984.
16. Boyd, A. M. *PC-DOS/MS-DOS*. New York: Bantam Books, 1985.
17. Issue Theme: "Storage Goes Optical." *Byte*, May 1986.
18. *HP-7475A Programming and Operation Manual*. Palo Alto, CA: Hewlett-Packard Co.
19. Knuth, W. A. *The Art of Computer Programming*, Vols 1-3. Reading, MA: Addison-Wesley, 1973.
20. Manna, Z., and K. Waldinger. *The Logical Basis For Computer Programming*. Reading, MA: Addison-Wesley, 1985.
21. Lafore, R. *Assembly Language Primer for the IBM PC and XT*. New York: New American Library, 1984.
22. Katzan, H. *Operating Systems: A Pragmatic Approach*, 2nd ed. New York: Van Nostrand Reinhold, 1986.
23. Presley, B. *A Guide To Programming the IBM Personal Computers*. Albany, NY: Lawrenceville Press, 1985.
24. Brainerd, W. S., J. L. Gross, and G. H. Goldberg. *FORTRAN 77: Fundamentals and Style*. Boston: Boyd & Fraser, 1985.
25. Pollack, L., and B. Cummings. *Programming in C on the IBM PC*. Englewood Cliffs, NJ: Prentice-Hall, 1984.
26. Zaks, R. *Introduction to PASCAL*. Berkeley, CA: Sybex, 1981.
27. Williams, G. "A Graphics Primer." *Byte*, Nov. 1982, 448-470.
28. Foley, J. D., and A. Van Dam. *Fundamentals of Interactive Computer Graphics*. Reading, MA: Addison-Wesley, 1984.

29. Flora, P. C. *The CAD Directory*, 2nd Ed. Blue Ridge Summit, PA: TAB Books, 1985.
30. Mark C. DeVecchio, 9067 Hillary Dr., San Diego, CA 92126.
31. da Cruz, F. *KERMIT.*, Bedford, MA. Digital Press, 1987.
32. Hawkins, D. T., and Levy, L. R. "Front End Software for Online Database Searching." *Online*, November 1985.
33. Dessy, R. E., "Local Area Networks," Parts I and II. *Anal. Chem.* 54(11), 1167-80A, 1982.
34. Archer, R. *The Practical Guide to LANs*. Berkeley, CA: Osborne/McGraw-Hill, 1986.
35. *Industrial PC Sourcebook*. San Diego, CA: ICS, 1986.
36. Wirth, N. *Programming in Modula-2*, 3rd., corrected ed., New York: Springer-Verlag, 1985.
37. *Reference Manual for the ADA Programming Language*. New York: Springer-Verlag, 1983.
38. "IGES, Version 3.0." PB 86-199759. Springfield, VA: NTIS, 1986.
39. EDIF Users Group, 2222 S. Dobson Rd., Bldg. 5, Mesa, AZ 85202.

3
Ac, Dc, and Transient Circuit Simulation Programs

Brian Biehl
Tektronix, Inc.
Beaverton, Oregon

3.1. INTRODUCTION

General-purpose circuit simulators can simulate any type of electronic circuit: analog, digital, mixed analog and digital. These circuits can contain many types of devices—resistors, capacitors, and transistors, to name a few. Circuit simulators can perform many types of analyses on these circuits, among which are DC operating points, small-signal frequency response, and large-signal transient response. The maximum frequency or fastest risetimes which can be simulated depend only on the accuracy of the device models used. The speed at which these simulators operate depends on the implementation of the numerical algorithms, compiler or interpreter efficiency, and the speed of the computer hardware.

This chapter describes general-purpose circuit simulators which can be used on personal computers. It begins by reviewing the history of personal computers relevant to circuit simulation.

Simulator architecture is described, including input parsing, setup procedures, matrix solution techniques, convergence criteria, and modeling. The analysis capabilities of these simulators are explained briefly, with examples of the more popular types of analyses. Device models for all the basic elements used in BIAS-D, PSpice, and MICROCAP II are described. Macro-modeling is described briefly.

Three simulator programs are featured in this chapter: BIAS-D,[1] a bare-bones

---

Programs similar to PSpice, which perform the same basic types of analyses, but are not featured in the examples presented in the book, are:

ALLSPICE: ACOTECH, 713 Santa Cruz Ave., Menlo Park, CA 94025.
IS_SPICE: intusoft, P.O. Box 6607, San Pedro, CA 90734.
Z/SPICE: ZTEC, 6745 Lindley, Reseda, CA 91335.
ECA: Tatum Labs, 1478 Mark Twain Court, Ann Arbor, MI 48103-9709.

simulator which run on the IBM PC's MS-DOS operating system as well as on the older CP/M systems; PSpice,[2] a version of SPICE[3] developed for the IBM PC; and MICROCAP II,[4] a third circuit simulation program which will also run on the IBM PC. A program listing of BIAS-D and Reference Manual are provided in the appendix at the end of this chapter.

Simulation programs[1,5] have been available for desk-top calculators since the early 1970s. Early on, the main disadvantage of these programs was not the programs themselves but the computer systems on which they ran. These computers had limited memory, slow execution speed, and supported few if any high-level languages. The largest memories available were 16 kilobytes of random access memory (RAM) with cassette tape as the storage media. Many trade-offs were required just to get a 10-node circuit to execute in a reasonable time.

The foregoing devices were the forerunners to "home computers," which first appeared in volume in the last 1970s. These early home computers contained 8-bit microprocessors with audio cassettes for program storage. The cassettes were soon replaced by $5\frac{1}{4}$-inch single-density floppy disk drives, with 80 to 100 kilobytes of storage per diskette. The maximum addressable RAM on these home computers was 64 kilobytes. These systems were again memory- and speed-limited, as also were the larger "desktops" of 10 years ago. However rather than $20,000 or more, they cost less than $3,000.

Today most personal computers are based on 16- or, recently, 32-bit microprocessors. These systems have dual- or quad-density $3\frac{1}{2}$-inch $5\frac{1}{4}$-inch floppies and, in many cases, 5-inch hard disks with 10 to 80 and more megabytes of storage; and 512 kilobytes or more of addressable RAM. They execute simulation programs 10 to 100 times faster than their predecessors, often still at costs of less than $3000. These systems are not language-limited as were their desktop predecessors, and having high-level language capabilities that include C, Pascal, Lisp, and Fortran. This is now an environment very attractive for circuit simulation.

3.2. THREE CIRCUIT SIMULATORS FOR PCs

This section briefly describes the capabilities and hardware requirements of the three aforementioned circuit simulator programs—BIAS-D, PSpice, and MICROCAP—for PCs. In addition, speed comparisons of several small computer systems are made with larger systems in terms of the simulator program BIAS-D.

3.2.1. BIAS-D

The circuit simulation program BIAS-D was originally written in BASIC for an HP 9830[1] desktop calculator. It was later converted into FORTRAN and executed on the HP 1000, DEC 11/45, and Prime 400 minicomputers, and on an IBM 370/168[6] mainframe computer. BIAS-D was then modified to run on a Cromemco Z80-based microcomputer system[7] with 64 kilobytes of RAM. BIAS-D presently executes on any MS-DOS or CP/M system with at least 64 kilobytes of memory. Program overlays on disk are not required.

A list of the capabilities and features of BIAS-D are given in Table 3-1. The 50-node circuit capability is based on a 90% sparse circuit with 75 elements or less. The node capabilities and element count can be adjusted depending on the available memory. The circuit elements are stored in a linked-list integer array with 16 bytes allocated to store resistors, current sources, and voltage sources; 24 bytes to store capacitors and inductors; 32 bytes for transistors; and 40 bytes for models. Details of the contents of each list can be found on page 77 of reference 8, end of this chapter. BIAS-D also uses sparse matrix storage and inversion techniques. Execution speed improvement techniques used in BIAS-D include optimal node reordering and sparse matrix decomposition.

Table 3-1. BIAS-D Capabilities

Nodes: 50 not including voltage sources and ground

Elements: 75 (approximate total) dynamically allocated

    Capacitors
    Inductors
    Voltage sources
    Current sources
    Bipolar transistors (*npn* and *pnp*)

Models (built-in)

    PUL—pulse
    SIN—sine wave
    EXT—Any arbitrary v(t)
    NPN—Ebers-Moll level 1
    PNP—Ebers-Moll level 1
    TEM—temperature (t.c. for R, C, and Beta)

Analysis types

    Dc operating points
    Dc transfer curves
    Transient
    Small signal frequency response

Interactive features

    Insert elements
    Alter elements (single value or sweep)
    Load circuit file to or from disk
    Save dc, transient or ac analyses data to disk
    Print circuit data at any time
    Graphics interface built-in

AC, DC, AND TRANSIENT CIRCUIT SIMULATION PROGRAMS

The *npn* and *pnp* transistor models in BIAS-D are equivalent to that of SPICE Level 1 BJT models without the voltage-dependent junction capacitors and the series collector, base, and emitter resistors. The model does have a low current beta roll-off parameter not included in the SPICE level 1 model. These models are described in more detail in Section 3.3.

Data is input using a semi-free format similar to SPICE. Element names are limited to two characters and node numbers to a maximum of 99. An example of the input format of BIAS-D is given in Figure 3-1. This is the input representation of the 24-node 5-transistor amplifier circuit shown in Figure 3-2.

```
* TEST CIRCUIT CKT13 (24 NODES)
*** INTEGRATED PREAMPLIFIER ***
* RESISTORS
R1 6 1 12K
R2 7 3 7.5K
R3 4 0 680
R4 7 6 9K
R5 8 0 5K
* TRANSISTORS
Q1 32 11 23 M2
Q2 34 21 43 M2
Q3 62 51 44 M2
Q4 64 61 53 M2
Q5 72 31 83 M2
* VOLTAGE SOURCES
VB  7 0 6.1
VS 9 0 1 M1
CS 9 1 1U
* BASE RESISTORS
RB1 1 11 100
RB2 2 21 100
RB3 5 51 100
RB4 6 61 100
RB5 3 31 100
* COLLECTOR RESISTORS
RC1 3 32 100
RC2 3 34 100
RC3 6 62 100
RC4 6 64 100
RC5 7 72 100
* EMITTER RESISTORS
RE5 83 8 10
RE4 53 5 10
RE3 44 4 10
RE2 43 4 10
RE1 23 2 10
* JUNCTION CAPACITORS
CE1 11 2 2P
CC1 11 3 2P
CE2 21 3 2P
CC2 21 4 2P
CE3 51 6 2P
CC3 51 4 2P
CE4 61 5 2P
CC4 61 6 2P
CE5 31 7 2P
CC5 31 8 2P
M1 PUL 0 -1 .5U .5U 5U .5U
M2 NPN 100 1 5E-15
END
```

Twenty-four Node Benchmark Circuit.

Fig. 3-1. Input format for BIAS-D using 24 node circuit.

32 ANALYSIS AND DESIGN OF ELECTRONIC CIRCUITS USING PCs

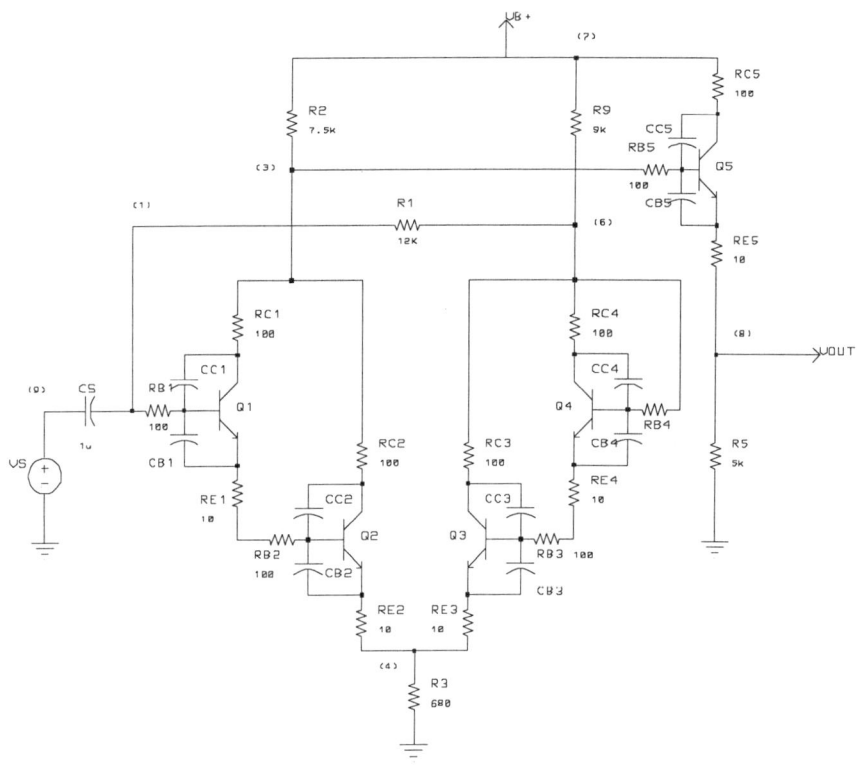

Fig. 3-2. Schematic of 24-node 5-transistor amplifier.

BIAS-D is an interactive program; its interactive features are included in Table 3-1.

Minimum Computer System for Executing BIAS-D:

1. A CPU with 64 kilobytes of RAM.
2. CP/M or any other operating system supporting FORTRAN.
3. A single disk drive, 81-kilobyte or larger.
4. A monitor or terminal.

3.2.2. PSpice

SPICE-2 has been available as a public-domain program from the University of California, Berkeley, since 1976.[3,9] SPICE is by far the most popular circuit simulation program in the world today. It was originally written in FORTRAN for a CDC-6600 mainframe computer, later converted to an IBM 360-series mainframe and eventually to the DEC VAX computer under both the DEC/VMS and Berkeley Unix operating systems.

AC, DC, AND TRANSIENT CIRCUIT SIMULATION PROGRAMS

PSpice is essentially a conversion of this popular simulator to run on the IBM PC, IBM PC XT and IBM PC AT and clones. Additional features have been added to the original program such as the gallium arsenide MOSFET models (GAAS) and Probe post-processing graphics. More recently, PSpice has been converted from FORTRAN into C. This has slightly increased the execution speed and greatly increased the maximum circuit size. A list of the capabilities and features of PSpice (SPICE) are given in Table 3-2.

SPICE uses a single large array to store all of the element and sparce matrix

Table 3-2. PSpice/SPICE Capabilities

General
  Eight character element names
  Maximum node number 9999
  Suffix abbreviations (eg., M = milli = 1E−3, K = kilo = 1E3)
  Number of devices limited only by available memory (practical limit of 120 transistors for IBM PCs)
  Subcircuit capability (nodes within a subcircuit are unique to subcircuit)
  Free format input

Built in Models
  Passive elements
    Resistors
    Capacitors
    Inductors
    Transformers
    Transmission lines
  Sources
    Independent voltage and current source
      Sinusoidal source
      Pulse source
      Modulated source
    Current-controlled current source
    Current-controlled voltage source
    Voltage-controlled current source
    Voltage-controlled voltage source
  Nonlinear active elements
    Bipolar junction transistors (BJT)
    Junction field effect transistors (JFET)
    Metal oxide field effect transistors (MOSFET)
    Diodes (germanium, silicon, and Schottkey)
  Analysis types
    Dc operating point
    Dc transfer curve
    Dc sensitivity
    Small signal transfer function (gain, input resistance and output resistance)
    Small signal frequency response
    Small signal noise
    Transient response
    Fourier analysis
    Analysis of above types at different temperatures

34 ANALYSIS AND DESIGN OF ELECTRONIC CIRCUITS USING PCs

data. This is managed internal to SPICE. Therefore the maximum number of elements and nodes are determined only by the memory available in the computer. For the IBM PC systems with 512 kilobytes of RAM, this limit is around 120 transistors (each transistor can represent 3 nodes. See Section 3.3.) for the FORTRAN version and twice that for the C version.

Minimum Computer System for PSpice:

1. IBM PC, PC XT or PC AT with 512 kilobytes of RAM and floating-point coprocessor.
2. MS-DOS 2.0 or later operating system.
3. A single $5\frac{1}{4}$-inch disk drive (IBM format compatable).
4. Graphics display terminal.

PSpice options include user-changeable Models and Probe. User-changeable models include the source code for the model-dependent routines and the object code necessary to link to the remaining PSpice modules. This option requires the MicroSoft FORTRAN Complier.[10]

Probe is a graphics postprocessor. It will run on the standard IBM PC monochrome monitor but with limited resolution. Probe is recommended for use with one of the following graphics options: (1) IBM Color Graphics Adaptor Card (CGA) (640 × 200), (2) the IBM Enhanced Graphics Adaptor (EGA) with the standard color display (40 × 350), (3) the Hercules Graphics Card (720 × 348) or (4) the Texas Instruments Professional graphics display (720 × 300).

3.2.3. MICROCAP II

MICROCAP II is a commercial program written by Spectrum Software of Sunnyvale, California. Capabilities of MICROCAP II are given in Table 3-3.

Minimum Computer System for Executing MICROCAP II:

1. IBM PC, PC XT, or PC AT with at least 256 kilobytes of RAM.
2. MS-DOS 1.1 or later operating system.
3. A single $5\frac{1}{4}$-inch disk drive (double-sided, double density).
4. A high-resolution monitor (*not* the IBM monochrome display monitor) with IBM Color Graphics Adapter Card (CGA) or equivalent. Color not required.

A printer with a parallel interface is optional (the standard IBM or MX80 printers are recommended). A schematic capture front end, which allows the entering of schematic symbols on the screen, and a graphics postprocessor are included in the basic price. Also included is a nice manual with many example circuits.

Table 3-3. MIRCOCAP II Capabilities

General	
Netlist input via schematic entry	
Interactive	
Nodes (transient and dc analysis	100
Nodes (ac analysis)	50
Built-in Models	
Voltage sources (batteries)	10
Voltage sources (sine)	6
Voltage sources (time)	6
User sources	1
Polynomial sources	25
Switches	50
Resistors	150
Capacitors	150
Inductors	50
Transformers	50
Diodes	100
Bipolar transistors	50
Mos transistors	50
Opamps	50
Analysis Types	
Dc operating point	
Transient	
Small signal frequency response	

3.2.4. Speed Comparisons of Computer Systems using BIAS-D

Since BIAS-D is written in standard FORTRAN, it is possible to execute almost the exact code (except for the timing and disk input/output (I/O) routines) on many different computer systems regardless of the CPU size (i.e., 8 bits or 32 bits). Table 3-4 lists eight computer systems used in execution-time comparisons. Each system was timed executing the 24-node circuit in Figure 3-2 for a transient analysis of 101 timepoints. This table gives the system, the time per iteration, and the relative execution speed (per iteration) compared to a VAX 11/780 minicomputer system running VMS FORTRAN. The total execution time for the VAX was 12 seconds compared to 82 minutes for the Cromemco and 3 minutes for the IBM PC. The systems are listed in approximate order of increasing execution speed. As could be expected, the 8-bit CPUs are slower than the 16-bit CPUs, which in turn are slower than the 32-bit CPUs.

Some items worth noting in Table 3-4 are:

1. Floating-point hardware or coprocessors increase execution times by a factor of 2 to 5.
2. Compiler efficiency can cause a significant difference in execution speed. This can be seen in the case of the VAX 11/780s. Both had identical

Table 3-4. Speed Comparison of Computer Systems Running BIAS-D (FORTRAN)

SYSTEM Operating System and Features	Ms per Iteration	Relative Speed†
8-Bit CPUs:		
1. Cromemco CR2D (CDOS, 4 MHz, Z80)	11s	410
2. TARBELL (CP/M, 4 MHz Z80)	11s	410
16-Bit CPUs:		
3. IBM PC XT (MS-DOS, 8086)	1.2s	45
(1) Linked without coprocessor library		
(2) Linked with 8087 coprocessor library		
(1) Without coprocessor	2.2s	83
(2) With coprocessor	410	15
4. Apollo (virtual memory, cache, 68000 10-MHz clock, accelerator)	140	5.2
7. HP-1000F (RTE VI, accelerator; 48-bit double precision)	61	2.2
32-Bit CPUs:		
8. VAX 11/750 (VMS 3.1; cache, virtual memory)		
(1) Without accelerator	61	2.2
(2) With accelerator	37	1.4
10. Prime 400 (PRIMOS; cache; virtual memory, accelerator)	64	2.4
11. VAX 11/780 (cache; virtual memory; accelerator)		
(1) Unix (Berkeley Version 4.1)	43	1.6
(2) VMS (rev. 3.2)	27	1.0
12. IBM 370/168 (MVS/TSO; cache accelerator)	4.5	.17
13. SPICE2G.6		
(1) VAX 11/780 VMS	43	1.6
(2) IBM PC XT (8086/87 coprocessor)	560	20.7

*Running test circuit in Figure 3-2 (24 nodes, 5 bipolar transistors)
†Relative to a VAX 11/780 VMS FORTRAN (with optimization)

hardware but one was using the Unix FORTRAN compiler and the other the VMS FORTRAN compiler.

3. SPICE-2 executed approximately 60% slower than BIAS-D on a VAX 11/780, whereas it executed (not shown) approximately twice as fast as BIAS-D on an IBM 370/168.[8]

The speed differences listed in Table 3-4 were caused by hardware or software limitations (or both). In the case of the 8-bit CPUs, the primary limitation is the double-precision multiply speed. The speed of a double-precision (64-bit) multiply on a 4-MHz Z80 (MicroSoft FORTRAN) is 10 ms, whereas on the VAX 11/780 VMS FORTRAN it is 2 μs—a factor of 5000 faster!

A compiler determines how efficiently hardware interfaces with software. Here efficiency is dependent on the machine instruction set, whether it takes advantage of the hardware, the programmer, and so forth. The differences in speed due to compilers can be more than the factor of 2 shown for the VAX in Table 3-4. In some cases, it may be a factor of 4 or 5.[8,11]

The size of the program affects its execution speed. Larger programs require more paging (disk I/O, which is slower than RAM). This is evidenced by the execution speeds of SPICE on a mini vs. a mainframe.

Interactive circuit simulation can be achieved at execution speeds of one second per iteration or faster. Slower than this and the user becomes bored. This effect can be evidenced by the 11 seconds per iteration for the 4-Mhz Z80 (Table 3-4) which took 82 minutes for a 101-point transient analysis—definitely not interactive. At 1 second per iteration, the same analysis would take 7.5 minutes—marginal but tolerable.

3.3. ARCHITECTURE OF A CIRCUIT SIMULATOR

Even though the programming languages or the instruction sets of circuit simulators differ, the internal structures are similar. The basic structure of a circuit simulator is shown in Figure 3-3. The input data parsing and circuit setup are preprocessing steps which are done before any analysis begins. The dc operating point analysis is usually done first, before any other analyses. The four primary analyses types are; dc operating point, transient, small-signal frequency response, and dc transfer curve. Other analysis types not shown in the figure are noise, dc sensitivity, and Fourier. Results of the analyses are sent to a printer, plotter, or disk file. If the results are sent to a disk file, postprocessing can be done at a later time.

3.3.1. Input Data Parsing

The input parser is the interface between the user and the computer. It determines the "friendliness" of the software. In general, an input parser reads a single line of data from a keyboard or file, checks for syntax errors, does limited preprocessing, and stores the information in a database within the simulator. Most simulators in use today allow free-format input data. That is, data fields need not conform to a rigid format; spaces usually denote the separation between data fields.

There are two possible modes of simulator operation: batch and interactive. A flow diagram of the batch-mode input parser is shown in Figure 3-4. Here input data which has been created on a text editor and stored in a file is read into the parser, a single line at a time. If this line contains a syntax error, an

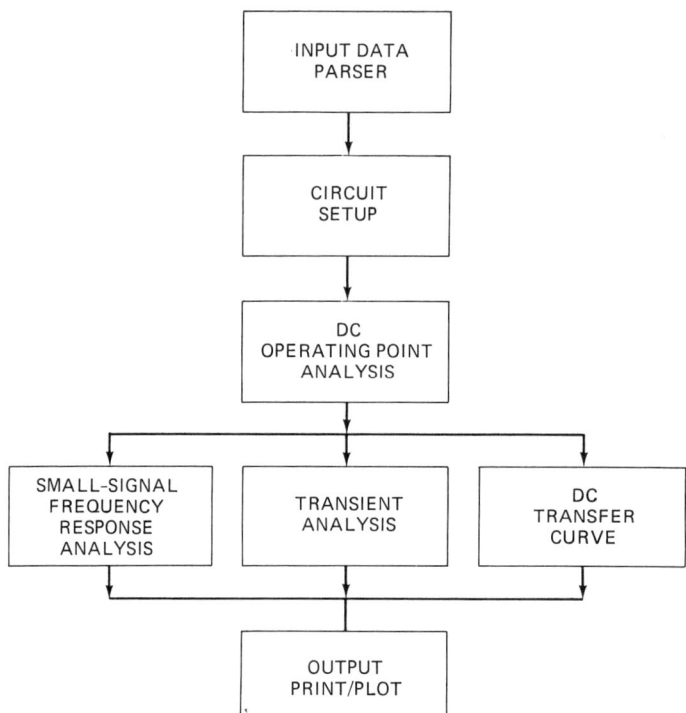

Fig. 3-3. Basic structure of a circuit simulator.

error flag is set and the error message is written to an output file. If, after all the input data has been processed, no errors are found, the simulator then goes to the next procedure: setup. If errors are found in the program, execution is terminated. The output file must then be read by the user to determine the error(s). Control commands such as "plot" or "print" are included in the input file.

In the interactive mode, the user enters the data via the keyboard, a single line at a time, or the user enters a previously generated input file. A flow diagram of the interactive mode interface is shown in Figure 3-5. As the data are entered, the data are parsed and checked for syntax errors. Errors are displayed immediately and can be corrected. After all the data are input, the program continues to the next procedure. The interactive mode parser should have provision to store the syntactically correct data in a file for later editing or resimulation. The control commands such as "plot" or "print" are input at the time of analysis.

Although an interactive circuit simulator is desirable, it is not always practical. If a computer system is so slow that the user must wait several minutes or even hours for the simulation results, and then respond to these results, an interactive

AC, DC, AND TRANSIENT CIRCUIT SIMULATION PROGRAMS 39

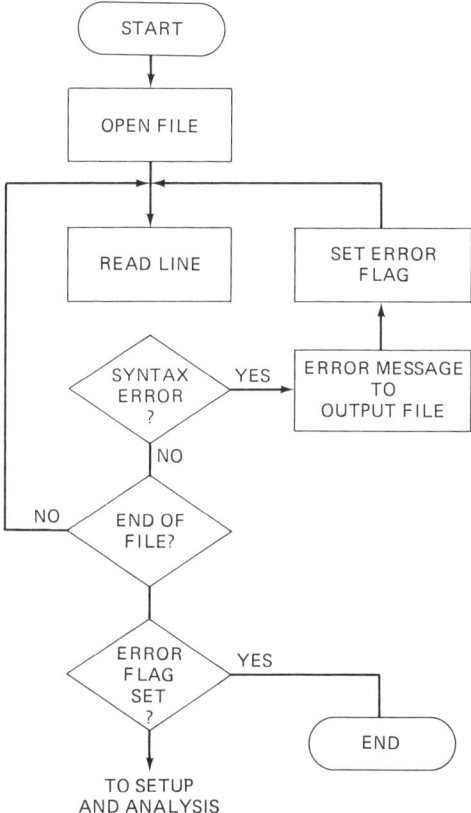

Fig. 3-4. Flow diagram of batch-mode input parser.

simulator should not be used; in this case, a batch simulator is best. Circuit simulators using the sparse matrix techniques described later in this chapter typically achieve solutions at times proportional to $n^{1.2}$ where n is the number of actual circuit nodes. This means that if an analysis of a 10-node circuit takes 2 seconds, a 100-node circuit will take approximately 32 seconds.

3.3.2. Circuit Setup

After the input data are stored, they must be restructured into a form suitable for analysis. Different simulator programs have different setup procedures; the setup process used in BIAS-D is explained here. The example circuit in Figure 3-6 is used to illustrate this restructuring process. This example is a six-node circuit with two transistors, five resistors, one capacitor, and two grounded voltage sources.

40 ANALYSIS AND DESIGN OF ELECTRONIC CIRCUITS USING PCs

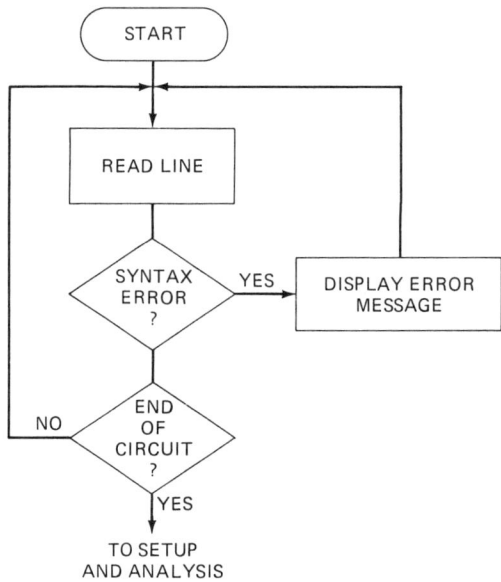

Fig. 3-5. Flow diagram of interactive mode input parser.

During the input parsing, a node vector, N_1, is generated containing all unique circuit node numbers. Since the input data node numbers are randomly assigned, the node numbers in this node vector are randomly ordered but the *number* of unique nodes, n, is known. Figure 3-6a shows the example circuit with the initial node numbering. Node vector N_1 is ordered into increasing numerical order. Another node vector, N_2, is created the same length as N_1 containing the sequence $\{1, 2, 3 \ldots n\}$ where n is the number of unique circuit nodes. At this point, the node vectors N_1 and N_2 are

$$N_1 = [1\ 2\ 3\ 5\ 6\ 10]$$
$$N_2 = [1\ 2\ 3\ 4\ 5\ 6] \tag{3-1}$$

As will be explained in Section 3.4, grounded voltage source nodes are handled separately. Nodes 10 and 1 in vector N_1 are the voltage source nodes. The equivalent location of the voltage nodes in vector N_2 are at positions 1 and 6. The voltage source nodes in N_2 are next moved to the right end of N_2 by exchanging positions with nonsource nodes. The rightmost position (6) is already a voltage source node, therefore only the (1) in the first position is exchanged with the (5) in position 5. N_1 and N_2 are now

$$N_1 = [1\ 2\ 3\ 5\ 6\ 10]$$
$$N_2 = [5\ 2\ 3\ 4\ 1\ 6] \tag{3-2}$$

AC, DC, AND TRANSIENT CIRCUIT SIMULATION PROGRAMS 41

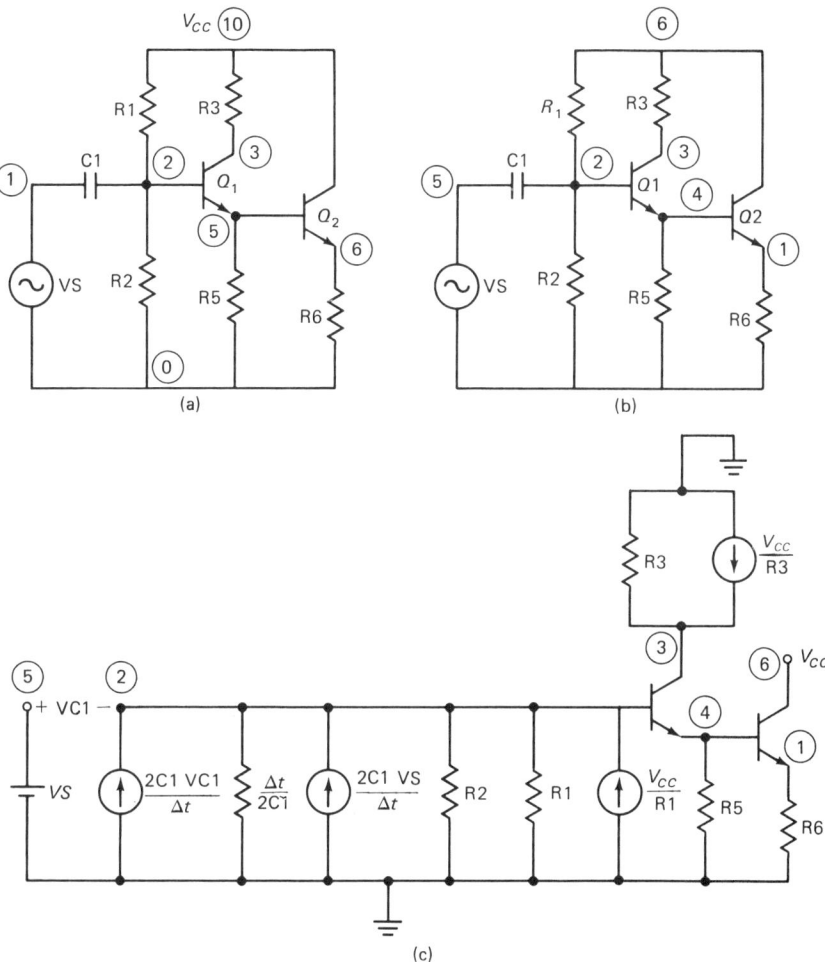

Fig. 3-6. Setup circuit: (a) with initial node numbering, (b) with renumbered nodes, (c) final restructured form.

Note that the equivalent source positions in N_2 are now 5 and 6. All circuit element nodes are now renumbered from the values in N_1 to the values in their equivalent location in N_2. That is, node 1 is renumbered to node 5; node 2 renumbered to node 2, and so forth. The results of this node renumbering is shown in Figure 3-6b.

The circuit is now restructured by converting elements connected to voltage sources into their Norton equivalents. For resistors, this is accomplished by grounding the node of each resistor connected to a voltage source and adding a Norton-equivalent current source from ground to the resistor's other node. Current source nodes connected to voltage sources are grounded. A capacitor is represented as a conductance in parallel with a voltage-dependent current source. A capacitor connected to a voltage source then is reconfigured by grounding the current source and treating the conductance as a resistor; that is, the resistor node connected to the voltage source is grounded and a Norton-equivalent current source from ground to the other resistor node is connected. Transistors are not restructured at this time.

Following these conventions, the example circuit shown in Figure 3-6b is restructured into that of Figure 3-6c. Three additional current sources are added, two for resistors R1 and R3 and one for capacitor C1. The values of these added current sources are stored symbolically, in the form of either a node number or element value location, since the resistor, the capacitor, or the voltage source value may be altered (BIAS-D is an interactive mode simulator) during an analysis session. The circuit is now in its final restructured form (Figure 3-6c). The known nodes, those of the grounded voltage sources, have been eliminated from the circuit. In the larger circuit-simulator programs such as SPICE-2, this elimination is done after the nodal-admittance matrix has been loaded.

3.3.3. Admittance Matrix Solution

At this point there are three possible paths to obtain a solution of the circuit node voltages: (1) direct solution of the admittance matrix equation

$$[Y][V] = [I] \qquad (3\text{-}3)$$

for the node voltage vector $[V]$ by taking the inverse of the admittance matrix $[Y]$ as

$$[V] = [Y]^{-1}[I] \qquad (3\text{-}4)$$

(2) using Gauss elimination[6] to reduce the matrix equation (Equation 3-3) and obtain the solution for the node voltages; and (3) using *LU* factorilization and sparce matrix techniques to reduce Equation 3-3 and solve for the node voltages.

Each of these paths has advantages and disadvantages. The direct method was used in earlier desktop simulators,[1,5] because it was programmed into ROM and therefore reduced the RAM memory requirements. For this reason, it also executed relatively fast. In todays personal computers, memory space is not as critical and the matrix inversion would be done in software. A matrix inversion requires significantly more computational effort (and therefore more time) than that of the other two methods. There is also a high probability that the large dynamic range of entries in the admittance matrix would result in a singular matrix and no solution would be obtained. For these reasons, it is not used in present-day circuit simulators. If the direct method is used, no further circuit processing would be required.

The Gauss elimination method does not require inverting the admittance matrix. The use of pivoting, either full or partial, greatly reduces the chance of obtaining a singular solution. Partial pivoting would allow swapping either the admittance matrix rows or columns.

There are two additional modification to the Gauss elimination procedure which greatly increase the node voltage solution speed. The first is *zero-checking*. The admittance matrix of a circuit is always sparse; that is, at least 50% of the entries in this matrix have a value of zero. Typically, the larger the circuit, the greater the number of zeros or sparsity of the admittance matrix. Since all matrix operations dealing with these entries are computed in double-precision arithmetic, at least 50% of the time, an operation is dealing with a value of zero. An operation in the Gauss elimination or *LU* factorialization procedure is of the form

$$A = A - B * C \qquad (3\text{-}5)$$

or

$$A = A/B \qquad (3\text{-}6)$$

A, B, and C can be checked for a value of zero; if a zero is found, either that operation or an entire row (or column) of operations can be eliminated. Hence the number of operations can be greatly reduced at the expense of a few zero-checking operations. Zeros towards the upper left of the matrix reduce the number of operations more than zeros at the lower right of the matrix. For this reason, a second modification to the Gauss elimination procedure is recommended.

The second modification requires the optimal reordering[13] of the circuit nodes to maximize the zero's toward the upper left of the admittance matrix. The reordering scheme implemented here does not always generate an optimal node order. This is because "fill-in's" caused by matrix "operations" are not counted in the optimization process. It is also possible that the original node order is already optimal. In this case, no speedup will be noticed. However, in general, the larger the circuit the greater the speed savings due to node reordering. There is some additional setup processing required due to this reordering process, however, as will be shown later in this section, but the effects are minimal.

In general, the Gauss elimination method is relatively easy to implement in software. For circuits of fewer than 50 nodes, this represents a viable method of solution for a circuit simulator. This method was used in several of the earlier Berkeley simulators[6,9] and was initially implemented in BIAS-D for speed and memory comparisons. Results of these comparisons will be given later in this section.

44 ANALYSIS AND DESIGN OF ELECTRONIC CIRCUITS USING PCs

The last but preferred method of matrix solution is LU factorialization. This method requires the "decomposition" of the admittance matrix, $[Y]$, into the product of an upper triangular matrix, U, and a lower triangular matrix, L. An upper triangular matrix has all the matrix positions below the diagonal filled with zeros. The admittance matrix equation is then given as

$$[Y][V] = [L][U][V] = [I] \tag{3-7}$$

The voltage vector solution is attained by a forward and backward substitution:

$$[U][V] = [L]^{-1}[I] = [I]' \tag{3-8}$$

and

$$[V] = [U]^{-1}[I]' \tag{3-9}$$

Since U and L are triangular matrices, the inverse L^{-1} or U^{-1} is not required to solve Equation 3-9.

The matrix LU decomposition process can be set up so that "pointers" indicate the matrix location of the next *nonzero* value for each operation. In this case, zero-checking as mentioned for the Gauss elimination process is eliminated. The pointer system permits storage of only the nonzero matrix terms. To set up the pointer structure, "mock decomposition" of the admittance matrix is required.

The mock decomposition process is performed with an incidence matrix, in which all entries are either a zero or a one. To generate an incidence matrix, each circuit element is added into the incidence matrix just as it would be added into the nodal-admittance matrix $[Y]$. The positions for each type of element is explained in Section 3.4.

The incidence matrix for the circuit in Figure 3-6 is:

$$[IY] = \begin{bmatrix} 1 & 0 & 0 & 1 \\ 0 & 1 & 1 & 0 \\ 0 & 1 & 1 & 0 \\ 1 & 1 & 0 & 1 \end{bmatrix} \tag{3-10}$$

Note that this is not optimal since the swapping of rows 1 and 3 will generate a matrix which is contains more zeros in the upper left of the matrix.

This decomposition process can include a Markowitz reordering algorithm similar to that used in the Gauss elimination method. However this reordering scheme results in a more efficient reordering because the matrix "fill-ins" are counted during the mock decomposition process. These fill-ins are not determined in the Gauss method since it does not require the mock decomposition process. The mock decomposition process, optimal reordering, and the pointer structure are all done during the initial setup procedure. Utilization of the pointer structures results in a significant speed improvement over the Gauss method. Some of this improvement can be attributed to the more efficient reordering method. These speed improvements will be shown later in this section.

As mentioned earlier, the same pointer structure used to locate the nonzero matrix entries in the nodal admittance matrix can also be used to store only the nonzero elements. The admittance matrix (which is always sparse) can then be stored as a linear array rather than as a matrix array. This results in a significant savings in memory. For example, a 50-node circuit (not counting grounded voltage sources) would require $50 \times 50 \times 8$ bytes of storage. Each position must be stored as a double-precision floating-point number which requires 8 bytes of storage. This is a total of 20,000 bytes of storage. Since the matrix is at least 50% zeros, a linear array would require a maximum of 10,000 bytes, giving a 10,000-byte memory savings. A 100-node circuit would result in at least a 40,000-byte savings. This is significant for small computer systems such as PCs. The speed overhead necessary to implement this sparse storage method is minimal since the pointers are already required for the LU decomposition process.

3.3.4. Circuit Convergence

If the circuit being analyzed is linear, then the node voltage solution attained in the previous section represents either the dc operating point voltages or the voltages at a single timepoint in a transient analysis. However, most of the circuits analyzed in circuit simulation programs contain active devices such as transistors. These circuits therefore must be linearized about an operating point and iterated until a solution is attained. One of the more difficult jobs of a circuit simulation program is to determine when the circuit has converged. Each nonlinear devices operating region has different characteristics with potentially wide variations in voltages or currents. Historically, circuit simulation programs have looked at node voltages or current variations to determine convergence. SPICE 2G looks only at branch currents for the last three iterations and determines the change in all branch currents. If these changes are below a predetermined threshold, then convergence is assumed. Earlier versions of SPICE looked at the node voltages for three consecutive iterations and if all node voltage changes (for three iterations) were less than 1 microvolt (this was the default value), then convergence was assumed. For some circuits, convergence cannot be attained within 100 iterations. In this case, the user must go in and change resistances or model parameters to slightly change the circuit conditions and therefore coax the circuit into convergence. Even swapping node numbers in some cases can cause a circuit to converge.

Convergence plays a significant role in the speed of the simulator. Each iteration requires the loading and solving of the nodal admittance equation. Typically 5 to 15 iterations are required for a dc operating point analysis. The transient analysis typically requires only 3 or 4 iterations because the time increments are usually small enough such that the node voltage differences between timepoints are small. For the transient analysis, the previous timepoint node voltages are used for the starting node voltages for the present timepoint solution.

3.3.5. Dc Operating Point Analysis

The Dc operating point analysis is the default analysis type in most circuit simulators. This analysis is essentially a debug analysis and is usually done as a means of getting to one of the other analysis types. The dc analysis solution ignores circuit capacitors and inductors either by setting the timestep to an infinite value (typically 10^{12} seconds) or by ignoring them altogether. This is explained in more detail later in this chapter. A dc transfer curve analysis is a dc operating point analysis in which the driving current or voltage is varied between two values. A flow diagram of the dc operating point analysis is given in Figure 3-7. The steps required to in each of the blocks of this figure have been described in the previous section. The dc operating point analysis in PSpice outputs all circuit node voltages, transistor and diode branch currents, voltage drops, and small signal ac parameters. The dc operating point analysis results from PSpice for the example circuit in Figure 3-1 is shown in Figure 3-8.

3.3.6. Transient Analysis

A transient analysis allows investigation of the time-dependent qualities of a circuit. Risetimes, falltimes, switching speeds, and oscillation frequency are examples of circuit parameters which can be determined (in a simulator) only by using a transient analysis. The flow diagram of the steps required to compute a transient analysis is shown in Figure 3-9.

The transient analysis for each timepoint is very similar to the dc operating point analysis but unlike the dc analysis the transient analysis adds capacitors and inductors into the nodal admittance matrix. The admittance values of these capacitors and inductors are dependent on the transient timestep. This is explained in more detail later in this chapter. Additionally, the transient analysis requires the timestep to be incremented, the voltage or current sources updated, and the capacitor and inductor charge incremented.

The timestep is estimated by determining the smallest critical time constant in the circuit. In early circuit simulators,[16] fixed timesteps required the user to be reasonably precise in determining them since they would ultimately determine the accuracy of the simulation results. For example, to get a 1% simulation accuracy for a sine wave requires approximately 20 points per cycle (trapezoidal integration method).

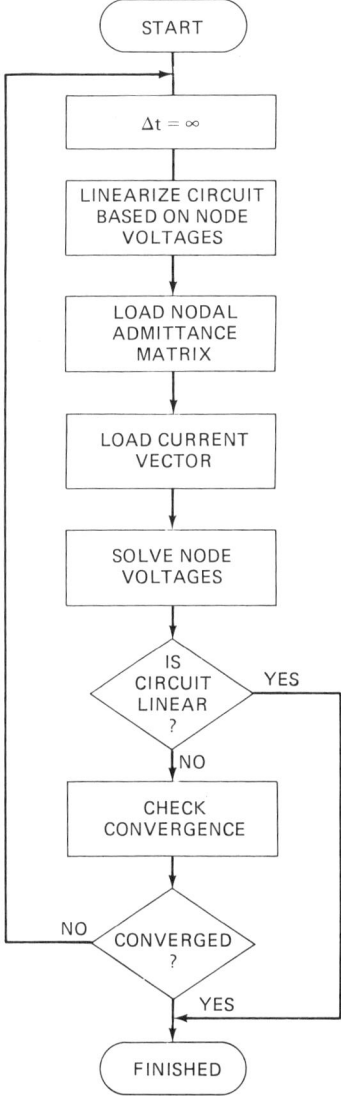

Fig. 3-7. Flow diagram of a DC operating point analysis.

Later simulation programs[17,18] allowed specifying subintervals in which the total time could be divided into subintervals, each of which could have its own timestep. This required the user to have some knowledge of circuit dynamics. From these programs evolved timestep control based on iteration count.[14,20] Today the most reliable timestep control is based on the truncation error of the integration method. In most analog circuits, the overhead in computing and

48 ANALYSIS AND DESIGN OF ELECTRONIC CIRCUITS USING PCs

```
****************** 22-AUG-86 *******            ***************************

* TEST CIRCUIT CKT13 (24 NODES)

****      INITIAL TRANSIENT SOLUTION        TEMPERATURE =    27.000 DEG C
******************************************************************************

NODE  VOLTAGE   NODE  VOLTAGE   NODE  VOLTAGE   NODE  VOLTAGE   NODE  VOLTAGE

( 1)  1.8337   ( 2)  1.2998   ( 3)  2.5520   ( 4)  0.6414   ( 5)  1.3001
( 6)  1.8342   ( 7)  6.1000   ( 8)  1.9000   ( 9)  0.0000   (11)  1.8337
(21)  1.2993   (23)  1.2998   (31)  2.5516   (32)  2.5515   (34)  2.5055
(43)  0.6461   (44)  0.6462   (51)  1.2996   (53)  1.3001   (61)  1.8342
(62)  1.7873   (64)  1.8338   (72)  6.0624   (83)  1.9038


      VOLTAGE SOURCE CURRENTS
      NAME        CURRENT
      VB+        -1.323D-03
      VS          0.000D+00


      TOTAL POWER DISSIPATION    8.07D-03  WATTS

************** 22-AUG-86 ********************             *****************

* TEST CIRCUIT CKT13 (24 NODES)

****   OPERATING POINT INFORMATION      TEMPERATURE =    27.000 DEG C
******************************************************************************


****  BIPOLAR JUNCTION TRANSISTORS


                 Q1          Q2          Q3          Q4          Q5
MODEL            M2          M2          M2          M2          M2
IB          4.60E-08    4.65E-06    4.69E-06    4.65E-08    3.76E-06
IC          4.60E-06    4.65E-04    4.69E-04    4.65E-06    3.76E-04
VBE            0.534       0.653       0.653       0.534       0.648
VBC           -0.718      -1.206      -0.488       0.000      -3.511
VCE            1.252       1.859       1.141       0.534       4.159
BETADC       100.002     100.000     100.000     100.000     100.000
```

Fig. 3-8. DC operation point analysis.

storing truncation errors and other parameters required for automatic timestep control negate the time saved in computing the fewer timepoints. However, many astable circuits such as multivibrators, Schmitt triggers, and so forth, will not converge using fixed timestep control methods.

A PSpice transient analysis simulation for the test circuit in Figure 3-1 is given in Figure 3-10. The driving source for this analysis is a 7-volt sine wave. As Figure 3-10 shows, this input voltage drives the circuit into saturation in the position direction but not in the negative direction.

3.3.7. Small-Signal Ac Frequency Response Analysis

The ability to compute small-signal frequency response is important in the design and analysis of analog circuits. The small-signal frequency response is found by determining the circuit voltage gain or transimpedance between the input and output nodes as:

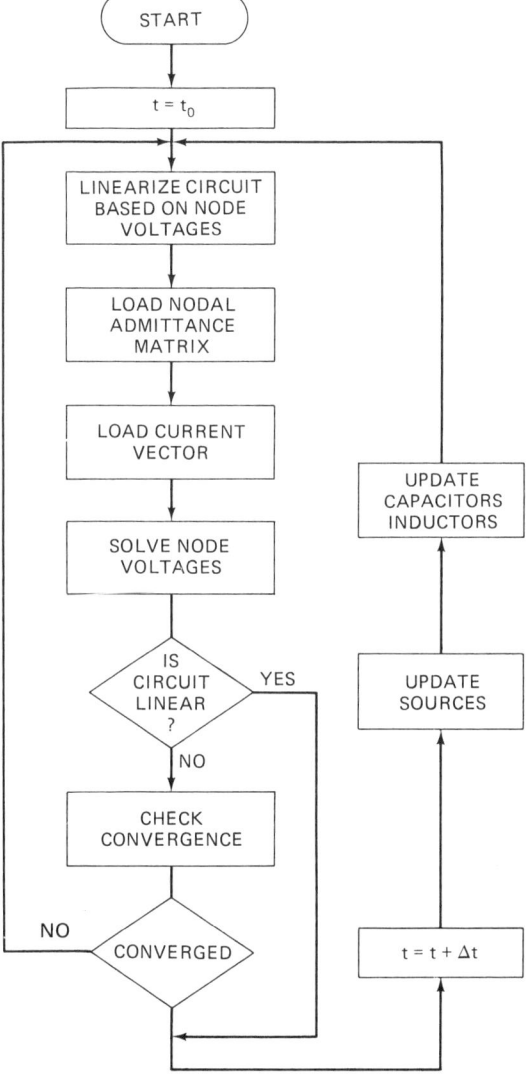

Fig. 3-9. Flow diagram for transient-analysis computation.

$$G_v = V_{out}/V_{in}$$

or

$$G_z = V_{out}/I_{in} \tag{3-11}$$

This gain is found by solving the complex admittance matrix equation (3-3) for the output node voltages and dividing by the complex input driving voltage or

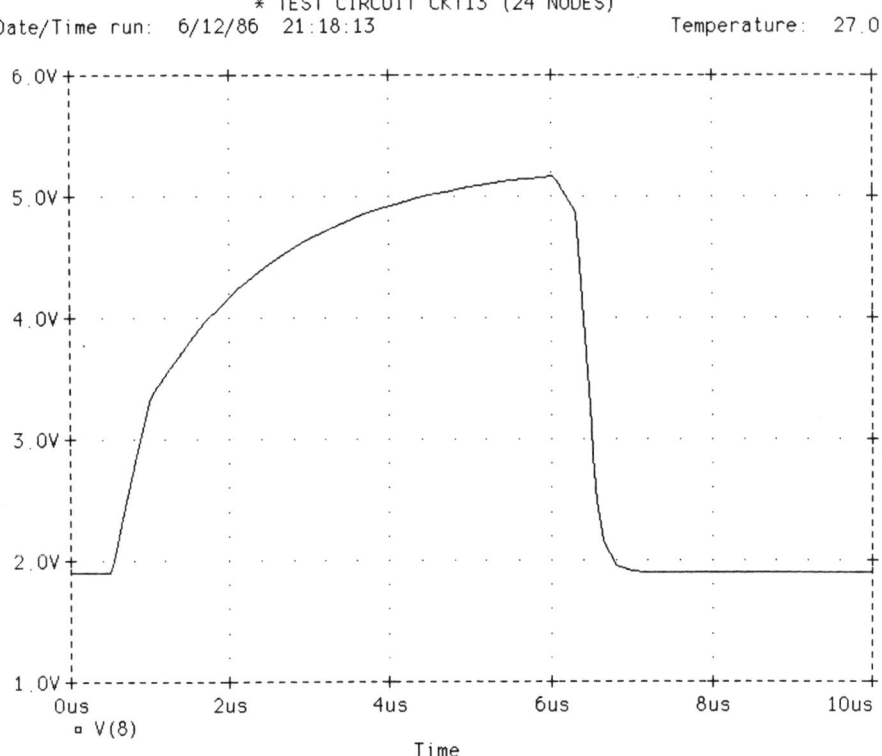

Fig. 3-10. Transient analysis output.

current. To do this the complex admittance matrix $[Y]$, is loaded with the real and imaginary equivalent conductances of the circuit elements. These elements are evaluated at each frequency of interest. For nonlinear or active devices such as transistors, these conductances are determined at the circuit's dc operating points. Therefore for these circuits, the dc analysis results must be available. The values of these linearized conductances for resistors, capacitors, inductors, and transistors are given in the next section. A complex driving current, given either as the input source or as a Norton equivalent of the driving voltage, is loaded into the complex current vector. This driving current or voltage is usually set to unity to simplify the calculation of Equation 3-11. This equation is then solved for the complex node voltage vector, $[V]$, at each frequency of interest. At this time either the output voltage magnitude, V_M, or phase, V_P is determined at the required node, n, as:

$$V_{M(n)} = \left\{ \left[V_{(n)\text{real}} \right]^2 + \left[V_{(n)\text{imaginary}} \right]^2 \right\}^{1/2} \qquad (3\text{-}12)$$

$$V_{P(n)} = \arctan \left[V_{(n)\text{real}} / V_{(n)\text{imaginary}} \right] \qquad (3\text{-}13)$$

AC, DC, AND TRANSIENT CIRCUIT SIMULATION PROGRAMS 51

This procedure is repeated for each frequency of interest (usually a sweep). A flow diagram of this procedure is shown in Figure 3-11. Since the frequency response analysis is computed using linear or linearized elements at a dc operating point, iterating to a solution is not required as in the dc and transient analyses. The same matrix pointer structure used for dc and transient analyses can also be used here. However, since the matrix is complex, all operations are also complex, requiring significantly more time to compute. This is especially true for computers with 8-bit microprocessors such as the Z80 and 8080 (See Section 2.3.). This complex matrix also requires twice the matrix storage space

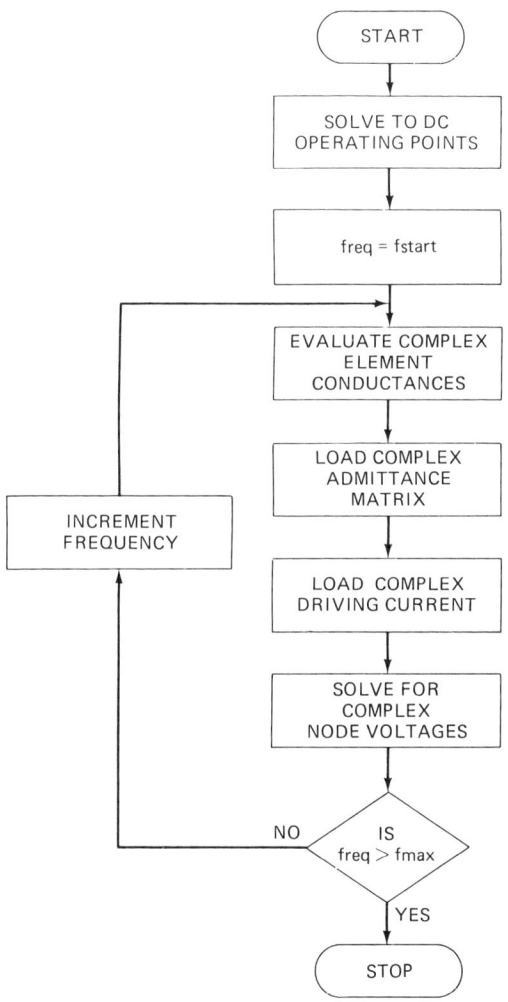

Fig. 3-11. Flow diagram for small-signal frequency response analysis.

52 ANALYSIS AND DESIGN OF ELECTRONIC CIRCUITS USING PCs

since the real and imaginary values must be stored in separate double-precision floating-point arrays.

A typical frequency response analysis would specify varying the analysis frequency in equal log increments from an initial frequency to a final frequency with 20 frequency points per frequency decade. An example circuit is shown in Figure 3-12. This is a passive high-Q notch filter. The PSpice input file for this circuit is given in Figure 3-13. The frequency response (in dB) for this circuit is shown in Figure 3-14 and the phase response is shown in Figure 3-15.

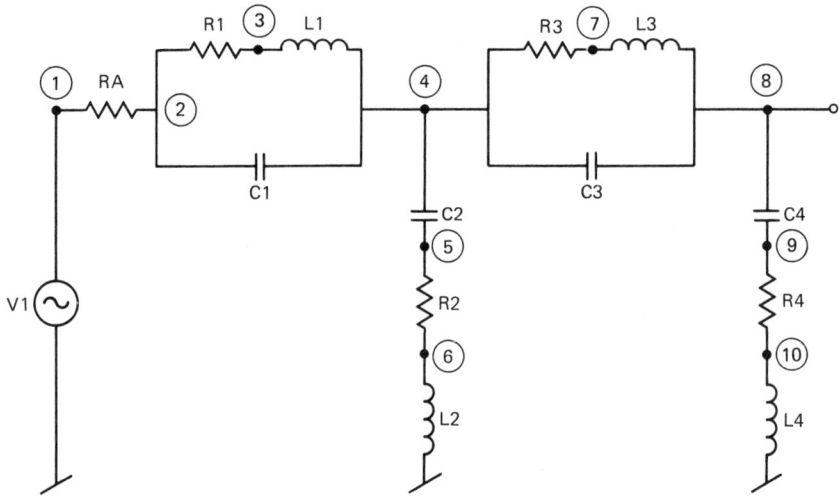

Fig. 3-12. Notch filter circuit diagram.

```
* Notch Filter Example Circuit
VIN 1 0 DC 0 AC 1
RA 1 2 1.032K
RB 8 0 1.032K
C1 2 4 .073UF
C2 4 5 294.9PF
C3 4 8 .073UF
C4 8 9 229.7PF
L1 3 4 157.5UH
L2 6 0 39MH
L3 7 8 122.8UH
L4 10 0 39MH
R1 2 3 .4
R2 5 6 100
R3 4 7 .3
R4 9 10 100
.AC DEC 100 40KHZ 70KHZ
.PLOT AC VDB(8) VP(8)
.END
```

Fig. 3-13. RSpice input file for notch filter circuit.

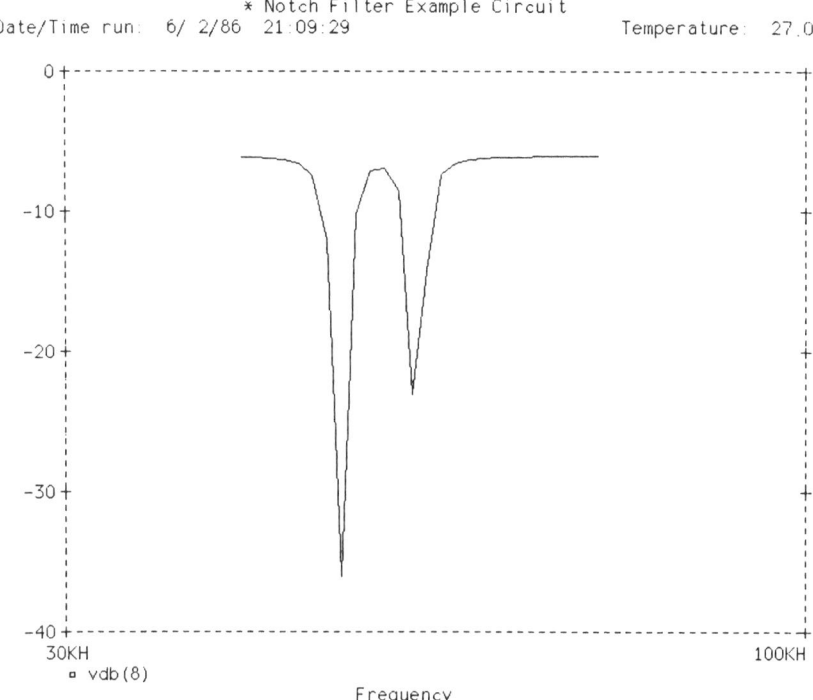

Fig. 3-14. Frequency response plot of notch filter circuit.

3.4. MODELING

Modeling is perhaps the most critical part of circuit simulation. Without proper models or without the proper implementation of these models, simulation results would be of little value.

There are many types of circuit devices or elements which can be modeled in a circuit simulator. The basic circuit elements are resistors, current sources, capacitors, inductors, and voltage sources. The models for these elements are described in some detail in this section. Diodes, transistors, and other nonlinear devices can also be modeled. These nonlinear circuit elements are also included in most circuit simulation programs and are briefly described in this section.

Circuit elements can be linear or nonlinear. The nonlinear elements must be "linearized" in order to be modeled by a digital computer. They may be linearized differently, depending on the operating region of the devices. They may also be different depending on the type of analysis being executed. Circuits which contain nonlinear elements models must be iterated to achieve convergence to the correct node voltages. After linearization, even the most complex models contain only two different types of circuit elements: (1) a resistor

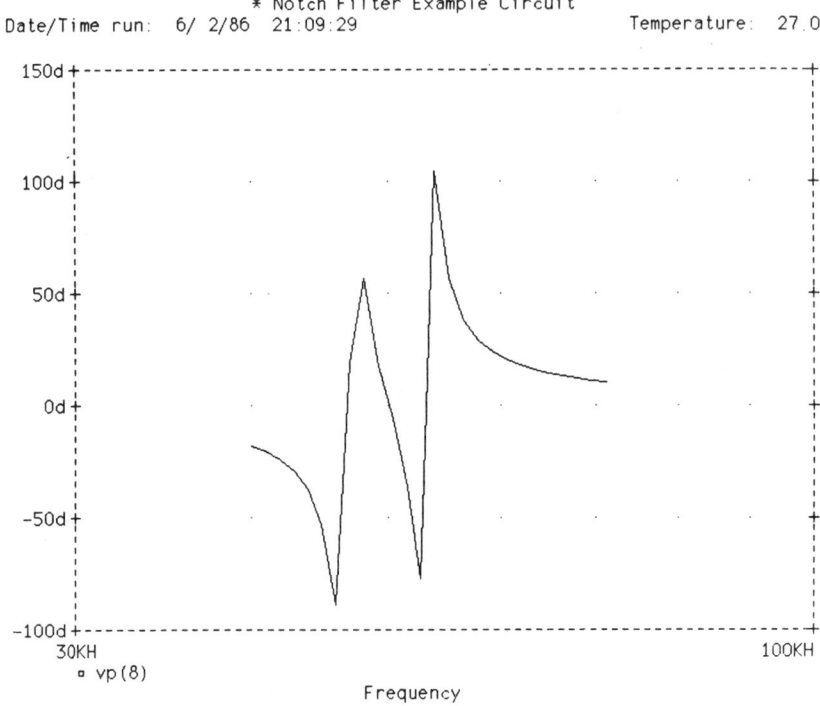

Fig. 3-15. Phase response plot of notch filter circuit.

(impedance) and (2) a current source, which may be voltage-dependent. The resistance or admittance elements of the model are added into the nodal admittance matrix, $[Y]$, of the matrix equation

$$[Y][V] = [I] \qquad (3\text{-}14)$$

The current portion is added into the current vector, $[I]$, of the matrix equation. This equation is then solved for the circuit node voltage vector, $[V]$.

3.4.1. Linear Resistor Model

A linear resistor is one of the basic circuit element models. The voltage-current equation for a resistor, R_A with a current $i(R_A)$ flowing from node 1 to node 2 due to a voltage drop $(v_1 - v_2)$ across R_A is given as (See Figure 3-16)

$$i_{R_A} = (v_1 - v_2)/R_A \qquad (3\text{-}15)$$

Fig. 3-16. Model of linear resistor.

In nodal admittance matrix form this equation is represented as:

$$\begin{bmatrix} 1/R_A & -1/R_A \\ -1/R_A & 1/R_A \end{bmatrix} \begin{bmatrix} v_1 \\ v_2 \end{bmatrix} = \begin{bmatrix} 0 \\ 0 \end{bmatrix} \quad (3\text{-}16)$$

The admittance submatrix on the left is added into the admittance matrix for the complete circuit in four locations as:

$$Y_{11} = Y_{11} + 1/R_A$$
$$Y_{12} = Y_{12} - 1/R_A$$
$$Y_{21} = Y_{21} - 1/R_A$$
$$Y_{22} = Y_{22} + 1/R_A \quad (3\text{-}17)$$

where the subscripts indicate the row and column position in the admittance matrix. There are no independent currents in the resistor model and therefore nothing is added into the current vector.

The linear resistor model is identical for dc analysis, small-signal frequency response (ac) analysis and transient analysis. For ac analysis, the resistance submatrix is added to the real part of a complex matrix.

3.4.2. Current Source Models

Another basic circuit element is the current source. Both the independent curent source and voltage-dependent current source models are keys to the implementation of many other models. Both current sources may be time-dependent, which is apparent only in the transient analysis. For example, PSpice and BIAS-D have several time-dependent source functions which include pulse and sinusoidal waveforms. These can be specified for the independent current source or an independent voltage source driving a dependent current source.

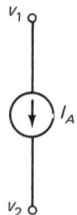

Fig. 3-17. Model of independent current source.

Independent Current Source. An independent current source, I_A, with a positive node (1) and a negative node (2) is shown in Figure 3-17. The current flow is assumed to enter the positive node and flow through the source to the negative node. The nodal admittance matrix equation for the current source is:

$$\begin{bmatrix} 0 & 0 \\ 0 & 0 \end{bmatrix} \begin{bmatrix} v_1 \\ v_2 \end{bmatrix} = \begin{bmatrix} I_A \\ -I_A \end{bmatrix} \quad (3\text{-}18)$$

The admittance matrix terms are zero since this is a pure current source. The current I_A is added to the existing current at node 1 and subtracted from the existing current at node 2 as shown by

$$I_1 = I_1 + I_A$$
$$I_2 = I_2 - I_A \quad (3\text{-}19)$$

The independent current source model is identical for dc and the transient analyses. For a dc analysis, the dc value of I_A would be specified or else it would default to the value at $t = 0$ for the transient analysis. For a time-dependent current source, the value of I_A added during the transient analysis will depend on the value of the waveform function at each timepoint. For ac analyses, all independent current sources are zeroed and therefore not added to the current vector

Voltage-Dependent Current Source. A voltage dependent current source I_B is shown in Figure 3-18. This is a four-terminal device with the value of the current I_B flowing from node 3 to node 4 depending on a constant, g_v, times the difference in voltage between node 1 and node 2. This is represented by the equation:

$$I_B = g_v(v_1 - v_2) \quad (3\text{-}20)$$

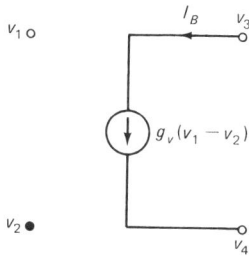

Fig. 3-18. Model of Voltage-Dependent current source.

Since this is a four-terminal device, it has a 4 × 4 nodal admittance submatrix representation:

$$\begin{bmatrix} 0 & 0 & 0 & 0 \\ 0 & 0 & 0 & 0 \\ g_v & -g_v & 0 & 0 \\ -g_v & g_v & 0 & 0 \end{bmatrix} \begin{bmatrix} v_1 \\ v_2 \\ v_3 \\ v_4 \end{bmatrix} = \begin{bmatrix} 0 \\ 0 \\ 0 \\ 0 \end{bmatrix} \qquad (3\text{-}21)$$

There are four locations in the admittance matrix where the value of g_v is added. These are added as:

$$Y_{31} = Y_{31} + g_v$$
$$Y_{32} = Y_{32} - g_v$$
$$Y_{41} = Y_{41} - g_v$$
$$Y_{42} = Y_{42} + g_v \qquad (3\text{-}22)$$

The admittance subscripts indicate the row and column position in the matrix. Interestingly enough, there are no independent currents added to the current vector. The voltage-dependent current source has the same model (and value) for dc operation point, transient, and small-signal ac analyses.

3.4.3. Linear Capacitor Model

The time-dependent voltage-current relationship of a linear capacitor is given by the integral equation

$$V_C(t) = 1/C \int i_C(t)\, dt \qquad (3\text{-}23)$$

For computer simulation this equation is discretized into the equation

$$V_C(t) = 1/C \int_{t-dt}^{t} i_C(t)\, dt + V_C(t - dt) \qquad (3\text{-}24)$$

The present voltage, $V_C(t)$ (Equation 3-24) is obtained by integrating the capacitor current from the last timepoint $t - dt$, to the present timepoint, t, and adding this result to the capacitor voltage at the last timepoint, $v_C t - dt$.

The capacitor model used for the transient response analysis is dependent on the type of integration method used, in other words, the method with which Equation 3-23 is approximated in the computer. The two most popular integration methods are Euler's[3,21] and the trapezoidal,[22] with the trapezoidal being the more accurate for the same number of timepoints per period. (Both PSpice and BIAS-D use the trapezoidal integration method for linear capacitors.) The difference between Euler's and the trapezoidal integration methods is shown in Figure 3-19; where A_1 and A_2 represent the integrated area for two consecutive timepoints.

If the trapezoidal method is used, Equation 3-24 can be rewritten as

$$i_C(t) = 2C/\Delta t * v_C(t) + I_0 \qquad (3\text{-}25)$$

where I_0 is known from the last timepoint and is given by

$$I_0 = -i_C(t - \Delta t) - 2C/\Delta t * v_C(t - \Delta t) \qquad (3\text{-}26)$$

Equations 3-25 and 3-26 are modeled by a resistance in parallel with a voltage-dependent current source as shown in Figure 3-20.

The value of the resistor, R, is $\Delta t/2C$ where C is the capacitance value and

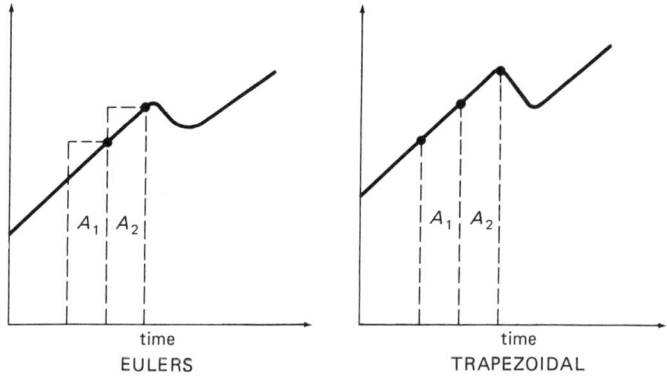

Fig. 3-19. Difference between Euler's and trapezoidal integration methods.

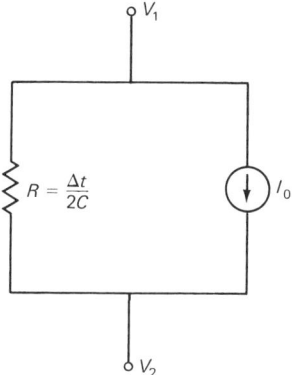

Fig. 3-20. Linear capacitor model.

Δt is the transient timestep increment. This timestep can be either fixed or variable (timestep control.)[3,21,23] The current source value, I_0, is a function of the capacitor current at the previous timepoint and the voltage across the capacitor at the present timepoint. This is the value given in Equation 3-26.

The nodal admittance matrix equation for the capacitor is then

$$\begin{bmatrix} 2C/\Delta t & -2C/\Delta t \\ -2C/\Delta t & 2C/\Delta t \end{bmatrix} \begin{bmatrix} v_1 \\ v_2 \end{bmatrix} = \begin{bmatrix} I_0 \\ -I_0 \end{bmatrix} \quad (3\text{-}27)$$

This requires the addition of the resistance $2C/\Delta t$ into four locations in the admittance matrix and the addition of current I_0 into two locations in the current vector. The value of I_0 in the current vector is updated for each timepoint in the transient analysis but remains constant for the iteration process in a nonlinear circuit. The resistance value $2C/\Delta t$ is changed only if a variable timestep is used and the timestep has changed.

This admittance matrix equation is the same for dc operating point, transient, and ac frequency response analysis. In the dc operation point analysis, the current source is set to 0 amps and the resistance is set to a large value, typically 1×10^{12} ohms. For a small-signal frequency response analysis, the current source is again set to zero amps. The resistance becomes an admittance with a pure imaginary value, $2\pi fC$, where f is the frequency of interest.

3.4.4. Linear Inductor Model

The model for a linear inductor is similar to that of the linear capacitor. The time-dependent voltage-current relationship of a linear inductor is given by the integral equation

$$i_L(t) = 1/L \int v_L(t)\, dt \tag{3-28}$$

For computer simulation, this equation is discretized into

$$i_L(t) = i_L(t - dt) + 1/L \int_{t-dt}^{t} v_L(t)\, dt \tag{3-29}$$

The present current, $i_L(t)$, is obtained by starting with the inductor current from the last timepoint, $i_L(t - dt)$, and adding the integrated inductor voltage from the last timepoint, $t - dt$, to the present timepoint, t.

Using the trapezoidal method of integration (as was done with the capacitor), Equation 3-29 can be rewritten as

$$i_L(t) = \Delta t/2L * v_L(t) + I_0 \tag{3-30}$$

where I_0 is again known from the last timepoint and is given by

$$I_0 = i_L(t - \Delta t) + \Delta t/2L * v_L(t - \Delta t) \tag{3-31}$$

These equations (3-30 and 3-31) are modeled by the same equivalent circuit as the linear capacitor, that is, a resistance in parallel with a voltage-dependent current source. This model is shown in Figure 3-21. The value of the resistor, R, is $2L/\Delta t$ where L is the inductance value and Δt is the transient timestep increment. The current source value, I_0, is a function of the inductor current at the previous timepoint and the voltage across the inductor at the previous timepoint. This is the value of I_0 given by Equation 3-31.

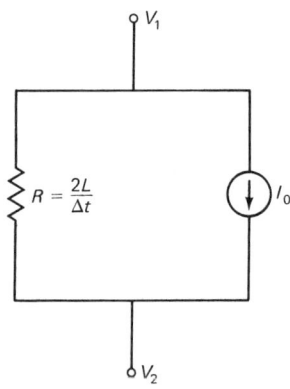

Fig. 3-21. Linear inductor model.

AC, DC, AND TRANSIENT CIRCUIT SIMULATION PROGRAMS 61

The nodal admittance matrix equation for the linear inductor is then

$$\begin{bmatrix} \Delta t/2L & -\Delta t/2L \\ -\Delta t/2L & \Delta t/2L \end{bmatrix} \begin{bmatrix} v_1 \\ v_2 \end{bmatrix} = \begin{bmatrix} I_0 \\ -I_0 \end{bmatrix} \quad (3\text{-}32)$$

This requires the addition of the resistance $\Delta t/2L$ into four locations in the admittance matrix and the addition of current I_0 into two locations in the current vector, the same as for the capacitor. The value of I_0 is updated for each timepoint in the transient analysis; the resistance value is changed only if a variable timestep is used and the timestep has changed.

This matrix equation is the same for dc operating point, transient, and ac frequency response analyses. For dc operation point analysis, the current source is set to 0 amps and the resistance set to a small value, typically 1×10^{12} ohms. For ac frequency response analysis, the current source is again set to 0 amps and the resistor becomes an admittance with a pure imaginary value $2\pi f/L$ where f is the frequency of interest.

3.4.5. Voltage Source Model

Many of the early circuit simulator programs such as ECAP[18] used a simple form of nodal analysis for solving the circuit equations. This form required all branch relations to be defined in terms of currents or voltage-controlled current sources. Voltage-defined branches were not allowed. Therefore voltage sources were modeled as a voltage source in series with a small resistor, usually 1 ohm. During setup, the voltage source and resistor were converted to their Norton equivalent of a current source in parallel with a resistor. This model was inconvenient and also added additional computation time to the simulation.

It is possible to modify the nodal analysis method to allow grounded voltage sources. If the voltage source is grounded, then the nonground node voltage is known and does not need to be part of the admittance matrix computations. This requires restructuring the circuit to reflect the removal of this source node. Elements connected to these voltage source nodes are changed to their Norton equivalents. An example of a single resistor R_1 connected to a grounded voltage source is shown in Figure 3-22. After restructuring, the resistor node connected to the source is grounded and a current source representing the value of the current through R_1 due to the presence of the voltage source is added in parallel with R_1, shown in Figure 3-23. This seems more complicated than just replacing the voltage source by its Norton equivalent. However, such is done only once during the setup procedure and requires just seconds even for large circuits. A more detailed description of this proceedure is given in Section 3.3.2.

An advantage of this technique is that it allows the use of a voltage source model with no series resistance. When this form of nodal analysis is used, there

62 ANALYSIS AND DESIGN OF ELECTRONIC CIRCUITS USING PCs

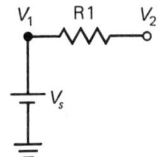

Fig. 3-22. Resistor connected to voltage source.

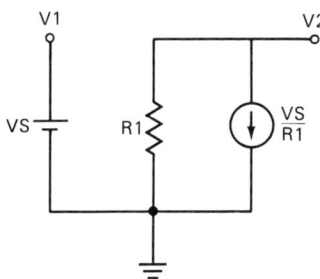

Fig. 3-23. Equivalent circuit of resistor connected to voltage source.

is not a simple representation of the voltage source in the nodal admittance equations; it is distributed among the elements that were connected to the voltage sources. This technique was used in some of the early simulators at Berkeley such as CANCER,[18] SLIC,[24] and SPICE1[26]; and is also in BIAS-D.[1]

If floating voltage sources are required, they must be modeled as current sources in parallel with a resistor, as in the original nodal analysis method. But floating voltage sources can also be modeled by additional modifications of the nodal analysis method. A modified nodal analysis (MNA) method developed by C. W. Ho et al.[24] is used in SPICE 2. The MNA method requires considerably more computational effort during the setup proceedure. If only grounded voltage sources are used in the circuit simulated, there is also considerable additional computational effort during the matrix solution as compared with the method used for grounded voltage sources.

3.4.6. Diode Model

A diode is a two-terminal nonlinear circuit element as shown in Figure 3.24. If a circuit contains nonlinear elements such as diodes or transistors, then the node voltages are determined by iteratively solving a set of linearized equations. The Newton-Raphson algorithm is the most popular method for linearizing these equations. Both PSpice and BIAS-D use the Newton-Raphson linearization method. The Newton-Raphson method requires an initial guess as to the operating point and a first derivative of the current-voltage equation evaluated

AC, DC, AND TRANSIENT CIRCUIT SIMULATION PROGRAMS   63

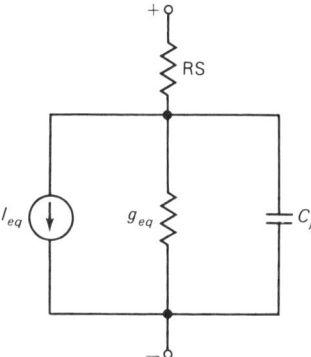

Fig. 3-24. Complete diode model.

at this operating point. An ideal diode has a voltage-current relationship given by

$$I_D = I_S(\exp(qV_D/nKT - 1)) \text{ amps} \qquad (3\text{-}33)$$

where I_S is the saturation current (amps), q is the electron charge (1.6×10^{-19} coulombs), V_D the anode-to-cathode voltage, k is Boltzmann's constant (1.38×10^{-23} joule/K), n is the emission coefficient (dimensionless, usually between 1 and 2), and T the junction temperature in degrees Kelvin (273 K = 0°C).

The saturation current, I_S, is sometimes confused with the diode leakage current specified in manufacturers' data sheets. The saturation current is proportional to the device junction area. Higher-power devices will have larger junction areas and therefore larger values of I_S. Typical values for I_S range from 1×10^{-16} for a low-power integrated device to 1×10^{-9} for a 50-watt discrete device. The saturation current cannot be measured directly; it is determined by a series of measurements with curve-fitting. A useful text for determining model parameters such as I_S is *Modeling the Bipolar Transistor* by Getreu.[25]

Diode leakage currents are typically 1000 times larger than I_S. These are caused by package leakage and surface contamination. They can be measured directly. This leakage current is modeled by a resistance, R_P, in parallel with the diode as shown later in the complete diode model (Figure 3-27.)

A linear current-voltage plot of the diode Equation 3-33 is shown in Figure 3-25. Shown in this plot is the Newton-Raphson "linearized" approximation of the diode in the forward biased region at an operating point I_0, V_0. In the forward biased region ($V_D \geq 0$) this approximation uses the familiar linear equation:

$$i_D = g_{eq}v_D + I_{eq} \qquad (3\text{-}34)$$

64 ANALYSIS AND DESIGN OF ELECTRONIC CIRCUITS USING PCs

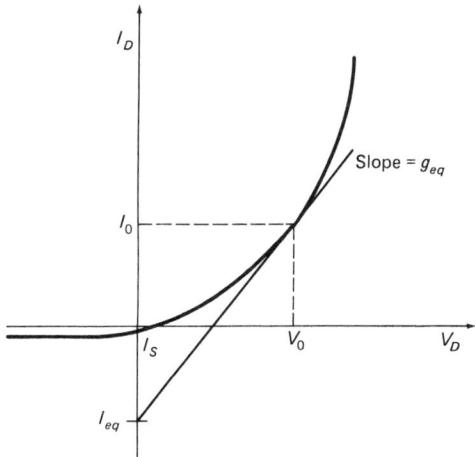

Fig. 3-25. Nonlinear and "linearized" V-I plot of diode equation.

G_{eq} is the slope of the diode equation at point V_0, I_0 and is obtained from the first derivative of the diode equation evaluated at V_0, I_0 as

$$g_{eq} = dI_D/dV_D \big|_{V_0, I_0} = q/KT * I_S[\exp(qv_D/KT)] \quad (3\text{-}35)$$

I_{eq} is the current intercept (y-axis) at $v_D = 0$.

In the reverse-biasd region ($v_D < 0$), the above approximation is still valid but represents computations that may be unnecessary. In fact, for values of v_D less than 1 volt, numerical underflow may occur in a digital computer. A model which uses the values established for g_{eq} and I_{eq} at $v_D = 0$ for all $v_D < 0$ is usually adequate.

The linearized model shown in Figure 3-26 represents Equation 3-35. This model is represented in the nodal admittance matrix equation form as

$$\begin{bmatrix} g_{eq} & -g_{eq} \\ -g_{eq} & g_{eq} \end{bmatrix} \begin{bmatrix} v_1 \\ v_2 \end{bmatrix} = \begin{bmatrix} -I_{eq} \\ I_{eq} \end{bmatrix} \quad (3\text{-}36)$$

which requires loading a conductance into four locations in the admittance matrix and a current into two locations in the current vector. This model is identical for the dc operating point analysis, transient analysis, and ac analysis. For ac analysis, the value of the equivalent current source I_{eq} is set to zero and therefore not loaded into the current vector.

Each time the circuit is iterated, a new linearized value for g_{eq} and I_{eq} is

AC, DC, AND TRANSIENT CIRCUIT SIMULATION PROGRAMS 65

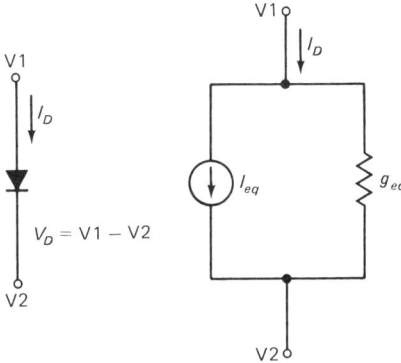

Fig. 3-26. Linearized diode model.

created. The solution is achieved when the values for V_0 and I_0 do not change more than, say, 0.1% for two or three iterations.

The model above represents an "ideal" diode. An actual diode has junction capacitance, transient time, series resistance, and breakdown voltage. The Spice 2G diode model is specified with 14 parameters.[3]

A complete diode model is shown in Figure 3-27. Where g_{eq} and I_{eq} are as described above, R_s is a series resistance primarily due to ohmic contact resis-

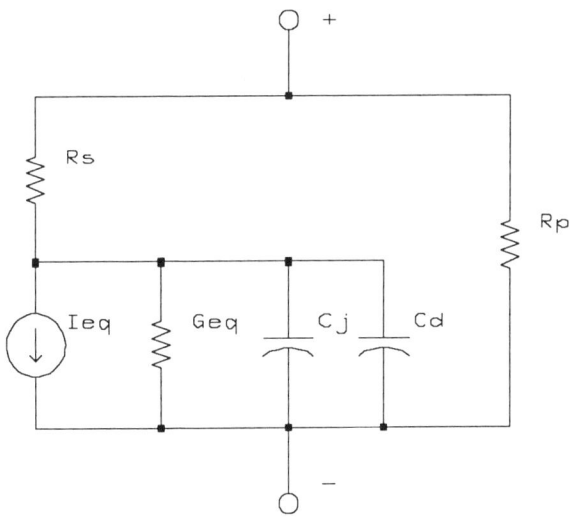

Fig. 3-27. Complete diode model.

tance, R_p is a parallel resistance due to the surface leakage mentioned earlier, C_j is a voltage-dependent junction capacitor evaluated at the operating point (V_0, I_0). The equation for C_j is given by:

$$C_j = C_{j0}[1 - v_D/\phi_b^{-m}] \qquad (3\text{-}37)$$

where C_{j0} is the capacitance at zero bias, ϕ_b is the contact potential and m a dimensionless coefficient with a value of between 0.3 and 0.5.

C_d is the capacitance due to minority carrier injection. The equation for C_d is given by

$$C_d = \left(\frac{q\tau I_S}{nKT}\right) \exp\left(qv_d/nKT\right) \qquad (3\text{-}38)$$

3.4.7. Transistor Models

Transistors are divided into two primary categories: bipolar junction (BJT) and field effect (FET). BJTs and FETs have both n and p polarity devices. Both polarities are usually modeled as identical devices with appropriate sign changes in the model equations.

Field effect transistors are further divided into junction (JFET) and metal oxide (MOSFET) devices. A listing of the different categories and types of transistor models available in PSpice is given in Table 3-5. Only the NPN and PNP BJT models are available in BIAS-D.

Category	Type	Description
BJT	NPN	n-type BJT
	PNP	p-type BJT
JFET	NJF	n-channel JFET
	PJF	p-channel JFET
MOSFET	NMOS	n-channel MOSFET
	PMOS	p-channel MOSFET

Table 3-5 Transistor Models Available in PSpice

Brief descriptions of the three different transistor models—BJT, JFET, and MOSFET—are given here.

Bipolar Junction Transistor. There are two levels of BJT models in SPICE 2: a modified Ebers-Moll model,[27] which is a level 1 model; or a modified Gummel-Poon model[28] which is a more advanced model. There are forty SPICE BJT model parameters which can be specified for the Gummel-Poon model.[7] If certain parameters are not specified, then the model defaults to the Ebers-Moll model. The basic Ebers-Moll *npn* transistor equations are given as

$$I_C = I_S[\exp(qV_{BE}/kT) - 1] - I_S(1 + 1/\beta_R)$$
$$\cdot [\exp(qV_{BC}/kT) - 1] \tag{3-39}$$

$$I_B = I_S/\beta_F[\exp(qV_{BE}/kT) - 1] + I_S/\beta_R[\exp(qV_{BC}/kT) - 1] \tag{3-40}$$

$$I_E = I_B + I_C \tag{3-41}$$

where I_S is the transistor saturation current, q the electronic charge, k is Boltzmann's constant, T is absolute temperature in degrees Kelvin, β_F the transistor dc forward current gain, and β_R is the transistor dc reverse current gain. For a *pnp* transistor, the equations are identical but the signs of I_C, I_B, and I_E are reversed. These equations can be more clearly understood if the transistor is drawn as a forward transistor (*npn*) superimposed on a reverse transistor as shown in Figure 3-28. Here each of the terms in Equations 3-38 through 3-40 can be represented as currents through either the forward or the reverse transistor. Parameters such as reverse beta, β_R, apply to parameters measured on the reverse transistor.

The large-signal model as represented by these Ebers-Moll equations is shown in Figure 3-29. where

$$I_{CC} = I_S[\exp(qV_{BE}/KT - 1] \tag{3-42}$$

$$I_{EC} = I_S[\exp(qV_{BC}/kT) - 1] \tag{3-43}$$

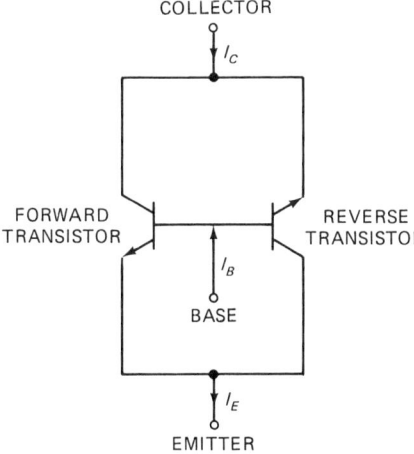

Fig. 3-28. Forward and reverse transistor representation.

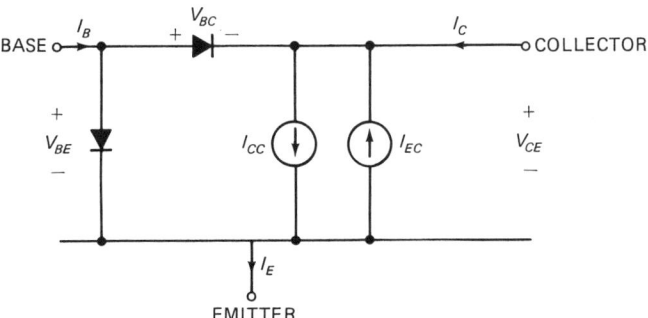

Fig. 3-29. Large signal BJT model.

The elements from the forward and reverse transistors can again be seen in this large-signal model.

The "linearized" equivalent of this large signal model is shown in Figure 3-30 where each junction diode is modeled as a resistance in parallel with a voltage-dependent current as was done with the diode model. The conductances g_m and g_π for the forward and reverse transistors are given as

$$g_{mf} = dI_C/dV_{BE} = q/kT * [\exp(qV_{BE}/kT)] \tag{3-44}$$

$$g_{mr} = dI_C/dV_{BC} = q/kT * [\exp(qV_{BC}/kT)] \tag{3-45}$$

$$g_{\pi f} = dI_B/dV_{BE} = g_{mf}/\beta_F \tag{3-46}$$

$$g_{\pi r} = dI_B/dV_{BC} = g_{mr}/\beta_R \tag{3-47}$$

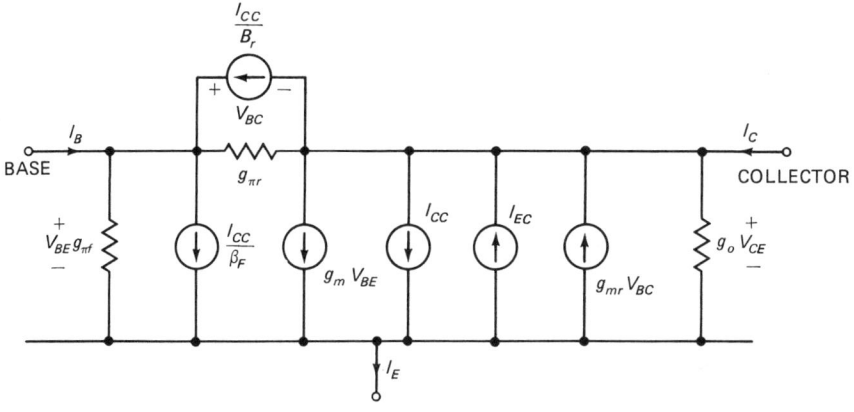

Fig. 3-30. Linearized BJT model.

AC, DC, AND TRANSIENT CIRCUIT SIMULATION PROGRAMS

The nodal admittance matrix equation for the BJT derived from the linearized model is given as:

$$\begin{bmatrix} (g_{\pi r} + g_{mr}) & (-g_{\pi r} - g_{mr} + g_{mf}) & (-g_{mf}) \\ (-g_{\pi r}) & (g_{\pi f} + g_{\pi r}) & (-g_{\pi f}) \\ (-g_{mr}) & (-g_{\pi f} - g_{mf} + g_{mr}) & (g_{mf} + g_{\pi f}) \end{bmatrix} \begin{bmatrix} V_C \\ V_B \\ V_E \end{bmatrix}$$

$$= \begin{bmatrix} (1 + 1/\beta_R) I_{EC} + I_{CC} \\ -I_{EC}/\beta_R - I_{CC}/\beta_R \\ (1 + 1/\beta_F) I_{CC} - I_{EC} \end{bmatrix} \quad (3\text{-}48)$$

This admittance matrix requires the addition of combinations of $g_{\pi f}$, $g_{\pi r}$, g_{mf} and g_{mr}, into nine locations in the admittance matrix for each transistor in the circuit. These are added for the dc operation point, transient, and small-signal ac analyses. The current vector requires loading combinations of I_{EC} and I_{CC} into three current vector positions for each transistor. This is done for dc and transient analyses. The current vector is zeroed for the ac analysis.

There are several additional elements included in the Ebers-Moll model in SPICE 2. Included are the collector base and base emitter junction capacitors, and series resistors R_C, R_B, and R_E in series with the collector, base and emitter of the transistor respectively. The complete model is shown in Figure 3-31. As

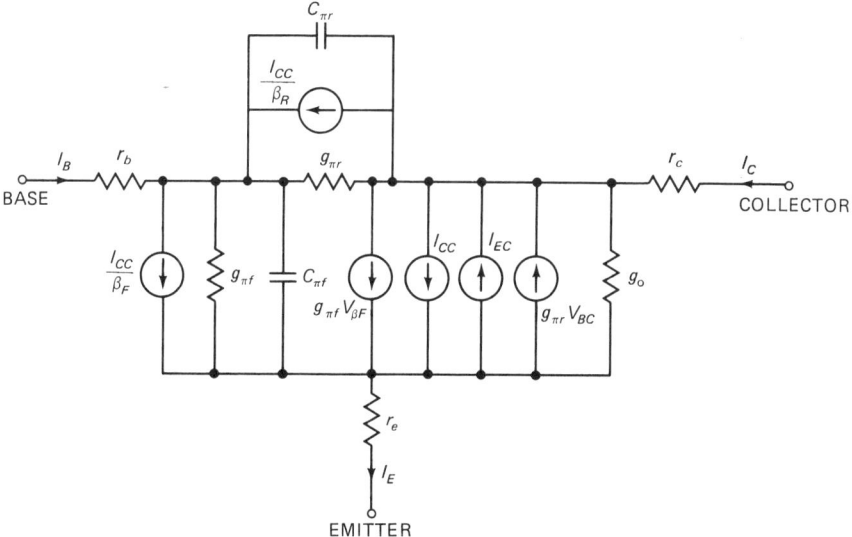

Fig. 3-31. Complete linearized BJT model.

can be seen in this figure, adding the collector base and emitter resistors effectively adds three nodes to the circuit for each transistor. This can increase the computation time significantly. The Gummel-Poon model is not described here but allows modeling effects such as current-dependent beta and f_T.

JFET Models. The junction field effect model used in SPICE is based on a modified Schichman-Hodges model.[29] The large signal Schichman-Hodges model for an *n*-channel JFET is shown in Figure 3-32. For the *p*-channel JFET model, the diodes and current source directions are reversed. The JFET is basically a symmetric device with the source and drain interchangeable. There are three regions of operation of the JFET: the cutoff region, resistance region, and the linear region. A plot of the drain-source curves for different gate voltages of an *n*-channel JFET is given in Figure 3-33. The large-signal equations for operation in these regions are given by the following equations:

Cutoff region:

$$I_D = 0 \qquad V_{GS} - V_{TO} < 0 \qquad (3\text{-}49)$$

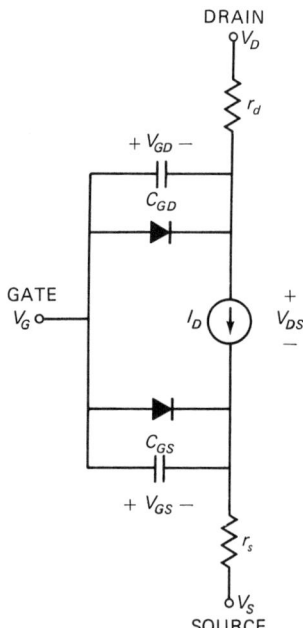

Fig. 3-32. Large signal JFET model.

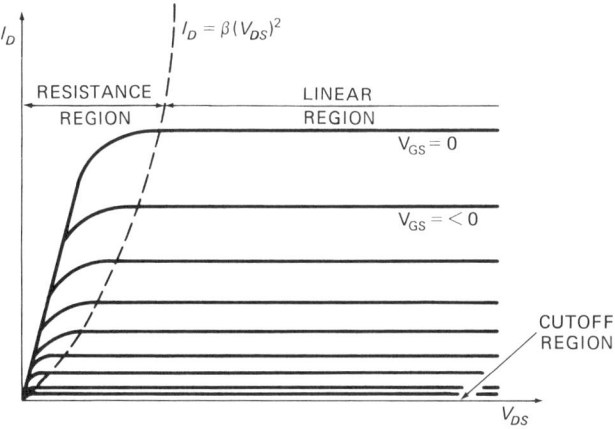

Fig. 3-33. Characteristic curves of n-channel JFET.

Resistance region:

$$I_D = \beta(V_{GS} - V_{TO})2 * (1 - \lambda * V_{DS}) \qquad 0 < V_{GS} - V_{TO} < V_{DS} \qquad (3\text{-}50)$$

Linear region:

$$I_D = \beta * V_{DS}[2 * (V_{GS} - V_{TO}) - V_{DS}] * (1 - \lambda * V_{DS}) \qquad (3\text{-}51)$$
$$0 < V_{DS} < V_{GS} - V_{TO}$$

Where β is transconductance, V_{TO} is the pinch-off voltage, and λ is the basewidth modulation term equivalent to the Early Voltage in BJT's. The gate current, I_G, is given by the current through the two-gate junction diodes as

$$I_G = I_S[\exp(qV_{GS}/kT) - 1] + I_S[\exp(qV_{GD}/kT) - 1] \qquad (3\text{-}52)$$

where I_S, k, and T are as defined for the BJT.

These diodes are normally reverse-biased and therefore only the I_S term is present in the model. These equations are presented here in the first quadrant, and they are also valid in the third quadrant.

The linearized JFET model which represents these equations is shown in Figure 3-34. The linearized model is identical for each of the three regions (See

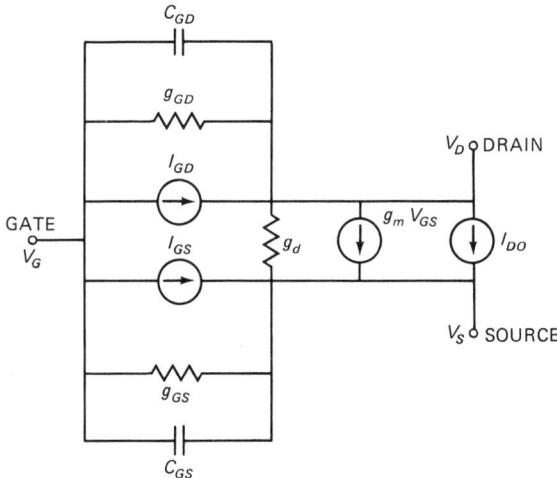

Fig. 3-34. Linearized JFET model.

Figure 3-33). However the values of g_m, g_d and I_{DO} are different in each region. For the resistance and linear regions these are given as

Resistance:

$$g_m = \left.\frac{\partial I_D}{\partial V_{GS}}\right|_{V_{GSO}, V_{DSO}} = 2 * \beta * V_{DSO} * (1 - \lambda * V_{DSO}) \quad (3\text{-}53)$$

$$g_d = \left.\frac{\partial I_D}{V_{DS}}\right|_{V_{GSO}, V_{DSO}} = 2 * \beta * \big[(V_{GSO} - V_{TO} - V_{DSO})$$
$$* (1 - \lambda * V_{DSO})$$
$$- \lambda * (V_{GSO} - V_{TO}) + 3/2 * \lambda * (V_{DSO})^2 \quad (3\text{-}54)$$

$$I_{DO} = I_D(V_{GSO}, V_{DSO}) - g_m * V_{GSO} - g_d * V_{DSO} \quad (3\text{-}55)$$

Linear Region: $\quad (3\text{-}56)$

$$g_m = \left.\frac{\partial I_D}{\partial V_{GS}}\right|_{V_{GSO}} = 2 * \beta * (V_{GSO} - V_{TO}) * (1 - \lambda * V_{DS}) \quad (3\text{-}57)$$

$$g_d = \left.\frac{\partial I_D}{\partial V_{DS}}\right|_{V_{GSO}} = \lambda * \beta * (V_{TO} - V_{GSO})^2$$

$$\begin{aligned} I_{DO} &= I_D(V_{GSO}, V_{DSO}) - g_m * V_{GSO} - g_d * V_{DSO} \\ &= \beta * (V_{TO})^2 - (V_{GSO})^2 \end{aligned} \quad (3\text{-}58)$$

The nodal admittance matrix equation representation of the JFET model, derived from this figure is given as:

$$\begin{bmatrix} (g_D + g_{GS} + g_m) & & -g_{GS} & -(g_m + g_D) \\ -g_{GS} & g_{GS} & & -g_{GD} \\ -g_D & -(g_m + g_{GD}) & (g_m + g_D + g_{GD}) \end{bmatrix} \begin{bmatrix} V_S \\ V_G \\ V_D \end{bmatrix}$$

$$= \begin{bmatrix} -(I_{DO} + I_{GG}) \\ -(I_{GS} + I_{GO}) \\ I_{GD} + I_{DO} \end{bmatrix} \quad (3\text{-}59)$$

Where g_{GD} and g_{GS} are the linearized gate diode resistors for the gate drain and gate source respectively. For dc and transient analyses, the admittance matrix in Equation 3-59 is added to the circuit nodal admittance matrix equation. The current vector in Equation 3-59 is also added to this equation. For a small signal frequency response, the current vector is zeroed and therefore only the admittance matrix portion is added. The equations used for the evaluation of g_m, I_{DO}, etc., in Equation 3-59 are determined by the region of operation of each device.

MOSFET Models. The MOSFET models in general are implemented as four terminal models. In addition to the gate, drain, and source terminals there is a substrate or bulk terminal which is present in all integrated devices. This terminal is more important in MOSFETs because a voltage bias between the bulk and source terminals affects the threshold voltage of the MOSFET. Three levels of MOSFET models are implemented in SPICE 2; 42 model parameters can be specified. Only the first order, or "level 1," model is described here. The level 1 MOSFET model implemented in SPICE 2 is a modified Schichman-

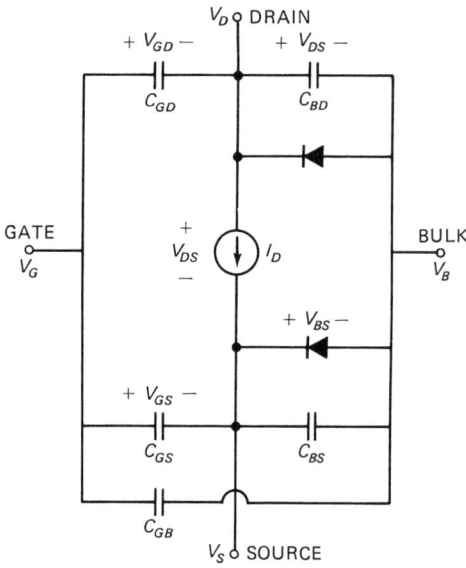

Fig. 3-35. Large signal MOSFET model.

Hodges model.[29] The large-signal representation of this model is shown in Figure 3-35. Three overlap capacitors, C_{GD}, G_{GS}, and C_{GB}, are modeled as linear capacitors in SPICE 2. The values of these capacitors are determined by the device geometries and oxide thicknesses. There are two junction diodes, one from bulk to drain and the other from bulk to source. The currents I_{BS} and I_{BD} through these diodes are given as

$$I_{BS} = I_S[\exp(qV_{BS}/kT) - 1] \tag{3-60}$$

$$I_{BD} = I_S[\exp(qV_{BD}/kT) - 1] \tag{3-61}$$

There is also a voltage dependent junction capacitor associated with each of these diodes; C_{BS} and C_{BD} given by

$$C_{BS} = C_{BDO}[1 - V_{BD}/\phi_b]^{-1/2} \tag{3-62}$$

$$C_{BD} = C_{BSO}[1 - V_{BS}/\phi_b]^{-1/2} \tag{3-63}$$

where C_{BDO} and C_{BSO} are the capacitances at zero junction bias voltage, and ϕ_b is the bulk junction potential.

Like the JFET, the MOSFET has three regions of operation: cutoff, resistance and linear. The drain-to-source current, I_D, in each region is given by the following equations.

Cutoff region:

$$I_D = 0 \quad V_{GS} - V_{TE} < 0 \qquad (3\text{-}64)$$

Resistance region:

$$I_D = \beta(V_{GS} - V_{TE})^2 * (1 - \lambda * V_{DS}) \quad 0 < V_{GS} - V_{TE} < V_{DS} \qquad (3\text{-}65)$$

Linear region:

$$I_D = \beta * V_{DS}[2 * (V_{GS} - V_{TE}) - V_{DS}] * (1 - \lambda * V_{DS})$$
$$0 < V_{DS} < V_{GS} - V_{TE} \qquad (3\text{-}66)$$

where V_{TE} is given by

$$V_{TE} = V_{T0} + \gamma[(\phi_b - V_{BS})^{1/2} - (\phi_b)^{1/2}] \qquad (3\text{-}67)$$

where γ is the bulk threshold parameter; ϕ_b, the bulk junction potential and V_{T0} the threshold voltage at $V_{BS} = 0$. Note that if the bulk terminal is shorted to the source, these equations (3-64 through 3-66) are identical to the JFET Equations 3-49 through 3-51. The linearized MOSFET model is shown in Figure 3-36. The equations for g_m and g_{DS} in the resistance and linear region are identical to the JFET Equations 3-53 through 3-57. g_{mBS} is given by

Resistance region:

$$g_{mBS} = d_{ID}/d_{VBS}\big|_{V_{GSO}, V_{DSO}, V_{BSO}} \qquad (3\text{-}68)$$

Linear region:

$$g_{mBS} = d_{ID}/d_{VBS}\big|_{V_{GSO}, V_{DSO}, V_{BSO}} \qquad (3\text{-}69)$$

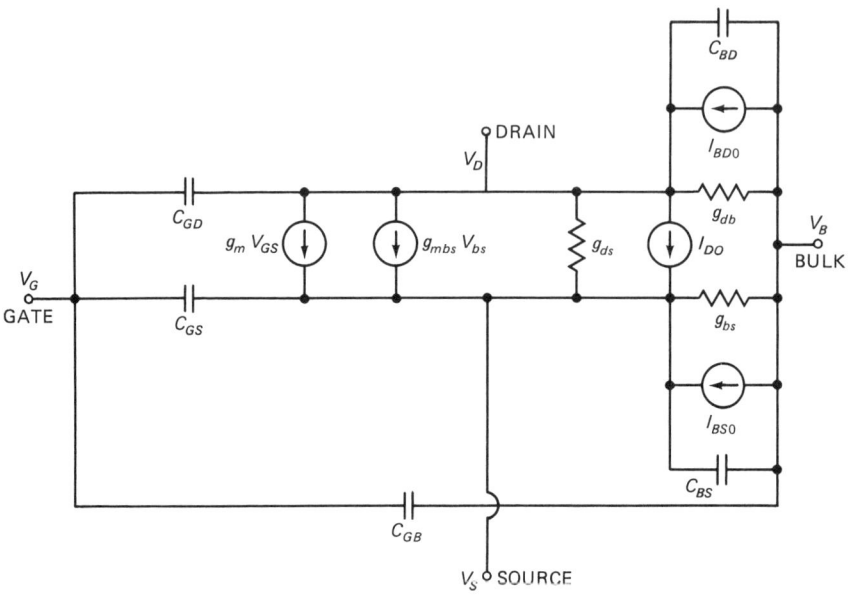

Fig. 3-36. Linearized MOSFET model.

The nodel admittance matrix equation for the MOSFET is not given here but is similar to that for the JFET given in Equation 3-59.

The admittance matrix for the MOSFET is the same for dc operating point, transient and small-signal frequency response analyses. For transient analyses the values of g_m, g_{DS} and g_{mBS} are dependent on the operating point voltages and therefore may vary for each timepoint. For small-signal frequency response the current vector is zeroed.

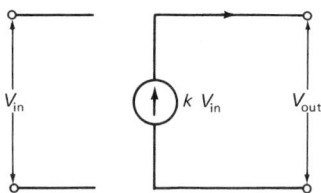

Fig. 3-37. Ideal operational amplifier macromodel.

AC, DC, AND TRANSIENT CIRCUIT SIMULATION PROGRAMS 77

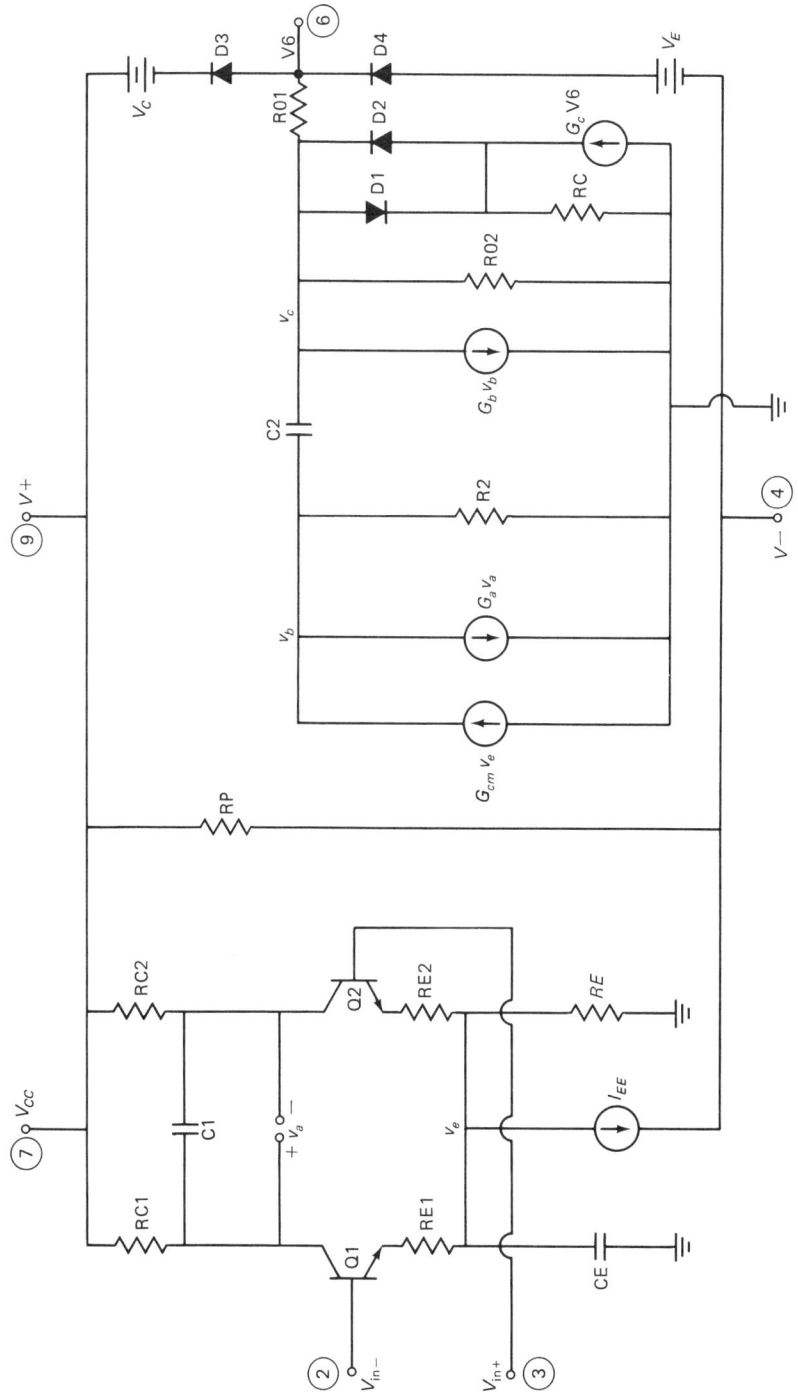

Fig. 3-38. Circuit diagram of a 741 operational amplifier macromodel.

78 ANALYSIS AND DESIGN OF ELECTRONIC CIRCUITS USING PCs

3.4.8. Macromodels

The term *macromodel* is most commonly used to represent a model which "simplifies" a more complex circuit function. The primary reason to create and use a macromodel is to decrease the circuit complexity and thereby increase the simulation speed.

Many macromodels have been proposed, both for analog and digital circuits;[30] some have been moderately successful but perhaps the most successful has been a macromodel for the operational amplifier by Pederson et al.[31]

The basic macromodel for the ideal operational amplifier is the voltage-controlled current source as shown in Figure 3-37. Where the output voltage

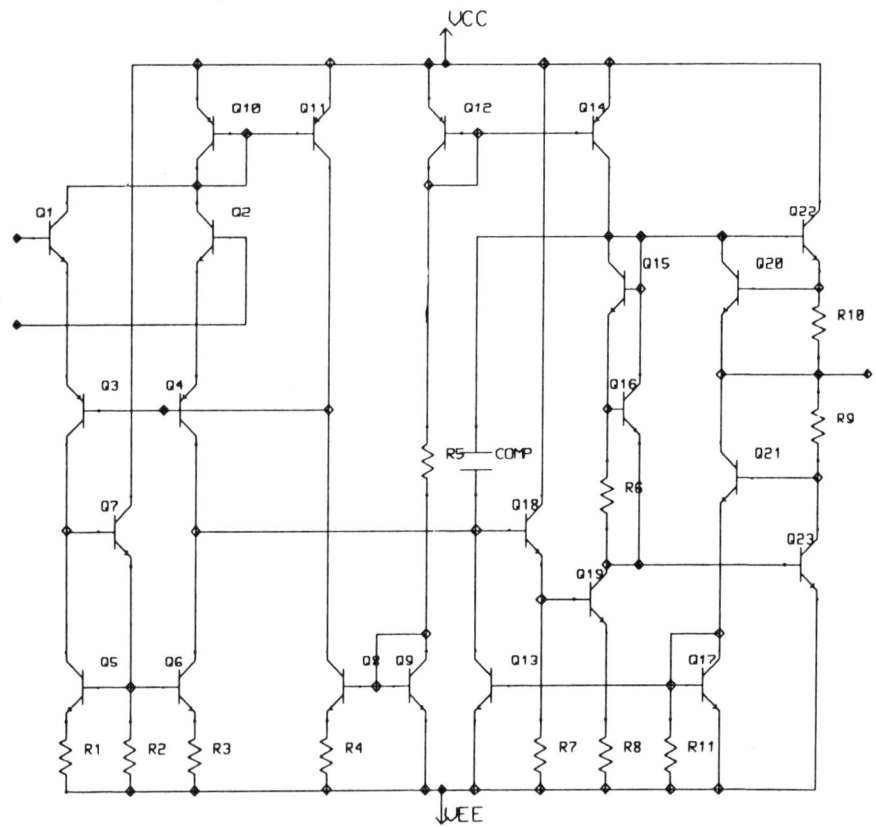

Fig. 3-39. Circuit diagram of a 741 operational amplifier.

AC, DC, AND TRANSIENT CIRCUIT SIMULATION PROGRAMS 79

V_{out} is proportional to a constant k times the difference in input node voltages ($V_1 - V_2$). Circuit elements can be added to model such parameters as finite input impedance and nonzero output impedance, but this basic model is still valid only for small-signal frequency response applications.

The macromodel proposed in Reference 31 is a near "black box" equivalent circuit of the 741 class of operational amplifiers (op amp). It models amplifier parameters such as high-frequency rolloff, slew rate, output limiting, and power dissipation. This op amp macromodel can be used for small-signal frequency response, dc operating point, and transient simulations. About the only limitation of this class of models is that it can only model operational amplifiers used with both positive and negative supplies. This is due to the internal ground references within the model.

A circuit diagram of this macromodel is given in Figure 3-38. Indicated in this figure are the input, interstage and output stages of the macromodel. Figure 3-39 shows a circuit diagram of the actual 741 operational amplifier. The circuit elements used in the macromodel are those available in most simulators, including PSpice. All the circuit element values and the transistor parameters

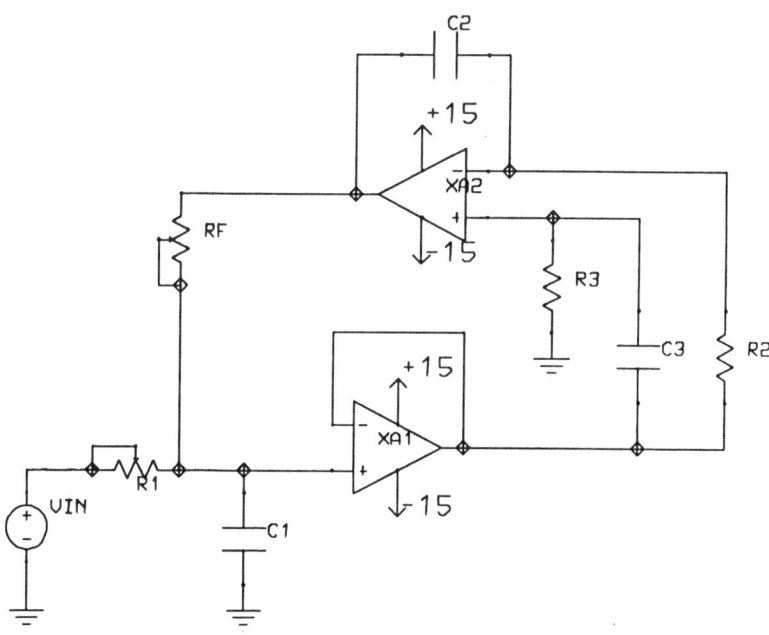

Fig. 3-40. Circuit diagram of an active filter example circuit.

80 ANALYSIS AND DESIGN OF ELECTRONIC CIRCUITS USING PCs

for this macromodel can be calculated from information obtained from manufacturers data sheets. Detailed equations with examples are given in Reference 31.

One advantage a computer simulator has over a breadboard is component selection. The circuit element values required for this macromodel in most cases

```
Jan 17 16:38 1986   ua741.dat  Page 1

* Active Filter Using UA 741 Macromodel
* Macromodel definition
.SUBCKT UA741   2  3  4  6  7
*     Node     Function    Node    Function
* ------------------------------------------
*       2      - Input      6      Output
*       3      + Input      7      V+
*       4      V-
* ------------------------------------------
Q1      17   2    1 M1
Q2      18   3    5 M2
RC1     17   7    5.305K
RC2     18   7    5.305K
C1      17  18    5.46PF
CE      16   0    2.41PF
RE1      1  16    2.712K
RE2      5  16    2.712K
RE      16   0    9.872MEG
IE      16   4    20.26UA
GCM      0  19   16  0  6.28N
GA      19   0   17 18  198.6U
R2      19   0    100K
C2      19  20    30PF
GB      20   0   19  0  247.49
RM      20   0    42.87
D1      20  21    DSD
RC      21   0    .02129MOHM
D2      21  20    DSD
GC       0  21    6  0  46.964K
RO      20   6    32.13
D4      23   6    DSD
D3       6  22    DSD
VE      23   4    2.303
VC       7  22    1.803
.MODEL M1 NPN BF=111.67 IS=8.000E-16
.MODEL M2 NPN BF=143.57 IS=8.309E-16
.MODEL DSD D IS=8.0E-16
* END OF MACROMODEL
.ENDS UA741
V+      12   0   DC  15
V-      10   0   DC -15
* Bandwidth adjust (R1 Maximum = 318K Ohm )
R1       1   2   10K
* Frequency adjust (RF Maximum = 31.8K Ohm)
RF       2   3   25K
C1       2   0   .1UF
C2       4   3   .001UF
R3       5   0   7.96K
XA2      4   5  10   3 12 UA741
R2       4   6   7.96K
C3       5   6   .001UF
XA1      7   2  10   6 12 UA741
RSHORT   6   7   .01
VIN      1   0   DC 0 AC 1
.PLOT AC VM(6) VP(6)
.AC DEC 20 100 10KHZ
.END
```

Fig. 3-41. SPICE input listing for active filter example circuit.

AC, DC, AND TRANSIENT CIRCUIT SIMULATION PROGRAMS   81

could not be obtained from a parts box. As an example, for a UA741 macromodel, transistor Q1 requires a beta of 111.67 and transistor Q2 a beta of 143.57. Different values of betas would affect the op amps input bias and input offset currents.

The complexity of this macromodel has reduced an 81-node circuit (including internal transistor nodes) with 193 branches to a 16-node circuit with 28 branches. The speed improvement is between 6 and 10 to 1.

A circuit diagram of an active bandpass filter using a UA741 macromodel is shown in Figure 3-40. A SPICE listing for this filter example is given in Figure 3-41. This listing includes the SPICE subcircuit for the UA741 operational amplifier macromodel. The frequency vs. magnitude small-signal frequency response for this circuit is shown in Figure 3-42. This was plotted using the PROBE feature of PSpice. Figure 3-43 shows the phase response for the same circuit.

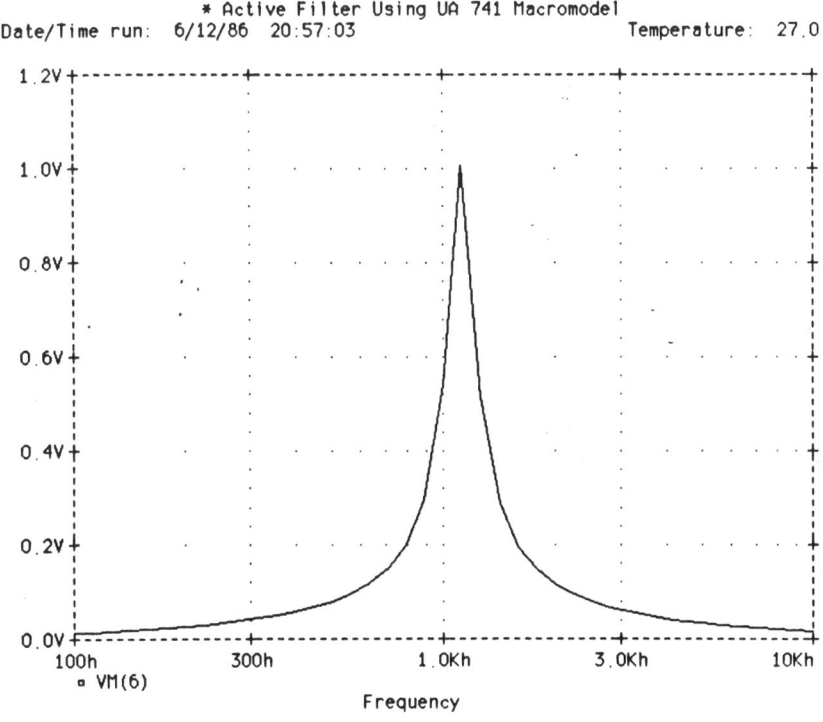

Fig. 3-42. Magnitude vs. frequency response of active filter using PSpice.

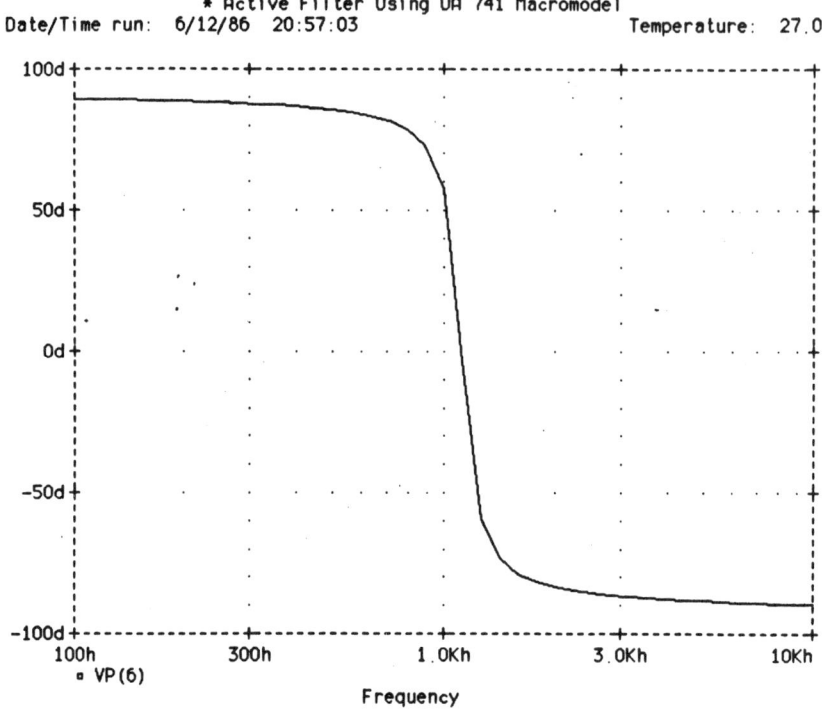

Fig. 3-43. Phase vs. frequency response of active filter using PSpice.

REFERENCES AND NOTES

3.1. Introduction

1. Biehl, B. L. "BIAS-D: A Semi-Interactive Circuit Analysis Program for Desktop Calculators and Minicomputers." Eighth Annual Asilomar Conference on Circuits, *Systems and Computers*, Dec. 1974.
2. MicroSim Corporation, 23175 La Cadena Drive, Laguna Hills, CA 92653. Attn: Paul W. Tuinenga, Vice President.
3. Nagel, W. "SPICE 2: A Computer Program to Simulate Semiconductor Circuits," Electronics Research Lab, ERL-M520, University of California, Berkeley, May 1975.
4. Thompson, A. V. "CAE Tools for Personal Computers." *Proceedings IEEE International Symposium on Circuits and Systems*, May 1984.
5. Newton, A. R., and G. L. Taylor, "BIASL.25, A MOS Circuit Simulator." Tenth Annual Asilomar Conference on Circuits, Systems and Computers, Nov. 22-24, 1976.

3.2. PC Capabilities

6. Biehl, B. L. "Circuit Simulation on Minicomputers." Tenth Annual Asilomar Conference on Circuits, Systems and Computers, Nov. 22-24, 1976.

7. Biehl, B. L., and C. W. Cairns, "BIAS-D as a Microcomputer-Based Circuit Simulator." Twelfth Annual Asilomar Conference on Circuits, Systems and Computers, Dec. 1978.
8. Biehl, B. L. *Interactive Electronic Circuit Simulation on Small Computer Systems*. Harry Diamond Labs, HDL-TM-79-30, 120 pp. Nov. 1979. Available from the Defense Documentation Center, Alexandria, VA.
9. Cohen, E. *Program Reference for SPICE 2*. Electronics Research Laboratory, ERL-M592, University of California, Berkeley, Jun. 1976.
10. Microsoft Corp., 16011 NE 36th Way, Box 97017, Redmond WA, 98073.
11. Tetewsky, A., "Benchmarking FORTRAN Compilers," *Byte*, 217, Feb. 1984.

3.3. Architecture

12. Ralston, A., *A First Course in Numerical Analysis*. New York: McGraw-Hill, 1965.
13. Markowitz, H. M. "The Elimination Form of the Inverse and Its Application to Linear Programming," *Management Sci.* 3, 255-269, Apr. 1952.
14. McCalla, W. J., and W. G., Howard, Jr., "BIAS-3—A Program for the Nonlinear Dc Analysis of Bipolar Transistor Circuits." *IEEE J. Solid-State Circuits*, SC-6, 14-19, Feb. 1971.
15. Idleman, T. E., F. S. Jenkins, W. J. McCalla, and D. O. Pederson. "SLIC—A Simulator for Linear Integrated Circuits." IEEE J. Solid-State Circuits, Vol. SC-6, 188-204, Aug. 1971.
16. *1620 Electronic Circuit Analysis Program (ECAP)*. 1620-EE-02x, IBM Application Program File H20-0170-1, 1965.
17. Nagel, L. W., and D. O. Pederson, "Simulation Program with Integrated Circuit Emphasis." Proc. Sixteenth Midwest Symposium on Circuit Theory, Waterloo, Canada, April 12, 1973; also Memorandum No. ERL M382, Electronics, Research Laboratory, University of California, Berkeley.
18. Nagel, L.W., and R. A. Rohrer. "Computer Analysis of Nonlinear Circuits, Excluding Radiation (CANCER)." *IEEE J. Solid-State Circuits*, SC-6, 166-182, Aug. 1971.
19. Johnson, E. D., C. T. Kleiner, L. R. McMurray, E. L. Steele, and F. A. Vassallo. "Transient Radiation Analysis by Computer Program (TRAC)." Autonetics Division, North American Rockwell Corp., Anaheim, CA, June 1968.
20. Jenkins, F. S., and S. P. Fan. "TIME—A Nonlinear Dc and Time-Domain Circuit Simulation Program." *IEEE J. Solid State Circuits*, Vol. SC-6, Aug. 1971, pp. 192-198.

3.4. Modeling

21. Chua, L. O., and Pen-Min. Lin. *Computer-Aided Analysis of Electronic Circuits*. Englewood Cliffs NJ: Prentice-Hall Inc., 1975.
22. Dommel, H. W., "Digital Computer Solution of Electromagnetic Transients in Single and Multiple Networks." *IEEE Transactions Power and Appar. Systems*, PAS-88, 378-385, Aug. 1970.
23. Bokhoven, V., "Linear Implicit Differentiation Formulas of Variable Step and Order", *IEEE Transactions on Circuits and Systems*, Vol. CAS-22, pp. 109-115.
24. Ho, C. W., A. E. Ruehli, and P. A. Brennan. "The Modified Nodal Approach to Network Analysis," Proc. 1974 International Symposium on Circuits and Systems, San Francisco, Apr. 1974.
25. Getreu, I., *Modeling The Bipolar Transistor*, Tektronix, Inc. Copyright 1976, Part No. 062-2841-00.
26. Vladimirescu, A., K. Zhang, A. R. Newton, D. O. Pederson, and A. Sangiovanni-Vincen-

telli, *SPICE Version 2G Users Guide*, Aug. 10, 1981, Dept of Electrical Engineering and Computer Science, University of California, Berkeley.
27. Ebers, J. J., and J. L. Moll, "Large Signal Behavior of Junction Transistors," *Proc. IRE*, 42, Dec. 1954.
28. Gummel, H. K. and H. C. Poon. "An Integral Charge Control Model for Bipolar Transistors." *Bell Systems Technical Journal*, Vol. 49, May/June 1970, pp. 827-852.
29. Schichman, H. and D. A. Hodges. "Modeling and Simulation of Insulated-Gate Field-Effect Transistor Switching Circuits." *IEEE J. Solid-State Circuits*, Vol. SC-3, Sept. 1968, pp. 285-289.
30. Greenbaum, J. R., "Digital-IC models for Computer-aided Design." *Electronics*, 46(25) (Dec. 6, 1973), pp. 121-125.
31. Boyle, G. R., B. M. Cohen, D. O. Pederson, and J. E. Soloman. "Macromodeling of Integrated Circuit Operational Amplifiers," *IEEE J. Solid-State Circuits*, Vol. SC-9, Dec. 1974, pp. 353-363.

4

Computer-Aided Design at Microwave Frequencies

Les Besser
Besser Associates, Inc.

4.1. HISTORICAL REVIEW

The computerization of the microwave design industry has been relatively slow compared to the other EE design disciplines. Traditionally, microwave designers came from the background of "artists and magicians" who achieved results by tuning circuits on the bench instead of using a systematic analytical approach. In the 1960s, the introduction of Hewlett-Packard's s-parameter test instruments revolutionized the industry, and soon related design techniques became available using measured data directly. The available computers of the 1960s were large and extremely awkward to use and, consequently, were rarely utilized in microwave circuit design.

By the late 1960s, the newly formed timeshare industry introduced more convenient operating systems and languages for engineers. Soon after, several software packages were also developed, and became commercially available to microwave designers. Unfortunately, due to relatively high expense and the resistance of management, progress was extremely slow. Circuits were still "designed" on the bench.

In the early 1970s, some of the software packages previously offered through timesharing became available for in-house computer installations. Several of the progressive companies began to recognize the advantages of the computer-aided approach and installed these programs on their own computers. However, in most cases the computers were initially bought by accounting or by production and were not suitable for scientific design. Availability was also a problem; the engineers had to take a second role in usage, and the early days of in-house computing were not as successful as most engineers had desired.

In the late 1970s, hardware manufacturers introduced the minicomputer and companies began to purchase them for dedicated scientific usage. Unfortunately, software development for microwave design did not keep up with this

86 ANALYSIS AND DESIGN OF ELECTRONIC CIRCUITS USING PCs

progress. The worldwide "software industry" consisted of a relatively small firm whose product, COMPACT, was the only program commercially available to microwave circuit designers. Still, progress was made and the concept of microwave CAD was slowly getting established.

By 1980, most microwave design managers recognized the importance of using computers; still the relatively large investment in computers and peripherals was extremely hard to justify. Very few companies offered engineers the luxury of individual computer terminals. In most cases, designers had to share terminals, often located elsewhere in their buildings. Despite all of these hardships, engineers began to rely more and more on CAD, and widespread usage was only a step away.

Table 4-1. Program Availability.

Program	Vendor and Address	Phone No.
AUTOART	Compact Software Inc. 483 McLean Blvd. Paterson, NJ 07504	201-881-1200
CLD	EEsof 5795 Lindero Canyon Rd. Westlake Village, CA 91362	818-991-7530
ECM	EEsof, as above.	
ESYN	EEsof, as above.	
FILSYN	DGS Associates 1353 Sarita Way Santa Clara, CA 95051	408-554-1469
IDD	EEsof, as above.	
LINECALC	EEsof, as above.	
MICAD	EEsof, as above.	
MONTE CARLO	EEsof, as above.	
MSTRIP+	Microwave Software P.O. Box 764 San Juan Capistrano, CA 92693	714-493-9501
MICROWAVESPICE	EEsof, as above.	
SYNMAT	RADOM Inc. 1035 Justin Place Meridian, ID 83642	208-323-0318
SUPERCOMPACT PC	Compact Software, as above.	
TOUCHSTONE TOUCHSTONE RF TOUCHSTONE SR	EEsof, as above.	

About the same time, personal computers began "popping up" in various parts of design departments; quite often engineers brought in their own personal machines so that they would have something available at all times. Again, software was not available. There were too many types of small computers and none of them had offered anything to microwave CAD. As in the case of the minis, engineers began to write their own programs, which of course was not the answer to the need. Generally, the in-house programs were undocumented and far from being user-friendly.

Finally IBM entered the market with a truly high-quality 16-bit personal computer and everything changed. The scientific community rapidly accepted the product and more and more engineers gained access to these machines. This was soon recognized by entrepreneurs and within a short time several CAD programs were offered for the IBM PCs and their clones. By then, even conservative microwave managers found it difficult to decline committing $10–20,000 to purchasing a PC and all the related software. Finally engineers had their dreams fulfilled—they could design circuits on their own computers.

The rest of this chapter will introduce several of these new computer programs and give examples for their applications. Some of these programs are also available for the HP 200 and 300 series desktop computers as well as the more popular portable machines, such as the Compaq and Corona. Table 4-1 indicates the sources of availability.

4.2. CIRCUIT ANALYSIS

Most of the pioneer microwave CAD programs were based on two-port manipulations, where the program created the equivalent two-port parameters of each circuit element, converted the two-port parameters to the most appropriate type for handling interconnections, (y-parameters for parallel, z-parameters for series, etc.) and finally printed results for the overall circuit. These programs had significant limitations, since they could only handle circuits allowed by the two-port type of interconnections. The next generation of programs offered in addition nodal or n-port analysis, with the capability of handling virtually unlimited circuit connections. This second approach generally required more computer time which became critical when the analysis was within an iterative loop. The optimum solution is a combination of both techniques, wherein the designer may choose the appropriate method dictated by needs. Outputs may be provided in tabular or in graphical form.

4.2.1. Capabilities

A high-frequency/microwave analysis program must include the following capabilities:

1. Accepting two-, three-, and four-port, measured s-, y-, and z-parameters (polar and rectangular forms); and measured noise parameters for various device configurations.
2. Conversion of the various types of parameters listed in item 1.
3. Flexible and efficient interconnection schemes.
4. A library of commonly used lumped and distributed circuit elements. The later ones may be defined either by electrical parameters (Z_0, θ) or by physical dimensions realized in several forms (microstrip, stripline, etc.)
5. Noise analysis with lossy and lossless matching and feedback components.
6. Computing stability.
7. Plotting rectangular and Smith Chart outputs, circles of stability, constant gain and constant noise.
8. Impedance mapping.
9. Element "tuning."
10. Accepting dependent parameters and functional inputs and element labels.
11. Various Q definitions and physical transmission line and dielectric losses.
12. Dispersive transmission line models and discontinuities.
13. Directional elements (couplers, circulators, etc.).
14. Transistor noise and s-parameter databanks.

4.2.2. Commercial Circuit Analysis Programs

The following will introduce the two most popular commercially available, general-purpose microwave CAD packages, TOUCHSTONE™ and SUPER-COMPACT™. Both of these programs are available for the IBM and HP personal computers, costing about $10,000, depending on the options selected.

TOUCHSTONE. This was the first major microwave CAD program offered on the IBM PC. It includes the capabilities outlined in Section 2.1 and accepts circuit files generated by its "sister" programs LINECALC™, ESYN™, and MICROWAVESPICE™. TOUCHSTONE can also pass files to MONTE CARLO™ and MICAD™ for tolerance analysis and circuit layout, respectively. A unique application can be seen through the use of MICROWAVESPICE™ to generate temperature and signal-level dependent s-parameters for a microwave transistor, which may then be used by TOUCHSTONE as part of a complex microwave network.

"Tuning" with TOUCHSTONE. Figure 4-1 illustrates the schematics of a feedback amplifier and the corresponding TOUCHSTONE input file, and Figure

```
! File: BJT0223A.CKT
! From Gonzalez book, page 223
! 0.01-1.5GHZ S-parameters
CKT
   S2PA  2 3 0   BJT0223
   RES   2 8   R\ 280
   IND   8 3   L\ 6.9
   DEF2P 2 3   FBA
   CAP   1 0   C\ 1.24
   IND   1 2   L\ 4.47
   FBA   2 3
   RES   3 0   R=300
   CAP   3 0   C\ 1.59
   IND   3 5   L\ 3.66
   DEF2P 1 5   FBAMP
OUT
   FBAMP DB[S21]  GR1
   FBAMP DB[S11]  GR2
   FBAMP DB[S22]  GR2
FREQ
   SWEEP 0.01  1.51  .05
GRID
   RANGE 0  1.5  .25
   GR1   7  13   1
   GR2  -30  0   5
```

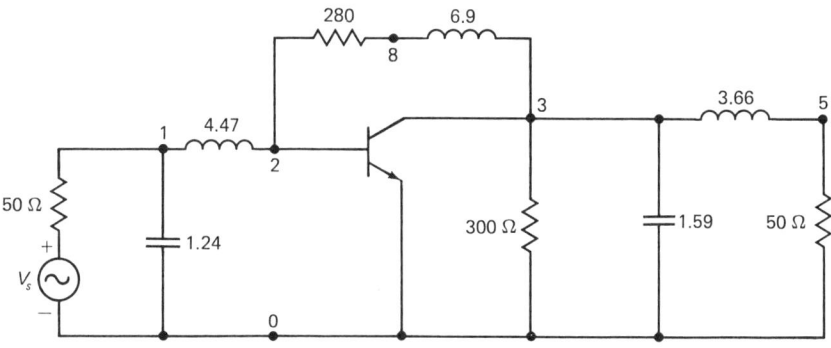

Fig. 4-1. Schematics and the corresponding TOUCHSTONE data file of a 10 KHz-1500 MHz feedback amplifier. Component values have been optimized for maximum gain flatness.

4-2 the resultant family of plots generated by the "Tune" mode of the program. In this example, the parallel feedback resistor value was changed from 180 to 480 ohms to examine the effect on the amplifier's gain. The optimum feedback resistor value is 280 ohms.

Circuit optimization, as seen later, offers simultaneous tuning of several components by the computer. Although optimization is faster and more practical, "manual" tuning often allows the designer to gain insight into component sensitivities that are otherwise buried inside the algorithms.

90 ANALYSIS AND DESIGN OF ELECTRONIC CIRCUITS USING PCs

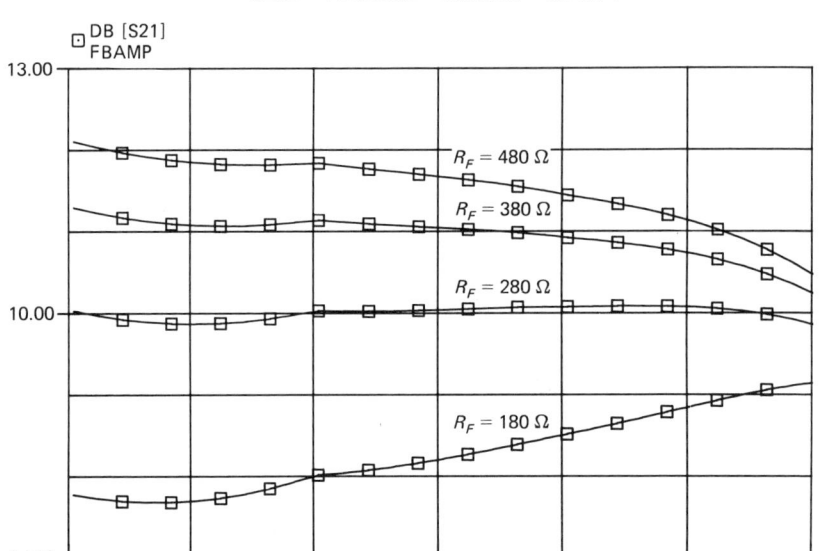

Fig. 4-2. Manual "tweaking" of the feedback resistor by the "TUNE" mode of TOUCHSTONE. Optimum gain response at the 10dB level is reached with a 280-ohm resistor.

```
* Fc=.8GHz
LAD
    TWO 1 2 Q1
* Stabilizing Resistor
* RES 2 0 R=500
AMP: 2POR 1 2
END
FREQ
    STEP .5GHZ 1.1GHZ .1GHZ
END
OUT
    PLO AMP S
END
DATA
    Q1: S
        .8GHZ   0.65 -95   0.035 40   5 115   0.8 -35
END
```

Fig. 4-3. SUPERCOMPACT data file of a transistor with a parallel resistor at the output. The 500-ohm resistor provides borderline stability factor at 800 MHz.

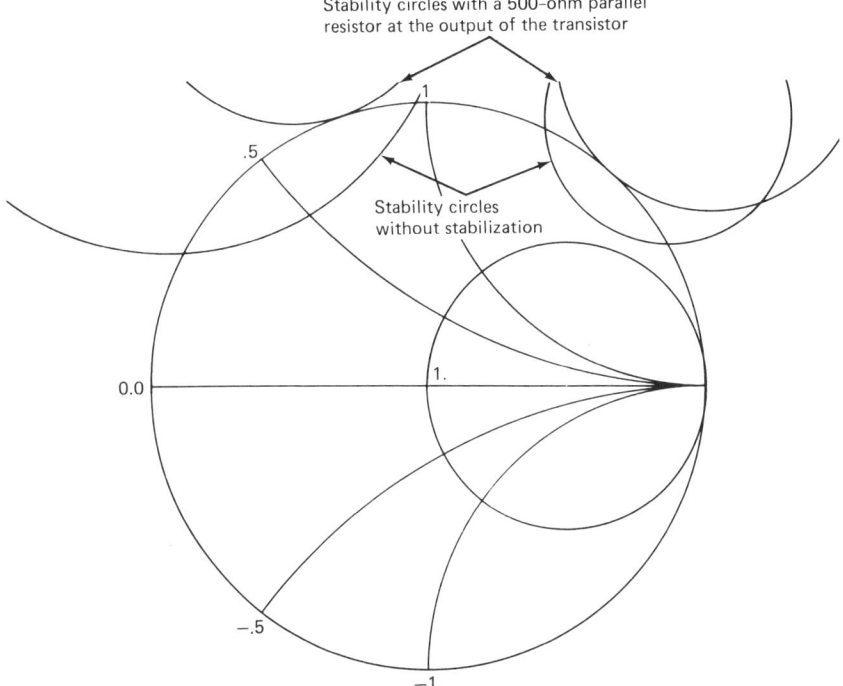

Fig. 4-4. Input and output stability circles with and without the stabilizing resistor. Note that the resistor moves *both* circles outside the Smith Chart.

SUPERCOMPACT PC.

The second major program is SUPERCOMPACT PC, a converted version of SUPERCOMPACT™, which has already been available for larger computers. SUPERCOMPACT PC includes subprograms for statistical analysis, synthesis, Lange-coupler design, and transistor modeling via the Negative Image design technique (Section 4.7). The program can pass data to AUTOART™ for circuit layout and mask generation.

Stability Analysis with SUPER-COMPACT PC.

Some of the capabilities

```
       SUPER COMPACT PC                              08/05/86

       FREQ        K         B1       MSG         K       B1      GMAX
       MHZ                             DB                          DB
                  WOSTB    WOSTB     WOSTB      WSTB     WSTB     WSTB
        800       0.55        +      21.55      1.02       +     20.77
```

Fig. 4-5. Adding a 500-ohm resistor to the transistor increases the stability factor to 1.02. When $K > 1$ and the B1 factor is positive, the transistor is unconditionally stable at a specific frequency.

of SUPER-COMPACT PC are introduced through the resistive stabilization of a potentially unstable transistor. The circuit file is shown in Figure 4-3 and the resultant stability circle plot in Figure 4-4, indicating that parts of the Smith Chart include potentially unstable terminations. After a 500-ohm parallel resistor is added to the output of the device (nodes 2 and 0), the stability circles move away from the Smith Chart and the tabulated output verifies unconditional stability ($k > 1$) within the 200–400-MHz band (Figure 4-5).

4.3. CIRCUIT SYNTHESIS

Synthesis can be viewed as a reverse procedure to analysis. The user inputs the electrical specifications, which are converted inside the program to resolve the circuit topology and the exact component values. Synthesis is a highly efficient design approach; however, it has severe limitations. In general, synthesis accepts only resistive source and load terminations, usually limits topologies to ladder connections and idealized (lossless) elements, and in most cases, the user must choose between the lumped and the distributed design approaches. Synthesis also requires a more sophisticated user, and so far has not gained much acceptance among designers. Still, it is a powerful design tool, and as soon as the programs become more simple and user-friendly, should be used more frequently.

One could differentiate between two basic types of microwave circuit synthesis: filter and matching network synthesis. The fundamental difference is that in the former, both passband and stopband specifications are given, while for matching networks, only the passband is specified. This difference seemingly makes matching network design much simpler; however, in the passband specification, additional requirements are present which do not exist in a typical filter design (gain-slope compensation, etc.). Also, terminations may be complex, uneven, and frequency dependent.

Since synthesis is restricted to resistive terminations, designers of active microwave circuits must first separate the impedances into real and imaginary parts. The real parts are used for terminations, and the reactive parts are temporarily placed into the circuit. At the end of the synthesis procedure, these reactive portions need to be transferred back to the appropriate terminations where they are physically present (Figure 4-6).

4.3.1. The FILSYN Program

FILSYN™ is perhaps the largest and most comprehensive commercially available filter design program for personal computers. The program can synthesize virtually all classes of filters, using active, passive digital, and microwave (distributed) realization. Its PLACER routine can find the optimum location of movable transmission zeros for multiple stopband specifications. Group delay

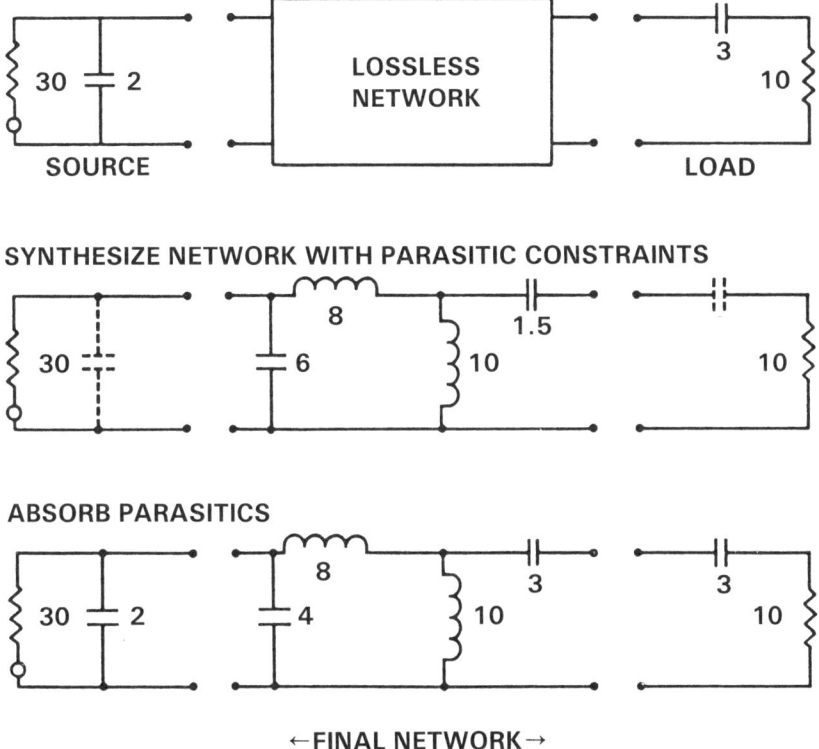

Fig. 4-6. Parasitics at the source and load are temporarily moved to the circuit, leaving purely resistive terminations for the synthesis.

equalization, Norton and Kuroda impedance transformations and predistortion are also included.

Two examples are shown to illustrated FILSYN's capabilities. The first one is realized with lumped elements; the second with commensurate-length transmission lines.

4.3.2. 70-MHz Bandpass Filter Synthesis with FILSYN

This example illustrates the design of a 70-MHz parametric bandpass filter, consisting of capacitively coupled parallel resonant circuits. Input and output impedances of the filter are 50 ohms. The required equal-ripple passband is 0.1 dB. The filters will have five resonant tank circuits; the inductors within the tank circuits should all have the same values.

The first task is to interactively enter the desired data by specifying the filter type, the bandwidth, the transmission zero types, and the input/output terminations, as shown in Figure 4-7.

At this point, FILSYN verifies the data and computes the overall network order, which is degree of 12. Next, the user requires the computer-generated topology as shown in Figure 4-8. The program has already scaled the circuit to the desired corner frequency and source impedance. However, impedance scaling has control over only the input termination. Generally, the user must reach the desired load termination by some form of impedance transformation. In this example, elements 12 and 13 offer Norton impedance transformation to a larger termination ratio, leading to a higher load impedance. A choice of T- or a Π-network is available to replace the two-capacitor section of elements 12 and 13, creating a new parallel branch between elements 11 and 12 (PI-option). The new parallel capacitor can be combined with element 10 and the impedance transform again with capacitor 9. This process can be repeated until the desired load impedance is reached. The circuit may also require some manipulation for

```
ENTER TITLE
> capacitively coupled resonator bandpass
FILTER KIND - LUMPED: 0, DIGITAL: 1 OR MICROWAVE: 2
> 0
FILTER TYPE - LOWPASS: 1, HIGHPASS: 2, LIN.-PHASE LOWPASS: 3, BANDPASS: 4
> 4
LOWER EDGE OF THE PASSBAND IN HZ
> 67.5e6
UPPER EDGE OF THE PASSBAND IN HZ
> 72.5e6
PASSBAND - MAX.-FLAT: 0, EQUAL-RIPPLE: 1, FUNCTIONAL INPUT: 2
> 1
WHAT IS THE BAND EDGE LOSS IN DB
> .1
BANDPASS - CONVENTIONAL: 1, PARAMETRIC: 2 OR MATCHING: 3
> 2
ENTER FREAL IN HZ. FOR DEFAULT, ENTER 0.
> 0
ENTER MULTIPLICITY OF TRANSMISSION ZERO AT ZERO
> 11
ENTER MULTIPLICITY OF TRANSMISSION ZERO AT INFINITY
> 1
ENTER NO. OF FINITE TRANSMISSION ZEROS
> 0
ENTER INPUT TERMINATION IN OHMS
> 50
ENTER OUTPUT TERMINATION (0. INDICATES OPEN OR SHORT)
> 50.
ENTER VALUE OF AVERAGE Q. IF NO PREDISTORTION, ENTER 0.
> 0
```

Fig. 4-7. Interactive input to FILSYN program, specifying an 12th-order bandpass filter with 50 ohm terminations. The carets indicate user inputs.

```
READ DATA FROM FILE: Y/N
> n
PLACER: P, SMAIN: S OR END: E
> s
ENTER TITLE
> capacitively coupled resonator bandpass
FILTER KIND - LUMPED: 0, DIGITAL: 1 OR MICROWAVE: 2
> 0
FILTER TYPE - LOWPASS: 1, HIGHPASS: 2, LIN.-PHASE LOWPASS: 3, BANDPASS: 4
> 4
LOWER EDGE OF THE PASSBAND IN HZ
> 67.5e6
UPPER EDGE OF THE PASSBAND IN HZ
> 72.5e6
PASSBAND - MAX.-FLAT: 0, EQUAL-RIPPLE: 1, FUNCTIONAL INPUT: 2
> 1
WHAT IS THE BAND EDGE LOSS IN DB
> .1
BANDPASS - CONVENTIONAL: 1, PARAMETRIC: 2 OR MATCHING: 3
> 2
ENTER FREAL IN HZ. FOR DEFAULT, ENTER 0.
> 0
ENTER MULTIPLICITY OF TRANSMISSION ZERO AT ZERO
> 11
ENTER MULTIPLICITY OF TRANSMISSION ZERO AT INFINITY
> 1
ENTER NO. OF FINITE TRANSMISSION ZEROS
> 0
ENTER INPUT TERMINATION IN OHMS
> 50
ENTER OUTPUT TERMINATION (0. INDICATES OPEN OR SHORT)
> 50.
ENTER VALUE OF AVERAGE Q. IF NO PREDISTORTION, ENTER 0.
> 0


GENERAL FILTER SYNTHESIS PROGRAM

capacitively coupled resonator bandpass                    26-DEC-86 11:11

    BAND-PASS FILTER
        EQUAL RIPPLE PASS BAND
            BANDEDGE LOSS                         =  .1000 DB.
            LOWER PASSBAND EDGE FREQUENCY         =  6.7500000D+07 HZ.
            UPPER PASSBAND EDGE FREQUENCY         =  7.2500000D+07 HZ.
        PARAMETRIC BANDPASS TYPE
        SPECIFIED STOP BAND
            MULTIPLICITY OF ZERO AT ZERO          = 11
            MULTIPLICITY OF ZERO AT INFINITY      =  1
            OVERALL FILTER DEGREE                 = 12
            VALUE OF FREAL                        =  6.9955343D+07 HZ.
            INPUT TERMINATION                     =  5.0000000D+01 OHMS
            OUTPUT TERMINATION                    =  5.0000000D+01 OHMS
            REQUESTED TERMINATION RATIO           =  1.0000000D+00
```

Fig. 4-7. (*Continued*)

** EVEN NUMBERED BRANCHES ARE SERIES, ODD ONES SHUNT **

capacitively coupled resonator bandpass

#	Element	Value
1	....R....	5.0000000D+01
2	C	4.8337949D-11
3	....L....	1.2974863D-08
4	C	3.7723026D-10
5	....L....	4.5762470D-11
6	C	1.1302407D-07
7	....L....	8.6537250D-14
8	C	5.9852780D-05
9	....L....	1.6318592D-16
10	C	3.1741442D-02
11	....L....	5.1391265D-19
12	C	1.0712671D+01
13	....C....	8.3323258D+01
15	....R....	2.5701851D-11

Fig. 4-8. The computer-generated filter topology has 11 transmission zeros at DC and one at infinity. The load resistance needs to be increased to 50 ohm by Norton impedance transformations.

physical realization, considering parasitics that may cause self-resonances at these frequencies.

The final circuit, impedance transformed to 50 ohm terminations, is shown in Figure 4-9. Note that the circuit is symmetrical and all the inductors are of the same value. The corresponding loss and delay plot are shown in Figure 4-10. Figure 4-10 displays the loss (A), details the ripple across the band, and shows the delay (B). With the use of ideal circuit elements, the filter has the expected 0.1-dB equal ripple response, but there is a significant delay distortion throughout the passband.

The next task is to evaluate the performance of the lossy filter, using element values with finite Qs. Specifying a Q-value of 150 for the components results in gain rolloff at the corner frequencies with an approximately 180-nanosecond group delay distortion in the passband.

CAPACITIVELY COUPLED RESONATOR BANDPASS

1	....R....	5.0000000D+01
3	....C....	2.4168974D-11
4	. C	2.4168974D-11
5	....C....	7.6622281D-11
7	....L....	5.1899454D-08
8	. C	5.6007974D-12
9	....C....	9.0057319D-11
11	....L....	5.1899454D-08
12	. C	4.3337557D-12
13	....C....	9.1319586D-11
15	....L....	5.1899454D-08
16	. C	4.3337557D-12
17	....C....	9.0057319D-11
19	....L....	5.1899454D-08
20	. C	5.6007974D-12
21	....L....	5.1899454D-08
23	....C....	7.6622281D-11
24	. C	2.4168974D-11
25	....C....	2.4168974D-11
27	....R....	5.0000000D+01

Fig. 4-9. The transformed versions of the filter shown in Figure 4-8. Both source and load terminations are now 50 ohms.

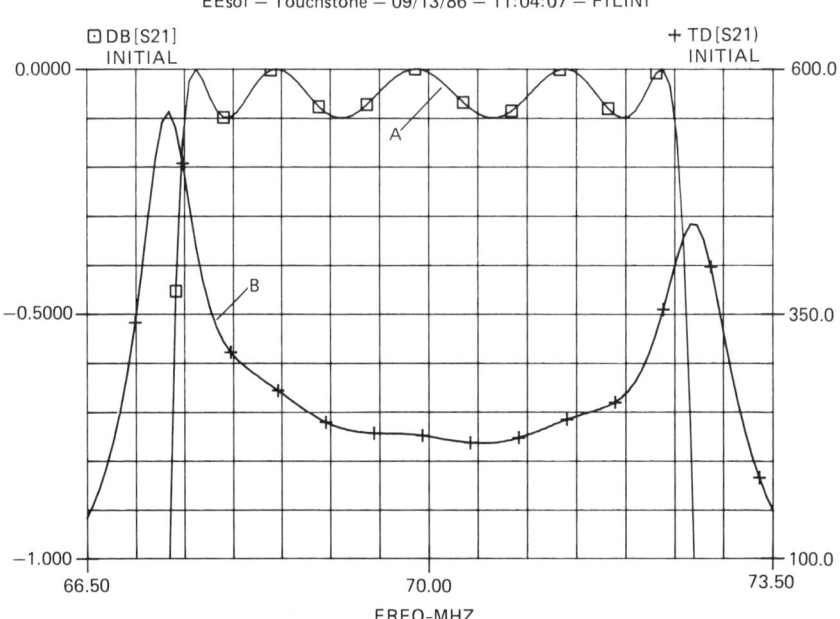

Fig. 4-10. Insertion loss (A) and group delay (B) response of the filter shown in Figure 4-9, using ideal components (infinite Q).

98 ANALYSIS AND DESIGN OF ELECTRONIC CIRCUITS USING PCs

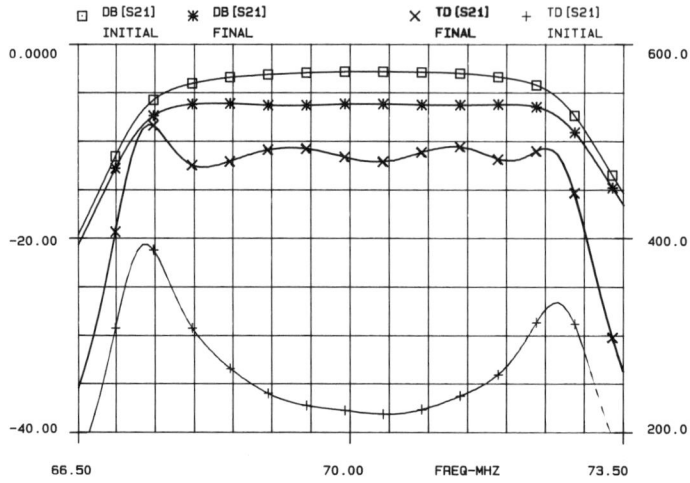

Fig. 4-11. Delay response before (+) and after (×) equalization, using all-pass two sections. The insertion loss increase is due to the loss of the equalizer sections.

The filter is now submitted to delay equalization. FILSYN determines that two second-order equalizer sections will be required to level the delay response. The final analysis of the filter/equalizer combination is shown on Figure 4-11. Notice that the program has reduced the initial delay error by almost an order of magnitude.

4.3.3. 5-GHz Lowpass Filter Design

This section illustrates direct distributed network synthesis with commensurate-length microstrip transmission lines. Specified stopband attenuation is 25 dB at 8 GHz. The desired topology includes three open-circuited parallel stubs separated by two unit elements (Figure 4-12). Depending on the initial specifications, this topology may be realized by either a third- or a fifth-order filter. The following procedure will start by designing the third-order network, followed by the fifth-order case and then compare the stopband responses of the two filters. Both circuits will be designed with 50-ohm terminations and 0.2-

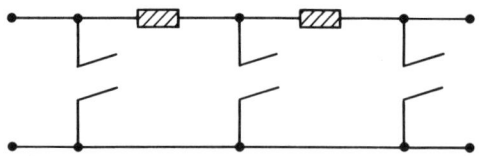

Fig. 4-12. Desired lowpass filter topology, using commensurate length of transmission lines.

```
ENTER TITLE
> 5GHz 3rd order lowpass filter
FILTER KIND - LUMPED: 0, DIGITAL: 1 OR MICROWAVE: 2
> 2
ENTER QUARTER WAVE FREQUENCY IN HZ
> 1e10
FILTER TYPE - LOWPASS: 1, HIGHPASS: 2 OR BANDPASS: 4
> 1
UPPER EDGE OF THE PASSBAND IN HZ
> 5e9
PASSBAND - MAX.-FLAT: 0, EQUAL-RIPPLE: 1, FUNCTIONAL INPUT: 2
> 1
WHAT IS THE BAND EDGE LOSS IN DB
> .2
STOPBAND - MONOTONIC: 0, EQUAL-MINIMA: 1 OR SPECIFIED: 2
> 2
ENTER NO. OF UNIT ELEMENTS
> 0
ENTER MULTIPLICITY OF TRANSMISSION ZERO AT QUARTER-WAVE FREQUENCY
> 3
ENTER NO. OF FINITE TRANSMISSION ZEROS
> 0
ENTER INPUT TERMINATION IN OHMS
> 50
ENTER OUTPUT TERMINATION (0. INDICATES OPEN OR SHORT)
> 50

GENERAL FILTER SYNTHESIS PROGRAM

5GHz 3rd order lowpass filter                          26-DEC-86 11:22

LOW-PASS FILTER
   EQUAL RIPPLE PASS BAND
      BANDEDGE LOSS                           =    .2000 DB.
      MAX. PASSBAND VSWR                      =   1.5386
      UPPER PASSBAND EDGE FREQUENCY           =   5.0000000D+09 HZ.
      QUARTER-WAVE FREQUENCY                  =   1.0000000D+10 HZ.
   SPECIFIED STOP BAND
      MULTIPLICITY OF ZERO AT QUARTER-WAVE FR. =  3
      NUMBER OF UNIT ELEMENTS                 =   0
      NUMBER OF FINITE TRANSMISSION ZERO PAIRS =  0
      OVERALL FILTER DEGREE                   =   3
      INPUT TERMINATION                       =   5.0000000D+01 OHMS
      OUTPUT TERMINATION                      =   5.0000000D+01 OHMS
      REQUESTED TERMINATION RATIO             =   1.0000000D+00
      NEAREST AVAILABLE TERMINATION RATIO     =   1.0000000D+00
```

Fig. 4-13. FILSYN specifications of the third-order transmission-line filter. Desired passband is 0-5G Hz.

dB equal ripple passband response from zero to 5 GHz. The commensurate quarter wavelength frequency of all transmission lines is set to 10 GHz, meaning that the lines will represent 45 degrees electrical length at the upper end of the passband (5 GHz).

Third-Order Filter Synthesis. The interactive response of FILSYN is shown in Figure 4-13. Since the desired topology cannot be synthesized directly (due to redundant unit elements), the third-order network is specified by entering three transmission zeros at quarter-wavelength frequency. Since in this specification, there are no redundant unit elements or any finite transmission zeros, the network order is equal to three.

The initial synthesized network is given in Figure 4-14. The symbol R refers to the terminations, while the Ls and Cs indicate short- or open-circuited transmission line stubs, with quarter-wavelength frequencies of 10 GHz.

At this point, the designer inserts two 50-ohm unit elements, one on each side of the circuit. Inserting these elements does not change the order of the network, since the magnitude of the gain response remains the same. The next two steps are consecutive lowpass Kuroda transformations, resulting in a circuit with three short-circuited series stubs and two unit elements. At this point, the dual circuit is requested, yielding the desired final topology as indicated in Figure 4-15. This is the final form of the third-order network. Note that all circuit elements are within the 20–120-ohm impedance range; therefore they can be easily realized on microstrip.

Fifth-Order Synthesis. In this case, the interactive FILSYN input response

```
5GHz 3rd order lowpass filter

**** ALL VALUES ARE IMPEDANCES ****
    1     ....R....        5.0000000D+01
              .   .
    3     ....C....        4.0731707D+01
              .   .
    4     .       L        5.7627161D+01
              .   .
    5     ....C....        4.0731707D+01
              .   .
    7     ....R....        5.0000000D+01
              .   .
```

Fig. 4-14. Computer generated topology by FILSYN provides two open circuited parallel stubs and one short circuited series stub. Designed topology can be reached through the Kuroda transformation capabilities of the program.

```
5GHz 3rd order lowpass filter

**** ALL VALUES ARE IMPEDANCES ****
     1      ....R....        5.0000000D+01
            .      .
     5      ....C....        9.0731707D+01
            .      .
            *      *
     7      *  U E *         1.1137725D+02
            *      *
            .      .
    11      ....C....        4.3382321D+01
            .      .
            *      *
    15      *  U E *         1.1137725D+02
            *      *
            .      .
    17      ....C....        9.0731707D+01
            .      .
    19      ....R....        5.0000000D+01
            .      .
```

Fig. 4-15. Final transformed version of the circuit from Figure 4-14, consisting of three open stubs and two cascade transmission lines (Unit Elements).

is identical to the one shown in Figure 4-16, with one exception. Two nonredundant unit elements are also specified, in addition to the three quarter-wavelength transmission zeros, represented by the two parallel open stubs and the series short stub. Therefore, the resultant filter will have an order of five.

The initial synthesized network is shown in Figure 4-17. Note that the unit elements have impedance values other than 50 ohms; consequently, they do contribute to the order of the network. This time, three consecutive lowpass Kuroda transformations are required to get to the final form shown in Figure 4-18. Notice that the topology of the third- and the fifth-order filters are identical; only the transmission line impedances are different. As one would suspect, the two filters will have significant differences in their stopband attenuation as shown in Figure 4-19; the fifth-order filter has approximately 14 dB greater attenuation at 8 GHz.

Although FILSYN can handle predistortion, both of the above illustrations used lossless (infinite-Q) components. In the case of the transmission line filter, the substrate parameters and discontinuities may also affect the performance significantly. In general, the designer will convert the electrical transmission line parameters to physical dimensions, then include all losses, parasitics, and discontinuities before arriving at the final design. A convenient program to provide such conversion is LINECALC, which is illustrated later in Figure 4-28.

102 ANALYSIS AND DESIGN OF ELECTRONIC CIRCUITS USING PCs

```
sfilsyn
            ***** S/FILSYN *****
   RELEASE 1.1  VERSION 08   10/01/86
              ** ROOT SEGMENT **
Copyright (C) 1983 - 1986 Dr. George Szentirmai.
           All Rights Reserved.
READ DATA FROM FILE: Y/N
> n
PLACER: P, SMAIN: S OR END: E
> s
ENTER TITLE
> 5g GHz 5th order lowpass filter
FILTER KIND - LUMPED: 0, DIGITAL: 1 OR MICROWAVE: 2
> 2
ENTER QUARTER WAVE FREQUENCY IN HZ
> 1e10
FILTER TYPE - LOWPASS: 1, HIGHPASS: 2 OR BANDPASS: 4
> 1
UPPER EDGE OF THE PASSBAND IN HZ
> 5e9
PASSBAND - MAX.-FLAT: 0, EQUAL-RIPPLE: 1, FUNCTIONAL INPUT: 2
> 1
WHAT IS THE BAND EDGE LOSS IN DB
> .2
STOPBAND - MONOTONIC: 0, EQUAL-MINIMA: 1 OR SPECIFIED: 2
> 2
ENTER NO. OF UNIT ELEMENTS
> 2
ENTER MULTIPLICITY OF TRANSMISSION ZERO AT QUARTER-WAVE FREQUENCY
> 3
ENTER NO. OF FINITE TRANSMISSION ZEROS
> 0
ENTER INPUT TERMINATION IN OHMS
> 50
ENTER OUTPUT TERMINATION (0. INDICATES OPEN OR SHORT)
> 50

GENERAL FILTER SYNTHESIS PROGRAM

5GHz 5th order lowpass filter                          26-DEC-86 11:27


     LOW-PASS FILTER
        EQUAL RIPPLE PASS BAND
           BANDEDGE LOSS                        =    .2000 DB.
           MAX. PASSBAND VSWR                   =   1.5386
           UPPER PASSBAND EDGE FREQUENCY        =   5.0000000D+09 HZ.
           QUARTER-WAVE FREQUENCY               =   1.0000000D+10 HZ.
        SPECIFIED STOP BAND
           MULTIPLICITY OF ZERO AT QUARTER-WAVE FR. =   3
           NUMBER OF UNIT ELEMENTS              =   2
           NUMBER OF FINITE TRANSMISSION ZERO PAIRS =   0
           OVERALL FILTER DEGREE                =   5
           INPUT TERMINATION                    =   5.0000000D+01 OHMS
           OUTPUT TERMINATION                   =   5.0000000D+01 OHMS
           REQUESTED TERMINATION RATIO          =   1.0000000D+00
           NEAREST AVAILABLE TERMINATION RATIO  =   1.0000000D+00
```

Fig. 4-16. Specifications of the fifth-order filter are similar to those shown in Figure 4-13, except that the two extra unit elements add to the order of the circuit.

COMPUTER-AIDED DESIGN AT MICROWAVE FREQUENCIES   103

```
5GHz 5th order lowpass filter

**** ALL VALUES ARE IMPEDANCES ****

        1      ....R....      5.0000000D+01
               .       .
               *       *
        3      *  U E  *      2.6138410D+01
               *       *
               .       .
               *       *
        5      *  U E  *      7.6664142D+01
               *       *
               .       .
        7      ....C....      2.4899055D+01
               .       .
        8      .    L         6.5212859D+01
               .       .
        9      ....C....      3.7944610D+01
               .       .
       11      ....R....      5.0000000D+01
               .       .
```

Fig. 4-17. Initial circuit topology generated by FILSYN shows the two nonredundant unit elements.

```
5GHz 5th order lowpass filter

**** ALL VALUES ARE IMPEDANCES ****

        1      ....R....      5.0000000D+01
               .       .
        3      ....C....      3.7944610D+01
               .       .
               *       *
        5      *  U E  *      8.4007706D+01
               *       *
               .       .
       11      ....C....      2.4211666D+01
               .       .
               *       *
       13      *  U E  *      8.4007706D+01
               *       *
               .       .
       17      ....C....      3.7944610D+01
               .       .
       19      ....R....      5.0000000D+01
               .       .
```

Fig. 4-18. A casual observer may not be able to differentiate between this filter and the one shown in Figure 4-15 with the exception of impedance changes. Comparing stopband attenuations however, reveals significant differences (see Figure 4-19).

104 ANALYSIS AND DESIGN OF ELECTRONIC CIRCUITS USING PCs

```
******* COMPUTED PERFORMANCE *******
                 3RD ORDER      5TH ORDER
FREQUENCY        LOSS           LOSS
IN HZ            IN DB          IN DB
 .00000D+00       .000000        .000000
1.00000D+09       .042957        .128297
2.00000D+09       .141256        .179841
3.00000D+09       .199786        .004177
4.00000D+09       .084473        .181790
5.00000D+09       .200000        .200000
6.00000D+09      4.580231      12.162432
7.00000D+09     14.615803      26.740836
8.00000D+09     27.358962      41.395514
9.00000D+09     46.626395      61.631061
1.00000D+10    886.274499     901.585527
```

Fig. 4-19. Insertion loss responses of the filters shown in Figures 4-15 and 4-18. Although they have similar physical appearances, the fifth-order filter indicates superior stopband performance.

4.3.4. Filter Synthesis Combined with Optimization

A unique group of design programs combine classical filter synthesis with transmission line synthesis and optimization to provide physical circuit dimensions directly. ECM (Edge-Coupled Microstrip), CLD (Comb-Line Design), and IDD (Inter-Digital Design) are specialized filter design programs that directly determine the final microwave circuit dimensions from the electrical specifications. The programs are a combination of classical filter synthesis and numerical optimization that provide exact equal-ripple (Chebyshev) performance over a specified bandwidth while meeting required stopband rejection. Since the programs use numerical techniques, the effects of parasitic elements are included in the design rather than being considered as an afterthought, or neglected entirely.

As an example of the concepts used in the above programs, consider the microstrip edge-coupled filter form shown in Figure 4-20. The filter consists of 10 edge-coupled sections that are normally used to realize an $N = 9$ resonator response. Because of the nonhomogeneous nature of the microstrip medium, each coupled line section can be accurately modeled only with a relatively complex circuit. Constraints inherent in the correct equivalent circuit preclude the use of classical synthesis techniques. In addition, fringing capacitances that exist from the open-circuited ends of each coupled line section can be accurately considered only by modeling the coupled line sections as four port networks.

As an alternate viewpoint, however, if the circuit of Figure 4-21 is capable of providing a desired response, all that is really required to achieve that response are the correct line lengths, line widths, and gap spacings. The above physical dimensions are not all independent parameters, and some can be fixed while the others are used as variable parameters. ECM allows the designer to specify the coupled line widths and adjusts the line lengths and gap spacings to achieve

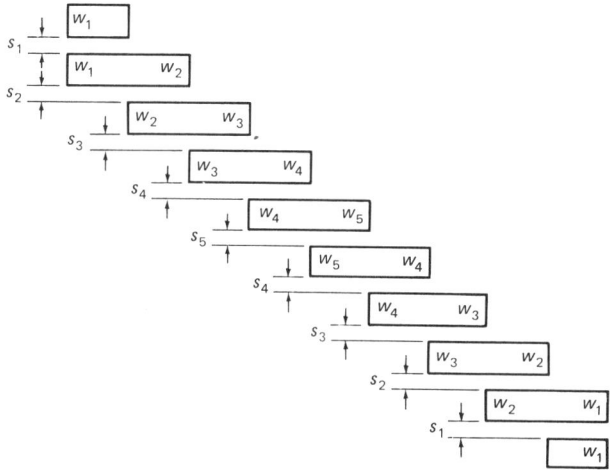

Fig. 4-20. General form of an edge-coupled microwave filter. Line widths may change between sections (i.e.: $w_1 \neq w_2$).

exact equal ripple response. Choosing wide line widths maximizes resonator unloaded Q (Q_u) and results in the lowest passband dissipation loss. However, wide lines result in narrow gap spacings for wide-bandwidth filters, and may be impractical for such cases. The choice of narrow line widths gives wider gap spacings and is attractive for wide-bandwidth designs. By doing several designs

```
A: >ECM

Enter FILENAME for design data and
S-parameter file /  <ECM>
EX1
For old design enter O / <new design>

For transformer input enter T /
<Edge coupled input>
H

    HELP MESSAGE

o--IIIIIIIIIIIIIIII   Transformer Input
       IIIIIIIIIIIIIII
              IIIIIIIIIIIIIII--o

o--IIIIIIII
       IIIIIIIIIIIIIIII   Edge-coupled Input
       IIIIIIIIIIIIIII
              IIIIIIIIIIIIIII
                     IIIIIIII--o

For transformer input enter T /
<Edge coupled input>
```

Fig. 4-21. Interactive inputs to the ECM program. User specifies the type of filter, desired attenuation, and the type of substrate used.

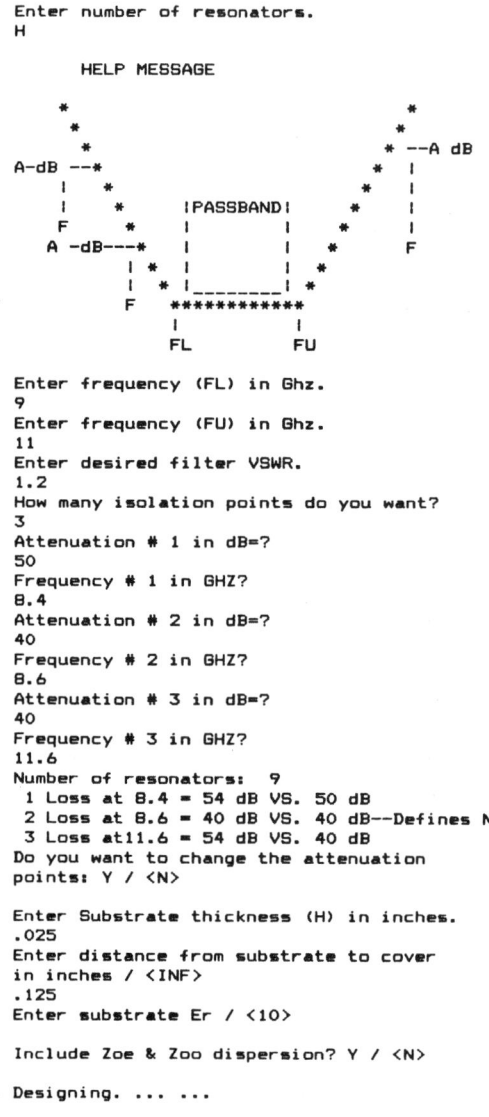

```
Enter number of resonators.
H
            HELP MESSAGE
                *                        *
              *                            *
            *                                *
A-dB --*                                       * --A dB
       |  *                                  *  |
       |    *         |PASSBAND|           *    |
       F      *       |        |         *      |
A -dB---*     |       |        |       *        F
        | *   |       |        |     *
        |   * |_____|   *
        F     ***********
              |                 |
              FL                FU

Enter frequency (FL) in Ghz.
9
Enter frequency (FU) in Ghz.
11
Enter desired filter VSWR.
1.2
How many isolation points do you want?
3
Attenuation # 1 in dB=?
50
Frequency # 1 in GHZ?
8.4
Attenuation # 2 in dB=?
40
Frequency # 2 in GHZ?
8.6
Attenuation # 3 in dB=?
40
Frequency # 3 in GHZ?
11.6
Number of resonators:  9
  1 Loss at 8.4 = 54 dB VS. 50 dB
  2 Loss at 8.6 = 40 dB VS. 40 dB--Defines N
  3 Loss at11.6 = 54 dB VS. 40 dB
Do you want to change the attenuation
points: Y / <N>

Enter Substrate thickness (H) in inches.
.025
Enter distance from substrate to cover
in inches / <INF>
.125
Enter substrate Er / <10>

Include Zoe & Zoo dispersion? Y / <N>

Designing. ... ...
```

Fig. 4-21. (*Continued*)

with different line-width choices, the designer is able to explore virtually all possible physical realizations without getting involved in equivalent circuit details.

To illustrate a typical design, consider the design of a 9–11 GHz equal ripple edge-coupled bandpass filter to provide the following rejection:

```
Analysis? N / <Y>
Input resonator Qu <program estimates Qu>

Program estimates Qu=231.

Enter analysis frequencies F1,F2,F3
8,12,.2

F(GHZ)   VSWR   LOSS(dB)   PHASE(DEG)   T-NS

 8.00    9.00    86.85        86.39      .17
 8.20    9.00    75.93        72.61      .21
 8.40    9.00    63.20        55.18      .28
 8.60    9.00    47.70        30.77      .42
 8.80    9.00    27.17       -12.20      .89
 9.00    1.20     4.02      -190.16     3.55
 9.2     1.20     2.10      -357.28     1.82
 9.4     1.18     1.74      -475.31     1.49
 9.6     1.03     1.56      -576.72     1.35
 9.8     1.17     1.49      -669.61     1.25
10.0     1.14     1.45      -758.40     1.21
10.2     1.10     1.45      -844.92     1.21
10.4     1.19     1.51      -932.46     1.22
10.6     1.03     1.60     -1023.54     1.32
10.8     1.10     1.87     -1123.83     1.51
11.0     1.20     3.08     -1254.00     2.45
11.2     9.00    18.10     -1407.31     1.00
11.4     9.00    33.92     -1453.31      .44
11.6     9.00    45.57     -1478.81      .29
11.8     9.00    54.77     -1497.02      .22
12.0     9.00    62.34     -1511.48      .18

Continue analysis? N / <Y>

Enter analysis frequencies F1,F2,F3
12.5,22,.5

F(GHZ)   VSWR   LOSS(dB)   PHASE(DEG)   T-NS

12.5     9.00    76.56     -1539.18      .13
13.0     9.00    86.37     -1560.58      .11
13.5     9.00    93.16     -1578.55      .09
14.0     9.00    97.61     -1594.31      .08
14.5     9.00   100.09     -1608.47      .08
15.0     9.00   100.77     -1621.44      .07
15.5     9.00    99.67     -1633.50      .06
16.0     9.00    96.68     -1644.88      .06
16.5     9.00    91.53     -1655.82      .06
17.0     9.00    83.62     -1666.60      .06
17.5     9.00    71.72     -1677.72      .06
18.0     9.00    52.34     -1690.61      .08
18.5     9.00    11.86     -1699.10      .73
19.0     9.00     6.40     -1737.22     1.19
19.5     1.24     4.56     -1790.11     2.73
20.0     6.47     5.76     -1850.80     1.45
20.5     9.00     9.87     -1895.76      .68
21.0     9.00    27.98     -1979.01     3.14
21.5     9.00    88.72     -2071.16      .12
22.0     9.00   125.49     -2086.98      .07

Continue analysis? N / <Y>
N

Computed loss based on Qu= 231

K    ZOE    ZOO   C(pf)  W(in.)  S(in.)  L(in.)

1   94.32  38.24  .020   .011    .0042   .1100
2   67.26  40.73  .029   .019    .0118   .1070
3   61.56  42.38  .031   .021    .0172   .1066
4   60.92  43.20  .031   .021    .0190   .1066
5   60.76  43.41  .031   .021    .0194   .1066

Enter 2 for new line width dimensions,
1 to continue using existing line widths
/ <end design>
```

Fig. 4-22. Output of ECM includes the dimensions of the physical circuit and the corresponding frequency response circuit. Additional modifications are also available if the circuit is not physically realizable.

$$IL > 50 \text{ dB @ } 8.4 \text{ GHz}$$
$$> 40 \text{ dB @ } 8.6 \text{ GHz}$$
$$> 40 \text{ dB @ } 11.6 \text{ GHz}$$

The filter is to be realized using a .025 inch (H) Alumina substrate ($Er = 10$). Cover height (CH) is to be .125 inch above the substrate. Desired maximum passband VSWR is to be 1.2 : 1.

The ECM design is shown in Figures 4-21 and 4-22. The program is capable of generating a TOUCHSTONE/MICAD-compatible output file for further processing.

4.3.5. Matching Network Synthesis

In active circuit design, it is frequently desired to begin the synthesis with a specific component type and value(s). In such cases, after specifying the topology, the user may find the desired component value by iteratively changing ripple (R) and minimum insertion loss (MIL) specifications. Generally, increased R or MIL results in larger parasitics.

Some of the matching network synthesis routines are capable of automating the above procedure. The SUPERCOMPACT PC synthesis routine is used here to illustrate the concept of "exact parasitic absorption" through the design of the transmission line matching network. The load is a 50-ohm resistor; the source may be modeled by a combination of a 10-ohm resistor and a series open stub.

The best approach in such a case is to begin synthesizing from the 10-ohm side and "force" the program to start with an element which is equal to the parasitic portion (series open stub) of the termination. The rest of the circuit

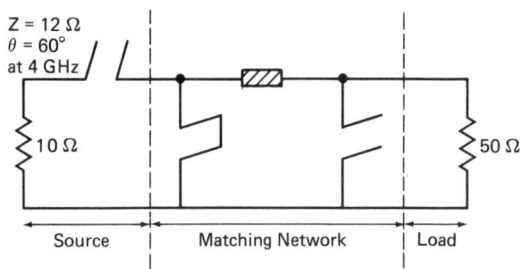

Fig. 4-23. One possible topology to match the complex source to the 50-ohm load. Initially, all transmissions lines will be synthesized with the same electrical lengths.

```
                       ***** SUPER-COMPACT PC SYNTHESIS *****
Do you want to specify a parasitic? (<Y>/N): Y
RL, F1 and F2 (in GHz)? (x,y,z): 50 2 4
Enter type of parasitic: OS
RS, Z0 and Angle at F2? (x,y,z): 10 12 60
Automatic Synthesis? (<Y>/N): N
List Topology Options? (Y/<N>): N
You have the following synthesis options:
      1) Structure with electrical length =  60.0000 DEGS
      2) Structure with specified length less than 90 DEGS
      3) Bandpass structure with quarter-wave length lines
Option? (x/< >): 1
Total elements in structure? (x/<4>): 4
#UE, #HPE & #LPE? (x,y,z/<1,2,1,>):
RL =     50.00000 Ohms      FREQ =    2.00000 to    4.00000 Ghz
RS =     10.00000 Ohms      Parasitic(s):
              OS =      12.00000 Ohms at   60.00000 Degs
Gain-Bandwidth Calculation
MIL? (x/< >):
RIP? (x/< >):
Ripple =    .18299 dB  MIL =    .11241 dB
    Another GBW calculation
    Change number of Elements
    Continue
Option? (A/E/<C>): C
dB Ripple? (x/<  .18299>):
dB MIL? (x/<  .11241>):
GBW Calculations Assume LHP Zeros.
All Zeros LHP? (<Y>/N): Y
SYN > EXT
The minimum structure has   2 HPE    1 LPE &   1 TRL
Lumped equivalent? (Y/<N>) N
Parasitic has been extracted:
   OST  SE    Z =   11.99605 Ohms     E =   60.00000    F = F2
Elements to be extracted are:   1 HPE,   1 LPE, & 1 TRL
EXT > EXT
 Select one of the following elements:
       SP    Z0=   31.3626 Ohms
       SS    Z0=    8.7294 Ohms
       UE    Z0=   17.3849 Ohms
Element? SP
EXT > EXT
 Select one of the following elements:
       SS    Z0=   12.0962 Ohms
       UE    Z0=   39.0073 Ohms
Element? UE
EXT > EXT
 Select one of the following elements:
       OP    Z0=   86.7813 Ohms
       UE    Z0=   31.2050 Ohms   (non-minimum)
Element? OP
Termination =    48.72605 Ohms
```

Fig. 4-24. SUPERCOMPACT synthesis specifications of the matching network shown in Figure 4-23. The program offers alternative components and configurations at each node, to synthesize other topologies if desired.

110 ANALYSIS AND DESIGN OF ELECTRONIC CIRCUITS USING PCs

```
                      SYNTHESIZED NETWORK
              .18299 dB Ripple         .11241 dB MIL

              R1=   10.00000  Ohms
  OST   SE    Z =   11.99605  Ohms   E = 60.00000    F = F2
  SST   PA    Z =   31.36263  Ohms   E = 60.00000    F = F2
  TRL   SE    Z =   39.00728  Ohms   E = 60.00000    F = F2
  OST   PA    Z =   86.78130  Ohms   E = 60.00000    F = F2
              R2=   48.72605  Ohms
EXT > SAV
Network saved.
SYN > PRI
                       SAVED NETWORK
          Ripple =   .18299 dB      MIL =    .11241 dB
          F1 =    2.00000 Ghz       F2 =   4.00000 Ghz

              R1=   10.00000  Ohms
* OST   SE    Z =   11.99605  Ohms   E = 60.00000    F =  4.0000 GHz (PARASITIC)
  SST   PA    Z =   31.36263  Ohms   E = 60.00000    F = F2
  TRL   SE    Z =   39.00728  Ohms   E = 60.00000    F = F2
  OST   PA    Z =   86.78130  Ohms   E = 60.00000    F = F2
              R2=   48.72605  Ohms
SYN > QUIT
```

Fig. 4-25. Final results of the SUPERCOMPACT matching network synthesis. All transmission line lengths are 60 degrees at 4GHz.

topology may be chosen from the Smith Chart—one possible form is the bandpass structure shown in Figure 4-23.

The step-by-step SUPERCOMPACT PC response is shown in Figure 4-24. The program has suggested several possible topologies, one of which is the form decided earlier. Since the resultant load is only 3% below 50 ohms, further transformation would not bring major improvement. The final circuit is shown in Figure 4-25.

4.4. TRANSMISSION LINE ANALYSIS AND SYNTHESIS

Programs of this type compute transmission line parameters either through numerical analysis or by approximations. The first type usually breaks a transmission line into conductor strips and requires extensive amounts of computational time. The MSTRIP program, introduced by Bryant and Weiss in the 60s, is an example of this type. Generally these routines handle only ideal transmission lines and only limited forms of realization are available.

The second category is often based on piece-wise approximations of closed-form equations. They offer more complete models, including the effects of losses and dispersion.

4.4.1. The MSTRIP+ Program

This program is a modified version of the original Bryant and Weiss routine. The program runs interactively on IBM PC and Apple II computers and includes dispersion effects, suspended substrate realization, and graphics.

Microstrip Analysis with MSTRIP+. An analysis of the impedance and the effective dielectric constant of a microstrip line of various w/h ratios on a 30-mil-thick Teflon-glass board with 1 oz. copper metallization is shown in Figure 4-26. The effects of dispersion on a 50-ohm line from 2 to 18 GHz is shown in Figure 4-27. Note that both the impedance and effective dielectric constants are frequency dependent.

4.4.2. The LINECALC Program

LINECALC is an interactive program representing the second group mentioned in the introduction to Section 4.4., using a spreadsheet-like format to convert transmission line parameters to physical dimensions and vice versa. In the case of synthesis, typical inputs are, in addition to the substrate parameters, the electrical length and impedance of the transmission line. The resultant output is the physical width and length of the line. The program can handle both single and coupled lines and various forms of physical realizations (microstrip, stripline, coplanar Waveguides, suspended substrate, and so forth). The transmission line analysis includes the effects of surface roughness, top cover, conductive and dielectric losses, and dispersion.

4.4.3. Transmission Line Synthesis with LINECALC

The electrical parameters of the third-order network, synthesized in Section 3.3.1, will now be converted to physical dimensions using LINECALC. Substrate parameters are the following:

Dielectric constant	9.9
Substrate thickness	25 mil
Metalization	.2-mil-thick gold
Dielectric loss factor	0.0005
Frequency	5 GHz
Realization type	Microstrip

The first element to be converted is the parallel open stub with 90 degrees electrical length at 10 GHz and 90.73 ohms impedance. The interactive response

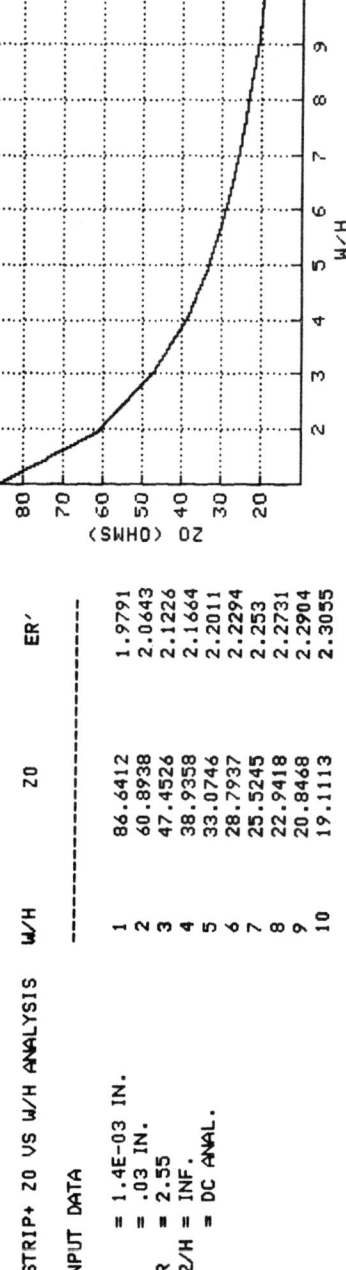

Fig. 4-26. Tabulated MSTRIP+ output and the related plot show the variation of microstrip transmission line impedance, as a function of the width/height ratio. Computations are performed at zero frequency.

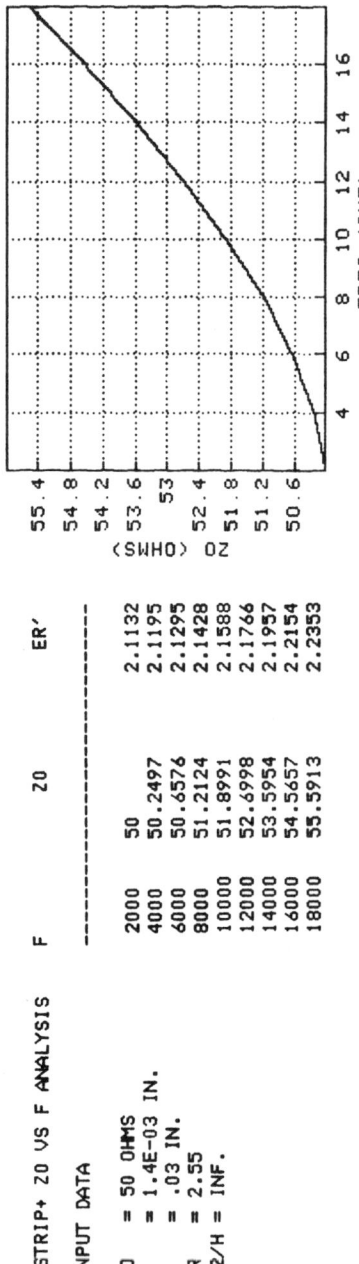

Fig. 4-27. Dispersive effects increase the microstrip line impedance from 50 ohms at 2 GHz to 55.5 ohms at 18 GHz. The effective dielectric constant also increases by about 5%.

COMPUTER-AIDED DESIGN AT MICROWAVE FREQUENCIES

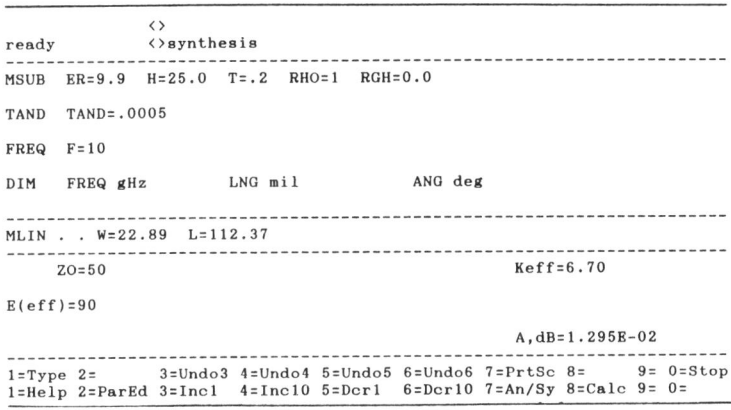

Fig. 4-28. Synthesis of the 50-ohm connecting microstrip lines by LINECALC. Using a 25-mil-thick substrate with 9.9 relative dielectric constant, the computed width is 22.89 mil. The length was arbitrarily chosen at 90 degrees at 10 GHz.

Z(OHMS)	W(MILS)	L(MILS)
90.731707	4.34	119.19
43.382321	30.41	110.63
111.37225	1.82	121.58
50.000000	22.89	112.37

Fig. 4-29. Tabulated results of the electrical to physical conversion of the three unique transmission lines of the filter shown in Figure 4-15 (note that two of the five elements are duplicated by symmetry). Dimensions for the 50-ohm lines are also given.

of LINECALC is shown in Figure 4-28. The tabulated summary of all three transmission line conversions are shown in Figure 4-29. The circuit will be analyzed and optimized in Section 4.5.1.

4.5. CIRCUIT OPTIMIZATION

When direct synthesis is not available, optimization may provide a life-saving function to the designer. Optimization requires the initial circuit topology and estimated initial values for the component values. The "quality" of the initial estimates will have a direct effect on the CPU time used and on the probability of convergence. Therefore, the designer should rely on a combination of engineering knowledge and past experience to come up with the best possible initial design.

114 ANALYSIS AND DESIGN OF ELECTRONIC CIRCUITS USING PCs

The optimization process converts the user's specification into the so-called error function. A typical form of the error function is

$$EF = \frac{1}{N} \sum_{FI}^{FF} W |\text{actual performance} - \text{goal}|^2$$

$$EF = \frac{1}{N} \sum_{FI}^{FF} W_1 |S_{11} - G_1|^2 + W_2 |S_{22} - G_2|^2 + W_3 |S_{21} - G_3|^2 + W_4 |NF - G_4|^2 \ldots$$

where the function is summed up through the desired frequency range using k discrete frequency steps. Different portions of the error function are weighed independently ($W_1 - W_4$) to emphasize specific requirements, and to compensate for numerical differences among them. G_1 through G_4 are the goals of the optimization.

For example, if the above function is used to optimize a low-noise amplifier with the following specifications through a given frequency range FI through FF

$$|S_{11}| = \text{not specified}$$

$$|S_{22}| < .31$$

Gain ripple < 1 dB

$$|NF_{act} - NF_{min}| < .1 \text{ dB}$$

$$EF = \frac{1}{N} \sum_{FI}^{FF} W_1 |S_{11}|^2 + W_2 |.31|^2 + W_3 |1|^2 + W_4 |.1|^2$$

then the weighting factors $W_1 - W_4$ should be assigned inversely proportional to the squared magnitudes of the error function components as shown in Table 4-2.

Table 4-2. Error Function Components

Parameter Type	Max. Acceptable Error Squared	Weight Selected
Gain ripple	1.00	×1
S_{22}	0.10	×10
NF	0.01	×100
S_{11}	Not specified	×0

Optimization is a multidimensional process wherein the number of dimensions is equal to the number of variables plus one. Simple problems will create "well-behaved" functions (surfaces) with a clearly defined single minima, to which the optimization can quickly converge. More complex circuits and requirements, particularly those for which conflicting specifications exist, will result in multimodal surfaces where the process may converge to any one of several low points, leaving the designer with an uncertainty about the true (global) optimum solution.

Once the error function has been defined, a particular search technique will be selected to minimize the error function in terms of the circuit variables. The search will begin at the initial component values specified by the user, then steered in the direction of the minimum by the optimizer.

A wide variety of search techniques have been published and applied successfully in microwave circuit design during the last 20 years. Directional search techniques, typically based on gradient directions, are generally very fast and efficient, but they have the tendency to converge to the local minima. Also, it may be difficult to compute the gradient analytically, and numerical approximations may be inaccurate and time-consuming.

If the function is suspected to be a multimodal type (such as the one shown in Figure 4-30), then a random search technique is preferred because it does not follow slope directions and has a much higher probability of converging to the global solution. Most modern CAD programs will include the combination

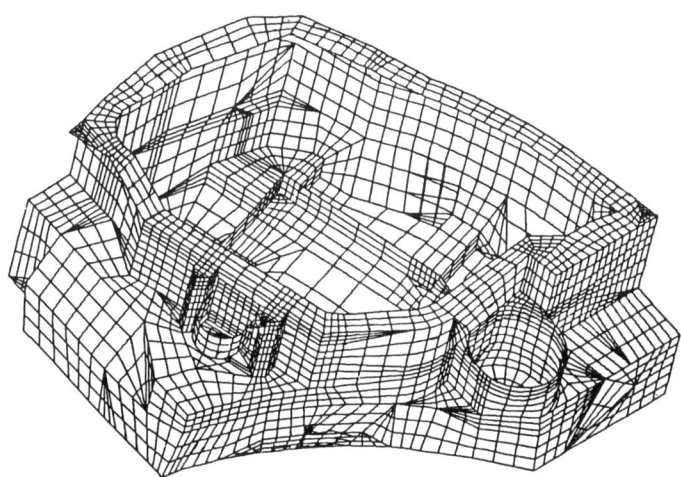

Fig. 4-30. Multi-modal error function surface displays several local minima to which optimization can erroneously converge.

of both search techniques and the user may interactively decide on the most appropriate usage.

In addition to the quality of the initial variable estimates, the number of variables, the number of frequencies at which the function is evaluated, and the complexity of the problem will have a direct effect on the outcome of the optimization. In the optimum design approach, the engineer will rely on graphical techniques and circuit synthesis to obtain an "intelligent estimate" and use optimization to fine-tune the circuit until the desired performance is reached.

Effective circuit optimization requires a computer that can handle the computational parts of the program in one piece instead of cutting it into several portions. If the program needs to be segmented due to memory limitations, then the various parts will be "swapped" in and out of memory a great number of times during an optimization and the response time may be painfully long. Optimization speed can be improved significantly by storing the nonvariable part of the evaluated portion during optimization. However, this further increases the memory requirements of the program and the tradeoffs need to be carefully examined.

A state-of-the-art microwave circuit optimization program such as SUPER-COMPACT or TOUCHSTONE requires at least 640 KBytes of main memory in a 16-bit personal computer. Although these programs are quite large, practical circuit optimization can be achieved in reasonable computer time. For example, a two-stage low-noise amplifier can be optimized for minimum noise and flat gain for an octave frequency range in 5 to 10 minutes of CPU time on the 8 MHz IBM PC/AT computer if the initial variables are within $\pm 50\%$ of the final values. Using a 32-bit PC results in major improvement.

Freq GHz	MS11 mag INITIAL	MS11 mag FINAL	MS21 dB INITIAL	MS21 dB FINAL
1.00000	0.067	0.042	-0.07	-0.04
2.00000	0.116	0.069	-0.13	-0.07
3.00000	0.134	0.063	-0.17	-0.08
4.00000	0.111	0.008	-0.17	-0.08
5.00000	0.033	0.093	-0.15	-0.14
6.00000	0.160	0.166	-0.32	-0.26
7.00000	0.589	0.200	-2.39	-0.46
8.00000	0.918	0.942	-11.52	-12.14
9.00000	0.961	0.983	-28.64	-33.89
10.00000	0.966	0.985	-82.19	-72.07

Fig. 4-31. The physical conversion and layout effects have deteriorated the filter response (INITIAL). Optimization by TOUCHSTONE have corrected the performance to an acceptable level (FINAL).

4.5.1. Circuit Optimization with TOUCHSTONE

The lowpass filter synthesized in Sections 4.3.3 and 4.4.3 is now analyzed and optimized, together with T-junctions to represent transmission line discontinuities. The junctions are created by the physical connections of the transmission lines and do affect the effective lengths of the lines. The performance change can be compensated by optimization; the initial and final responses are compared in Figure 4-31. It required 400 random trials and 4 minutes of CPU time to reach the final performance.

4.6. STATISTICAL ANALYSIS

In a survey of the attendees of continuing education microwave courses, it was found that fewer than 5% of practicing microwave circuit designers do any kind of tolerance or statistical analysis after the nominal design is completed. Lack of a realistic yield prediction generally results in a total redesign or significant production slowdown. At the other extreme, designers may use more expensive parts than needed by not knowing what tolerances are allowable.

The ideal approach would be to include tolerance effects from the very beginning and optimize the "toleranced" circuit. Unfortunately, this increases the requirements for computer capacity to the point of impracticality for most designers. Still, a simplified Monte Carlo type of tolerance analysis should be done routinely at the completion of the initial design to prevent surprises when the circuit is placed into production. Performing a day's worth of computer-aided tolerance analysis may prevent months of painful corrections.

In an analysis with a Monte Carlo program, the computer simulates the performance of a number of circuit samples; the components are randomly selected within their appropriate tolerance limits. The random selection may assume uniform distribution unless the user has sampled data available for the tolerance spread to be used directly for more accurate results (Figure 4-32). After a reasonable number of circuits are simulated (typically 100 to 500 random samples), the program can estimate the yield based on the computer-predicted performance. Results can be displayed in tabular form or on histograms (Figure 4-33). Reviewing the predictions, the designer can make cost decisions such as whether to tighten some of the tolerances or to allow tuning of the components.

4.6.1. Yield Analysis with SUPER-COMPACT

The 5 GHz lowpass filter, optimized in Section 5.1, is now submitted to a Monte Carlo analysis. All transmission line widths are assigned $\pm.3$-mil etching

118 ANALYSIS AND DESIGN OF ELECTRONIC CIRCUITS USING PCs

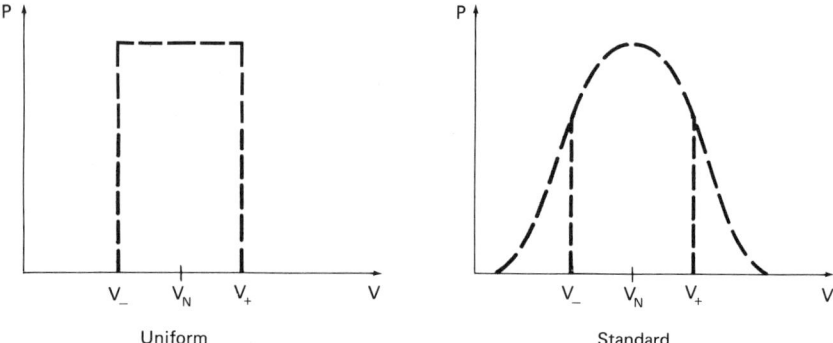

Fig. 4-32. Uniform and standard tolerance distributions for Monte Carlo analysis. Uniform distribution may be the best choice when sampled data is not available.

tolerances with the assumption that line lengths are not affected significantly. Substrate thickness and dielectric constant are varied within $\pm 5\%$. Maximum acceptable input/output VSWR and insertion loss are 1.2 : 1 and 0.3 dB respectively. The results of 500 random trials are shown in Figure 4-34, predicting a 90% yield without any tuning.

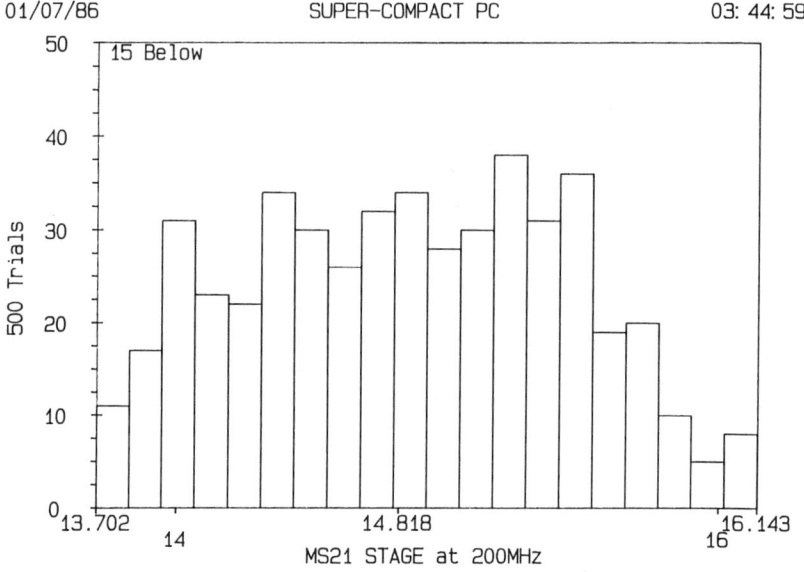

Fig. 4-33. Tolerance effects can be plotted in histograms. This plot displays the predicted gain-spread of a VHF amplifier production lot of 500 units. Nominal gain is 14.8dB; the acceptable range is 14-16 dB. About 52 circuits fail these specifications, resulting in an 89.6% yield.

```
 # Circuit    Frequency   Parm  Hist_Low   Nominal   Hist_High  Pass    Low     High
 1:DISCOPT    1.000GHz    MS21   -0.043    -0.041     -0.039   100.0%   0.0%    0.0%
 2:DISCOPT    1.000GHz    MS11    0.037     0.042      0.047   100.0%   0.0%    0.0%
 3:DISCOPT    2.000GHz    MS21   -0.077    -0.072     -0.067   100.0%   0.0%    0.0%
 4:DISCOPT    2.000GHz    MS11    0.06      0.069      0.077   100.0%   0.0%    0.0%
 5:DISCOPT    3.000GHz    MS21   -0.088    -0.083     -0.079   100.0%   0.0%    0.0%
 6:DISCOPT    3.000GHz    MS11    0.053     0.063      0.072   100.0%   0.0%    0.0%
 7:DISCOPT    4.000GHz    MS21   -0.085    -0.084     -0.083   100.0%   0.0%    0.0%
 8:DISCOPT    4.000GHz    MS11    0         0.008      0.021   100.0%   0.0%    0.0%
 9:DISCOPT    5.000GHz    MS21   -0.159    -0.144     -0.132   100.0%   0.0%    0.0%
10:DISCOPT    5.000GHz    MS11    0.078     0.093      0.109    90.0%   0.0%   10.0%
Yield =  90/100 (90.00000 %)
COMMAND> Yield Histogram Last Next Penplot
Yield: perform yield analysis on current circuit
```

Fig. 4-34. Tabulated results of the Monte Carlo analysis shows that all filters pass a .3-dB maximum insertion loss limit. However, 10% of the units fail the input match requirements at 5 GHz (|s11| < .10).

4.7. ACTIVE DEVICE MODELING

Active microwave circuit designers are achieving desired performance by creating matching networks that selectively improve or worsen existing impedance matches between devices and input/output terminations. Transistors are typically characterized by their measured, frequency dependent s-parameters, and noise parameters in case of a low-noise design. The task is then to modify the original measured parameters until they meet the overall circuit specifications by addition of the proper matching networks.

To allow the proper design of any two-port network, the source and load impedances must be known. In high-frequency amplifiers, the matching networks may be terminated at one or both sides by transistors, which complicates the design task. The actual impedances represented by the transistors in the circuit are generally not known, and cannot be determined analytically. This uncertainty is caused by the fact that transistors usually are characterized alone in 50-ohm systems (50-ohm load source and load terminations), while in the circuit the actual input and output impedances of the transistor can change drastically due to the input-output interaction. The circuit designer needs to know the actual transistor impedances because they represent the terminations for the matching circuits.

An alternative approach in designing the matching network is that, instead of computing the transistor impedances, one finds by optimization the actual frequency-dependent source and load terminations that result in the desired overall response with a given transistor. Simply stated, one takes the original transistor parameters, inserts them between the variable source and load, and changes the terminations until the desired performance is reached.

For example, for a transistor with measured 50-ohm s-parameters, it is possible to determine what kind of source and load will result in a desired transducer gain G_T through a given frequency range. It is necessary to solve the function

$$G_T = \frac{|s_{21}|^2 \left(1 - |\Gamma_S|^2\right)\left(1 - |\Gamma_L|^2\right)}{\left|(1 - s_{11}\Gamma_S)(1 - s_{22}\Gamma_L) - s_{12}s_{21}\Gamma_L\Gamma_S\right|^2}$$

The terminations Γ_S, Γ_L, and the S-parameters may all be frequency-dependent. For the generalized case, this problem can be solved only by optimization since there are two unknowns in one equation.

When the desired source and load are found, matching networks can be designed to transform the 50-ohm source and load terminations to the desired new source and load. This approach, called Negative Image Device Modeling, simply replaces the experimental approach in which the desired source and load terminations are determined by the use of triple-stub tuners cascaded to the 50-ohm terminations.

Practical microwave CAD programs will not allow direct changes of source and load terminations during optimization. Variable source and load terminations can, however, be simulated by cascading, to the 50-ohm terminations, ideal transformers with variable impedance ratios (to change the real part of the terminations) and variable reactive elements (to change the imaginary parts). In most cases, the resistive part and one or two reactive components will be sufficient to characterize the desired termination, even for broadband designs.

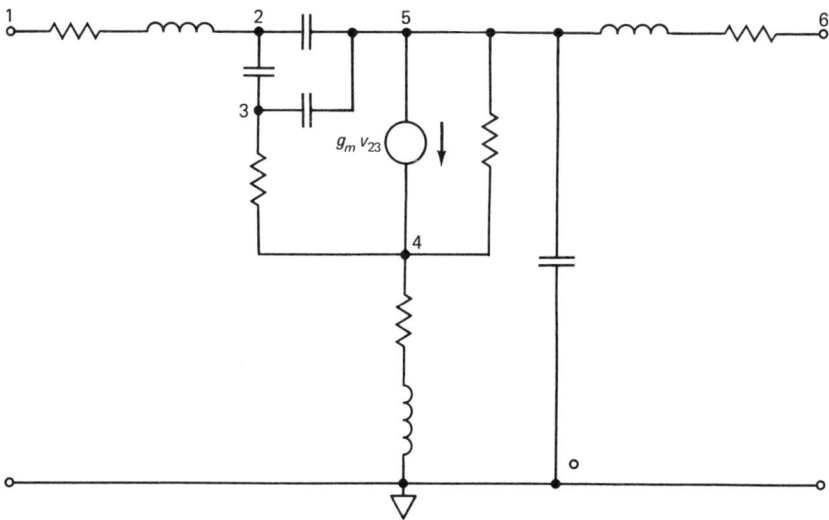

Fig. 4-35. Small-signal RF equivalent circuit of an NEC 700 GaAs FET. Component values are generally found by optimization.

Depending on the bandwidth, the matching networks can then be designed by Smith Chart manipulations, synthesis, or optimization. Since this modeling approach takes into consideration transistor gain rolloff, synthesis programs without gain-slope specifications become usable. SUPER-COMPACT includes a special routine, called PMS (Port Model Synthesis), to model equivalent circuits to approximate the desired source and load terminations for a specified amplifier performance.

4.7.1. Transistor Equivalent-Circuit Modeling

Device designers need to find equivalent circuits to describe the broadband behavior of transistors. For complete characterization, nonlinear signal-dependent models are needed. However, small-signal behavior can be approximated by linear models, such as the one shown in Figure 4-35 for a GaAs microwave FET. The TOUCHSTONE data file used in the optimization is displayed in Figure 4-36, while Figure 4-37 illustrates the procedure used to find the component values of the device model.

```
CKT
  SRL   1 2   R=3.5 L\0.01036
  CAP   2 3   C\ 0.34392
  RES   3 4   R\ 1.24561
  SRL   4 0   R\3.23321 L\0.06052
  CAP   2 5   C\ 0.02611
  SRL   5 6   R=2.0 L\0.17433
  CAP   5 0   C\ 0.01408
  VCCS  2 5 3 4  M=-0.06  A=0  R1=1E6  R2\345.31970  F=1E9  T=2
  CAP   3 5   C\0.02943
  DEF2P 1 6   A          ! A is the circuit model of a FET

  S2PA  1 2 0  NEC700
  DEF2P 1 2   B          ! B is the actual FET

FREQ
  SWEEP 2 18 4           ! USE THIS TO OPTIMIZE
! SWEEP 1 24 1           ! USE THIS FOR DISPLAY

OUT
! DISPLAY S11, S22, S21, S12 ON SC3
  A  S11
  A  S22
  A  S21
  A  S12
! DIRECT SPAR TO SCN
! A  SPAR

OPT
  A  MODEL  B            ! requests optimizer to adjust data to make A = B
```

Fig. 4-36. TOUCHSTONE data file to optimize the small signal FET model of Figure 4-35.

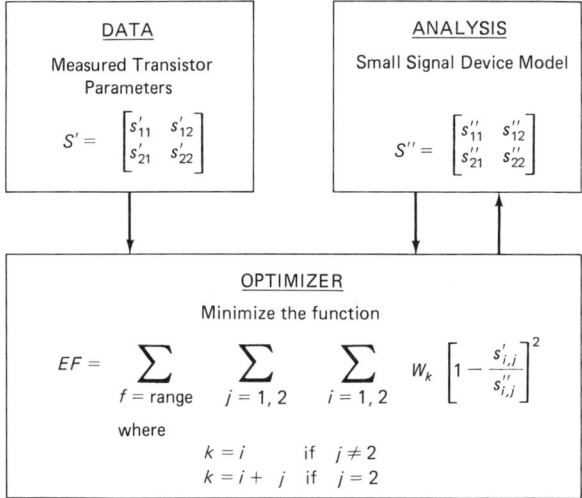

Fig. 4-37. The error function of the optimization compares the four frequency-dependent s-parameters of the model (S'') with the measured broadband data (S'). Components of the model are optimized until the two sets of s-parameters are equal ($S' = S''$).

4.8. CIRCUIT CONTENTS LAYOUT AND MASK DESIGN

Although the design part of the circuit development has received considerable attention for the last 20 years, the crucial part that follows up the circuit design, creation of the circuit layout mask, has been totally neglected. Until very recently, virtually all microwave manufacturers throughout the world, small or large, laid out circuits by tedious manual procedures which often required artistic ability beyond the capabilities of many engineers. Typically, the circuit designer arrived at an initial layout by moving templates around the drawing board until everything fit into the appropriate space. Then a critical review was often required by a more senior designer who "eyeballed" the possible coupling effects among the various components and checked that fundamental design rules, i.e., critical spacings, etc., had not been violated. At that point, the designer made a sketch of the layout and turned it over to the drafting department for a detailed drawing. Next, a ruby mask was cut from the drawing and the final mask was created by appropriate photo reduction. The time requirement from the completion of circuit design to the creation of the mask typically varied from one week to two months. Since the procedure involved several people and groups, the possibility of human error was very high.

Creating a computerized process to convert an electrical design to a physical layout requires the unusual combination of microwave, mechanical, and software engineering. Such programs need to solve the existing microwave problems, offer flexible mechanical drafting, and have a high level of user interaction. Consequently, the first commercial layout program, AUTOART, was introduced only in 1983, on a super-mini computer.

The convenient and high-resolution graphics now available on PCs have become attractive to the software developers of microwave circuit layouts, resulting in several new products that are now available at reasonable cost. Two commercially available programs, AUTOART and MICAD, fall into this category. Both programs are capable of automatic conversion from an electrical circuit file (of SUPER-COMPACT or TOUCHSTONE, respectively), offer interactive layout manipulations, and interface to mask generators. The programs accept all standard transmission line elements and are capable of creating additional spaces and forms necessary for component mounting and tuning.

In addition to converting TOUCHSTONE circuit files, MICAD can accept

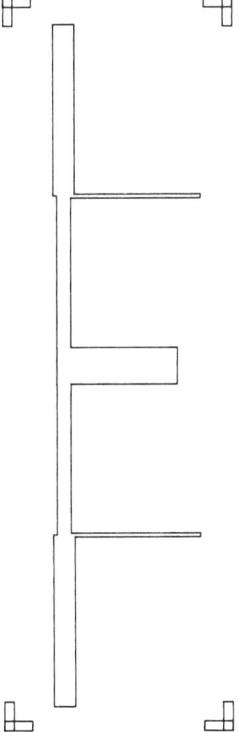

Fig. 4-38. Metalization mask of the 5G Hz lowpass filter, created with the MICAD program. Discontinuities are inserted in the form of "TEE-junctions" between the stubs and cascade transmission lines.

direct input either by element specifications or by geometrical references. The program is capable of generating multiple overlays (i.e., conductor, resistor), driving pattern generators and coordinatographs directly, or interfacing with mechanical CAD systems such as ComputerVision or CALMA.

4.8.1. Circuit Layout with MICAD

To illustrate the automated circuit layout, the 5-GHz lowpass filter optimized in Section 4.5.1 was submitted to MICAD. Figure 4-38 shows the resulting layout with mask alignment marks for microstrip realization.

4.9. NONLINEAR CIRCUIT DESIGN

While most of the small-signal, steady-state linear circuit design problems have been solved, nonlinear circuit design is in much worse condition. Although linear approximations may be used to get a ballpark solution, most high-frequency engineers rely on an empirical approach to design power amplifiers, oscillators, multipliers, and so forth. The fundamental problem is that while the linear circuit problems have been well researched and tied to measurable quantities, many of the nonlinear problems still need investigation and experimental verification. Most of those who use any CAD tool rely on SPICE or an equivalent program that is basically developed for low-frequency applications and does not have microwave circuit elements or capabilities. Some promising software development has been conducted at various universities, using the harmonic balance technique and commercially available versions will soon be available from software vendors. One SPICE derivative has also been modified for microwave usage and is further described in the next section.

4.9.1. The MICROWAVESPICE™ Program

MICROWAVESPICE is an enhanced, interactive, menu-driven version of the Berkeley SPICE program. This new version is available on the IBM and HP Vectra personal computers, taking full advantage of each system's graphics capabilities and using the TOUCHSTONE full-screen editor. Lossy TEM lines, single and coupled microstrip lines, microwave BJT, FET and diode models, and user-defined linear and nonlinear functions are also included. The program can generate s-parameters for active device models at various power levels and temperatures and store them in files for TOUCHSTONE processing. With such combinations, a designer can predict both dc and ac performance, taking advantage of transient, harmonic, and noise analysis.

S-parameter Generation with MICROWAVESPICE. S-parameters for a network are generated by first defining and using a subcircuit in a larger network,

COMPUTER-AIDED DESIGN AT MICROWAVE FREQUENCIES   125

```
CKT
    S2PA_A1      10   20   0         GASFET.S2P  [MODEL=GASFET]
    DEF2P        10   20              TRAN
    TRAN_X1      10   20
    CAP_C1        5   10   C=1E-6
    CAP_C2       20   30   C=1E-6
    DEF2P         5   30              FET
MODEL
    GASFET      GAS  MODEL=2  BETA=.01  VBI=0.8  RD=2  RG=2
         +           RS=2  CGDO=30F  CDS=.3P  FC=.5  TAU=5P  AF=1.0
         +           CGSO=0.3P  A0=.80  A1=.333  A2=-.0625
         +           A3=-0.205  VDSO=3.0  GAMMA=2.5
SOURCE
    FET         IVS_VDS     20    0        DC=5.0
    FET         IVS_VGS     10    0        DC=-1
    FET         IVS_VIN      5    0        AC=1
    FET         RES_RIN      5   10        R=50
    FET         RES_ROUT    30    0        R=50
CONTROL
    FET         DC          VDS 0 12 .25 VGS -2 0 0.5
    FET         AC          LIN 10 1G 10G
    FET         OPTIONS     LIMPTS=5000
SPICEOUT
    FET         DC          I(ALL)
    FET         DC          ALL
    FET         SP*         X1 GASFET.S2P  50
```

Fig. 4-39. MICROWAVESPICE input file, describing a nonlinear GaAs FET model. The last statement instructs the program to generate and store small-signal s-parameters in a file.

then requesting a linear ac sweep (specifying the number of frequency points, the starting frequency, and the ending frequency). After the ac analysis is complete, the data is readily available in table or graph form, on screen or on paper.

The MICROWAVESPICE circuit file shown in Figure 4-39 illustrates the

```
! GASFET CHARACTERISTIC CURVES AND SPARAMETERS
!    TEMPERATURE=  27.000 DEG C      12/6/85      8:39:43
# S  GHz  MA  R  50.0
! FREQ    M(S11)  A(S11)   M(S21)  A(S21)   M(S12)  A(S12)   M(S22)  A(S22)
  1.000   .9941  -10.39   3.9869  166.98   .0173   81.07    .8385  -13.22
  1.474   .9874  -15.23   3.9385  160.90   .0251   76.93    .8358  -19.36
  1.947   .9785  -19.97   3.8742  154.94   .0327   72.89    .8322  -25.37
  2.421   .9678  -24.61   3.7961  149.10   .0398   68.97    .8280  -31.21
  2.895   .9557  -29.10   3.7065  143.41   .0465   65.20    .8233  -36.87
  3.368   .9425  -33.45   3.6079  137.90   .0526   61.58    .8183  -42.31
  3.842   .9285  -37.65   3.5026  132.56   .0583   58.14    .8132  -47.52
  4.316   .9142  -41.69   3.3928  127.41   .0635   54.86    .8081  -52.50
  4.789   .8998  -45.56   3.2805  122.44   .0681   51.77    .8032  -57.24
  5.263   .8855  -49.28   3.1673  117.67   .0723   48.84    .7984  -61.75
  5.737   .8715  -52.84   3.0546  113.06   .0760   46.09    .7941  -66.02
  6.211   .8580  -56.25   2.9434  108.68   .0793   43.50    .7900  -70.07
  6.684   .8450  -59.52   2.8347  104.45   .0823   41.06    .7864  -73.90
  7.158   .8327  -62.65   2.7289  100.39   .0849   38.78    .7831  -77.52
  7.632   .8210  -65.64   2.6267   96.49   .0871   36.64    .7802  -80.94
  8.105   .8099  -68.52   2.5282   92.75   .0891   34.63    .7778  -84.18
  8.579   .7996  -71.27   2.4337   89.16   .0508   32.75    .7756  -87.24
  9.053   .7899  -73.91   2.3432   85.71   .0923   30.98    .7739  -90.14
  9.526   .7808  -76.45   2.2568   82.39   .0936   29.32    .7724  -92.88
 10.000   .7724  -78.89   2.1743   79.20   .0947   27.77    .7712  -95.48
```

Fig. 4-40. Computed 50-ohm s-parameters from 1 to 10 GHz, generated by MICROWAVESPICE using the data file shown in Figure 4-39.

method for creating and storing s-parameters. The resultant data is shown in Figure 4-40.

4.10. ILLUSTRATIVE EXAMPLES

To summarize the CAD functions described above, two design examples will be shown. The first one demonstrates impedance matching to a one-port load, using lumped elements. The second example is a low-noise amplifier design using microstrip transmission lines.

4.10.1. Impedance Matching to a Complex Load

When the real parts of two terminations to be matched have large ratios (i.e., 10 or more), synthesis is recommended even for relatively narrow frequency bands. If one or both terminations are complex, the real and reactive parts need to be separated because classical synthesis accepts only real terminations. Depending on the impedance response of the terminations, finding the real and reactive parts of simple equivalent circuit(s) may already require some computer assistance and familiarity with the Smith Chart. Generally, if the impedance or admittance plot of the termination follows the constant resistance or constant conductance circles, a simple series or parallel combination of a resistor and parasitic element is sufficient to model the termination. Synthesis can then be performed by temporarily moving the reactive part(s) of the termination(s) into the circuit.

The following example illustrates a procedure in which a 50-ohm source is matched to a complex load whose impedance does not follow either the constant resistance or constant conductance circles. Consequently, simple series or parallel modeling is not adequate; the load equivalent circuit must include both series and parallel components. A designer familiar with the Smith Chart can approximate the network for this narrow passband. The task, however, becomes more difficult as the bandwidth increases.

The procedure uses the SYNMAT and SUPER-COMPACT PC programs in designing the matching network. Since the latter has already been introduced, it is appropriate to describe the SYNMAT programs next.

The SYNMAT Program SYNMAT is an interactive matching network synthesis routine which works with lumped elements only, though it offers a lumped-to-distributed element conversion that may be suitable for narrowband applications. The program has a unique modeling section which evaluates all possible two- and three-element RLC combinations and selects the best ones to fit the response of a given termination, described by its measured reflection coefficient.

Step-by-Step Design Procedure. The procedure includes the following steps:

1. The SYNMAT program is used first to find the most suitable equivalent network to model the complex frequency-dependent load, shown in Figure 4-41. Based on the evaluation of the various impedance functions, a resistor and parallel LC or series L with a parallel C result in the lowest error (see Figure 4-42). Optimizing the later combination does not improve the SYNMAT solution significantly; the optimum values are $r = 2$ ohms, $L = 4.85$ nH and $C = 131$ pF.
2. The circuit design begins from the real part of the complex termination with the synthesis of a two-port network to match 2 to 50 ohms. "Exact parasitic absorption" is used to fully include the 4.85 nH parasitic inductance of the termination in the circuit. The next element may not be exact, but it will have to be a capacitor of at least 131 pF. The rest of the circuit is unrestricted, though for optimum performance equal amounts of highpass and lowpass elements are chosen (2 + 2). The synthesis requires

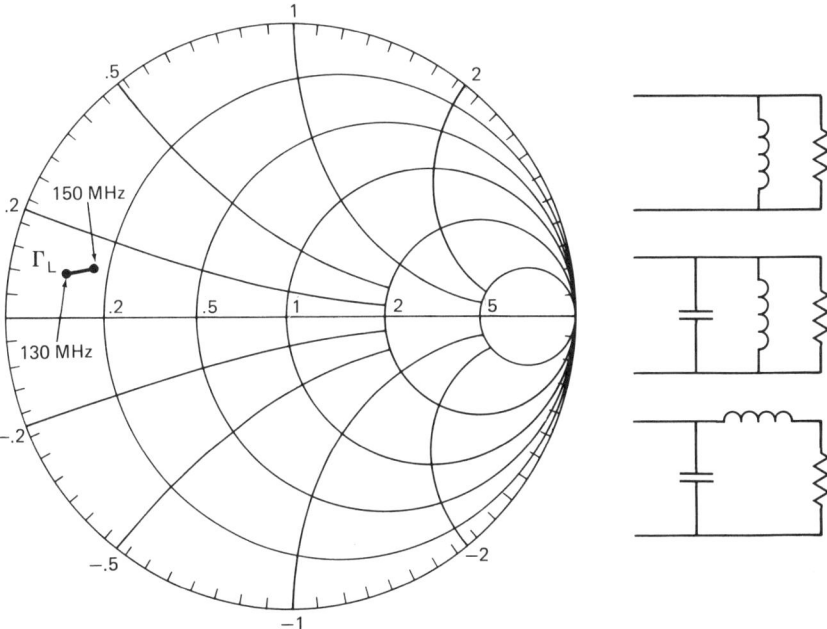

Fig. 4-41. Measured response of the one-port reflection coefficient with three possible equivalent circuits, found by the SYNMAT program.

128 ANALYSIS AND DESIGN OF ELECTRONIC CIRCUITS USING PCs

```
SPARAMETER DATA USED FOR MODEL :
   f(GHz)     S11MAG      S11ANG      S21MAG      S22MAG      S22ANG
   0.130      0.810       168.659     0.000       0.000       0.000
   0.150      0.728       166.031     0.000       0.000       0.000

ELEMENT VALUES ARE IN  Ohms   pF   nH

BEST 2-ELEMENT MODEL FOR S 1  1
    LP         R
   17.6665    12.4510
FIT IS EXACT AT Fupper. . . .ERROR AT Flower =           39.066    %
APPROXIMATE WORST CASE ERROR IN dB =                      1.432   dB

BEST THREE ELEMENT MODEL OF SERIES-SERIES or SHUNT-SHUNT TYPE FOR S 1  1
    LP         CP          R
   7.2520     91.5138     12.4510
FIT IS EXACT AT Fupper. . . .ERROR AT Flower =            0.422    %
APPROXIMATE WORST CASE ERROR IN dB =                      0.018   dB

BEST SERIES-SHUNT or SHUNT-SERIES THREE ELEMENT MODEL FOR S 1  1
    CP         LS          R
  131.5978    4.8421      1.9910
FIT IS EXACT AT Fupper. . . .ERROR AT Flower =            0.157    %
APPROXIMATE WORST CASE ERROR IN dB =                      0.007   dB
```

Fig. 4-42. SYNMAT approximation finds that the third combination (R, LS, CP) results in the lowest error (.157%) in modeling the measured reflection coefficients.

Norton impedance transformation to obtain a 50-ohm load resistance. At the end of the synthesis, the circuit is separated into two parts: (1) the 4.85-nH inductor and 131-pF of the parallel capacitor are moved back into the source termination, and (2) the rest forms the matching network (Figure 4-43).

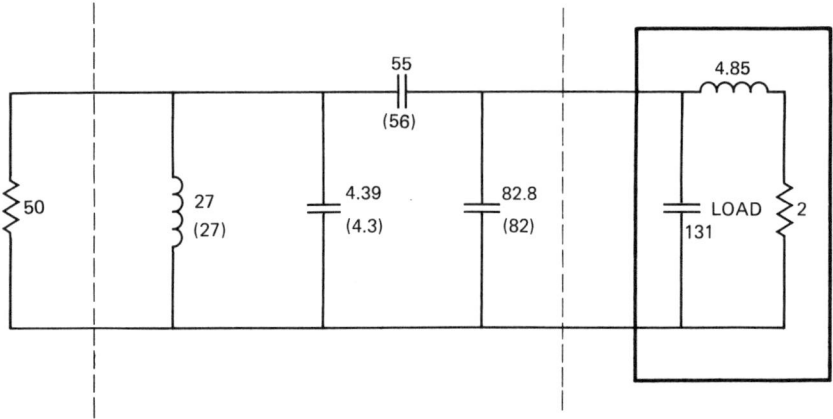

Fig. 4-43. Synthesized network to match 50 ohms to the complex load of Figure 4-42. Component values in parentheses indicate the nearest available ±5% standard components.

3. The synthesized circuit is analyzed between a 50-ohm source and the complex load, using idealized (lossless) components. The result is an equal-ripple response with nearly zero loss and ripple through the 130–150-Mhz passband. However, the synthesized values are not practical since they would have to be specially produced for the nominal values. Instead, all components are replaced by the nearest available standard 5% parts. At this point, finite losses are also added to the matching elements. The previously perfect response is now distorted, as shown in Figure 4-44.
4. Next, statistical tolerance (Monte Carlo) analysis is performed to predict yield under realistic production environment. All components, including the complex load, are assigned $\pm 5\%$ tolerances, assuming uniform distribution. Circuits with insertion loss less than .3 dB are acceptable.

The program has predicted a 84.6% yield, as shown in Figure 4-45. Since the normal response rolls off at the low end of the passband, one would expect poor yield at those frequencies. Examining the tolerance analysis output confirms this: 6.2% of the circuit fails at 130 MHz, and so forth. If this yield is not acceptable, a cost decision is needed to determine whether tighter tolerances or tuning will provide the best alternative.

A more logical way of predicting and controlling yield is the so-called "toleranced design," wherein the tolerance effects are included in the synthesis or

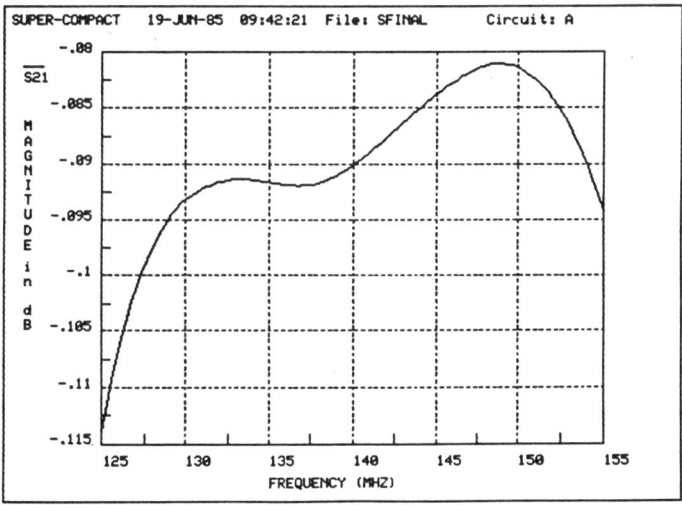

Fig. 4-44. The perfect initial equal-ripple response is changed by deviating from nominal to standard available component values.

130   ANALYSIS AND DESIGN OF ELECTRONIC CIRCUITS USING PCs

```
# Circuit  Frequency     Parm  Hist_Low  Nominal  Hist_High  Pass    Low   High
1:A        130.000MHz    MS21  -0.405    -0.093   -0.092     93.8%   6.2%  0.0%
2:A        132.500MHz    MS21  -0.359    -0.091   -0.093     97.0%   3.0%  0.0%
3:A        135.000MHz    MS21  -0.284    -0.092   -0.088     96.8%   3.2%  0.0%
4:A        137.500MHz    MS21  -0.372    -0.092   -0.086     98.2%   1.8%  0.0%
5:A        140.000MHz    MS21  -0.363    -0.09    -0.083     98.8%   1.2%  0.0%
6:A        142.500MHz    MS21  -0.307    -0.087   -0.087     99.4%   0.6%  0.0%
7:A        145.000MHz    MS21  -0.248    -0.084   -0.083    100.0%   0.0%  0.0%
8:A        147.500MHz    MS21  -0.22     -0.081   -0.079    100.0%   0.0%  0.0%
9:A        150.000MHz    MS21  -0.227    -0.081   -0.081    100.0%   0.0%  0.0%
------------------------------------------------------------------------------
Yield = 423/500 (84.599998 %)
```

Fig. 4-45. Tabulated results of the Monte Carlo analysis reveals that most of the failures are at the low-frequency end of the passband. Predicted yield is 84.6%.

optimization. Unfortunately, this process increases the dimension of the design and puts greater stress on the computer. CPU time requirements are easily increased by two or three orders of magnitude. However, with the proposed interface between PCs and the more powerful mainframe computers, this task should soon be possible.

4.10.2. Low-Noise Amplifier Design

>Design goal: 20-dB gain, noise figure less than 1.5 dB through the frequency range of 8–10 GHz.

An efficient and systematic method of designing a two-stage amplifier is to begin with individual stages to provide a matched interstage for the complete amplifier. Using the available power gain (G_a) method allows us to select the source impedance of the first stage for minimum noise and create a matched output impedance. The second stage is designed by the operating power gain (G_p) method, which results in a matched input. Then the two stages can be cascaded without changing the performance of the individual stages.

Typically the interstage matching network is designed as one network since the desired input and output impedances are known from the single-stage design procedures. Therefore the output impedance of the first stage should be transformed directly to the desired source impedance of the second stage. However, for illustrative purposes, the following outline treats each stage separately; the interstage network is made up by combining the output network of the first stage and the input network of the second stage. As mentioned, direct interstage network design will result, in general, in a simpler network.

Design Outline.   Since the required bandwidth is only about 20%, a simple design can be done by selecting the matching components at the center frequency

(9 GHz) and optimizing the performance of the complete circuit for the full bandwidth (8–10 GHz). This design procedure follows:

1. Plot the constant-noise, constant available gain, and input/output stability circles of the first stage at the center frequency. Minimum noise operation ($F = .95$ dB) can be realized at approximately 10.5-dB gain, inside the stable source impedance region. The load impedance is chosen at the conjugate of the resultant output impedance (Figure 4-46).
2. Select suitable input and output topologies using distributed circuit components. Since the desired source reflection is almost exactly on the constant conductance circle of one ($G = 1$), a parallel element to the 50-ohm source can be used at the input. A quarter-wave cascade transmission line (UE) and a parallel shorted stub provide appropriate impedance transformation at the output as shown in Figure 4-47. Optimization will extend the response of both circuits for the 8–10-GHz bandwidth. At this point, lossless elements are used; the effect of physical losses will be examined later.

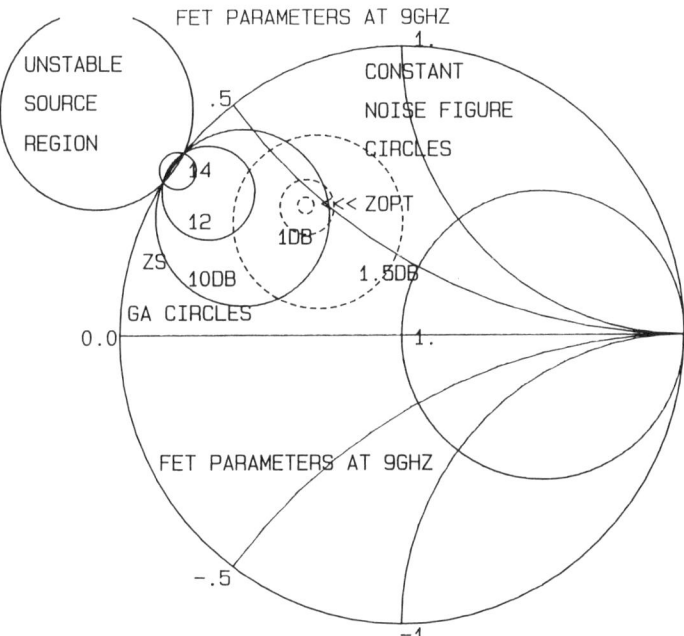

Fig. 4-46. Source impedance of the first stage is chosen for minimum-noise-figure operation at the 10.5-dB gain level. The output is then conjugate-matched to provide the highest available gain.

132 ANALYSIS AND DESIGN OF ELECTRONIC CIRCUITS USING PCs

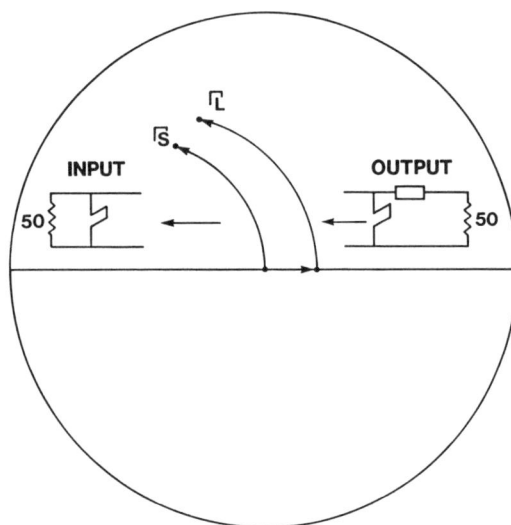

Fig. 4-47. Impedance transformations on the Smith Chart create the desired source and load terminations. The former can be realized by a simple shunt stub across the input, while the load requires a two-element transformer.

3. Analysis of the first stage shows 10.5-dB gain, .95-dB noise, and a virtually perfect output match at the center frequency. Since the input is tuned for minimum noise, the input reflection coefficient is quite high. This stage is potentially unstable, but since resistive stabilization would hurt the noise performance, resistance will not be added to the circuit. If the next stage is also designed for 10.5-dB gain, the cascaded ideal amplifier will have 21-dB gain, with a 1-dB margin to compensate for losses and tolerance effects.
4. Second-stage design begins with plotting constant operating power-gain circles and selecting a load point as far as possible from the unstable load region. After the load has been selected, the conjugate input match is chosen as the source (Figure 4-48). In this case, both the source and load are near the constant conductance circle of $G = 1$, so again, parallel shorted stubs can be used as matching networks. Most likely, the 20% bandwidth can be covered by optimizing these simple networks; wider passbands would require larger-order networks and synthesis in addition to Smith Chart approximation. Examination of the constant-noise circles predicts a second-stage noise figure of 1.5 dB. The source and load terminations are both in the stable regions. Analyzing the second stage verifies the expected gain and noise performance with a nearly perfect input match at 9 GHz. If this stage is cascaded to the matched output of the first stage,

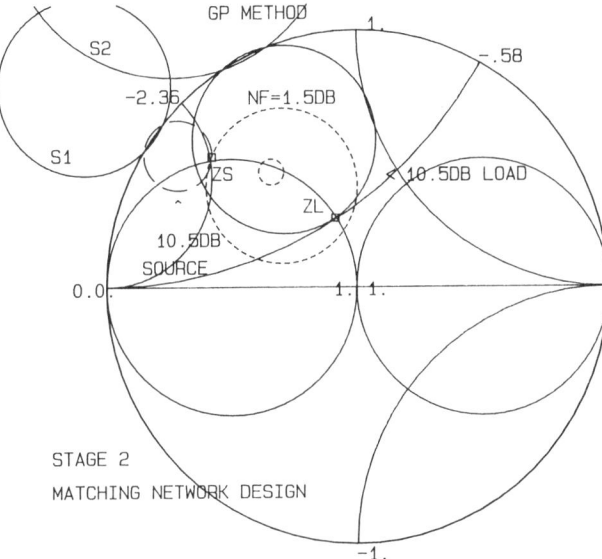

Fig. 4-48. The second stage requires only simple simple impedance transformations; single parallel stubs across the source and load terminations are sufficient.

the gains of the two stages should add directly at 9 GHz. The overall noise figure of the two-stage circuit is determined by the relationship

$$F = F_1 + \frac{F_2 - 1}{G_a}$$

where F_1 and F_2 are the noise figures of Stage 1 and Stage 2 respectively and G_a is the available gain of Stage 1. Substituting the appropriate power gain and noise figures predicts a .15-dB increase in overall noise figure due to second-stage contribution. Analyzing the two-stage cascade shows the expected performance at 9 GHz, where the interstage is matched. However, at other frequencies, where the match does not exist, the performance rapidly deteriorates and it is necessary to optimize the circuit. The impedances and electrical length of all transmission lines are declared to be variables.
5. Optimization of the amplifier sets the gain at 21 dB while maintaining an overall noise of less than 1.4 dB for the full bandwidth. The responses before and after optimization are shown in Figure 4-49.
6. After converting the ideal matching networks to actual transmission lines, resistive and dielectric losses, dispersion, discontinuities and other

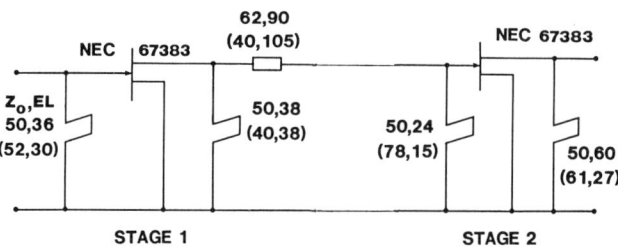

Fig. 4-49. Schematic of the two-stage low-noise amplifier, depicting the electrical transmission line parameters before and (in parentheses) after the optimization. Impedance values are in ohms; line lengths are given in degrees.

physical effects can be examined. The electrical-to-physical conversion is done again by the LINECALC program. Table 4-3 summarizes the electrical parameters and the corresponding physical dimensions of the transmission lines of the optimized amplifier. Amplifier noise increases due to input network losses, while other losses and parasitics mainly affect gain performance. Figure 4-50 compares the performances of the ideal (lossless elements) electrical and converted physical circuits. An additional optimization is necessary to compensate the effects of physical losses and parasitics. Figures 4-51 and 4-52 show the final physical circuit and its response.

7. Layout may also affect the performance, and additional optimization may again be needed to correct resulting changes. Dc bias may be fed to the devices through the parallel shorted stubs, in which case, the appropriate parasitics should be added to model dc decoupling, and so forth. The first

Table 4-3. Results of Transmission Line Parameter Conversions

Substrate Parameters
ER = 9.9 H = 15 mil T = .15 mil F = 9 GHZ
RHO = 1 RGH = 4 micron TAND = 0.0006

Electrical Parameters		Physical Dimensions	
Z_0 Ohm	E_{eff} Degree	Width (mil)	Length (mil)
52	30	12.8	42.4
40	38	21.5	52.3
40	105	21.5	144.4
78	15	4.4	22.0
61	27	8.9	38.7

COMPUTER-AIDED DESIGN AT MICROWAVE FREQUENCIES 135

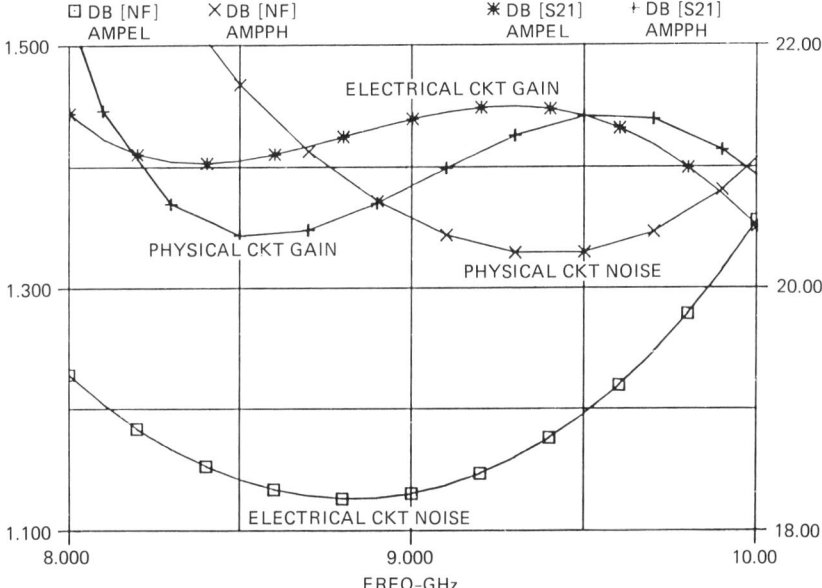

Fig. 4-50. Adding the discontinuities, parasitics, and physical losses changes the amplifier's performance; the most noticeable effect is the increase of noise. Performance can however be improved by additional optimization.

phase of the layout is performed automatically by the MICAD program, which can convert TOUCHSTONE circuit files into the simplest physical layout. Some of the circuit elements, i.e., transistors, capacitors, etc., will require additional inputs to specify the space needed to attach the physical components to the circuit.

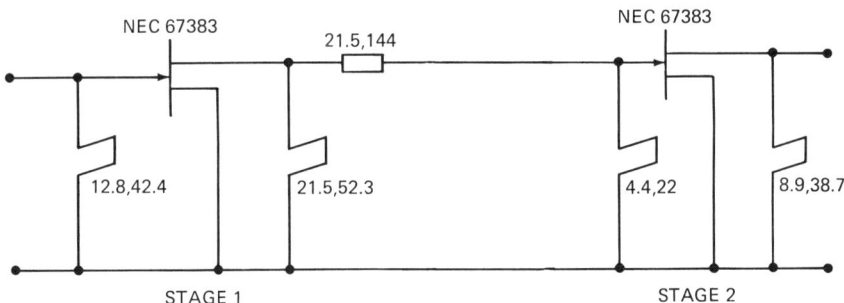

Fig. 4-51. Schematic of the physical circuit, after the changes of the final optimization. Transmission line widths and lengths are given in mils.

136 ANALYSIS AND DESIGN OF ELECTRONIC CIRCUITS USING PCs

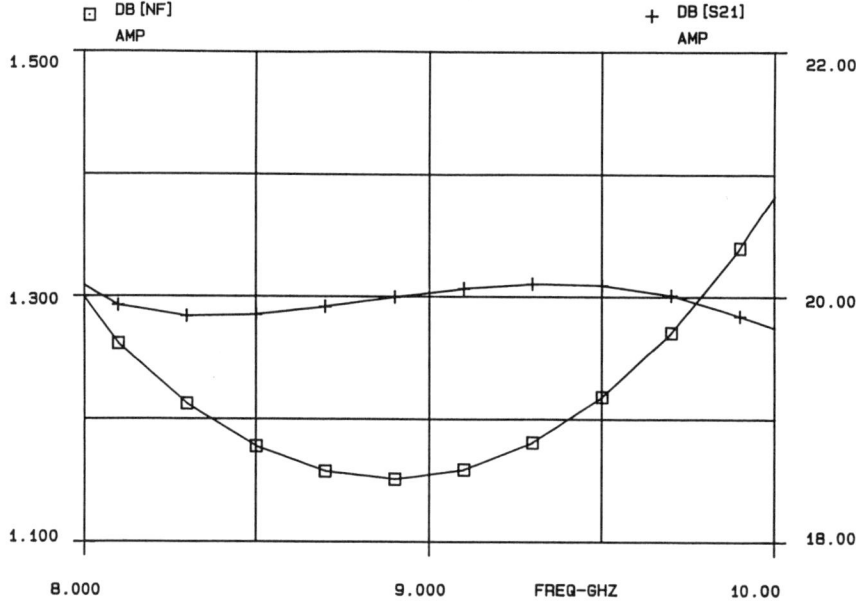

Fig. 4-52. Performance of the final circuit, ready for physical layout.

The autoprocessed layout, created by MICAD, is shown in Figure 4-53. At this point, the designer can further manipulate the layout to add bias connections and to provide grounding via "plated thru-holes." Some of the transmission lines may be "meandered" to minimize the size of the circuit or to provide tuning capabilities. Wiring and assembly diagrams can also be created as shown in Figure 4-54.

Fig. 4-53. Initial MICAD circuit layout made by converting the TOUCHSTONE file of the circuit shown in Figure 4-52. This "auto-layout" may be further manipulated by the designer to add active devices, bias inputs, and coupling and bypass capacitors.

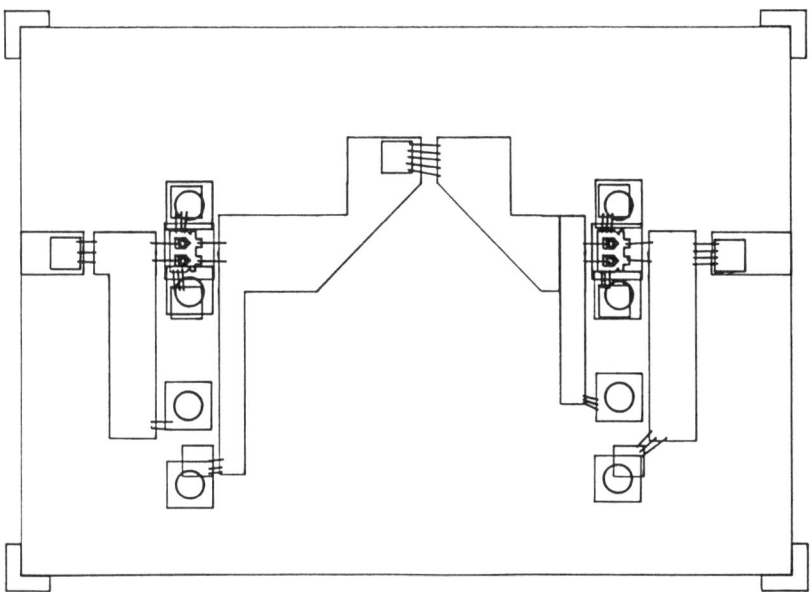

Fig. 4-54. Final assembly diagram of the low-noise amplifier, including chip components and bondwires.

4.11. THE FUTURE

If the trend of computer cost/performance continues to improve, the next generation of personal computers will be sufficient to handle most of the day-to-day design needs of microwave engineers. Seeing the rapid payoff, management is more willing to invest in capital equipment to further improve CAD facilities.

Nonlinear circuit design techniques need major R&D effort—beyond the means of any single business entity. A significant improvement is not likely without some form of government subsidy, or the entrance of a major corporation (i.e., IBM, HP) into the industry, because a meaningful approach requires a combination of measurement techniques, theoretical analysis, and software development that small organizations can not offer or afford. The recently introduced HP 8510 network analyzer system represents a major breakthrough in linear microwave measurement capabilities in both the time and frequency domains. Interfacing the network analyzer with CAD will establish a new link between design and manufacturing, and form the first step toward the concept of total automation.

5

Digital Design And Analysis

Robert Osann
Independent Consultant

5.1. INTRODUCTION

In the early 1980s, computer-aided engineering (CAE software) for digital systems was virtually nonexistent except on large computers with high prices. As graphic workstation prices gradually moved downward, programs for digital design and analysis began to see a growing popularity. For the most part, the applications were semiconductor IC design as opposed to general digital systems. In the last three years, however, a number of things have happened which will bring true CAE closer to reality for the majority of digital design engineers. The first of these is the appearance and acceptance of personal computers as a viable platform for low-cost CAE software. Of particular significance is the IBM personal computer, which currently enjoys the largest installed base and, for the first time, has the memory-addressing capability necessary for serious CAE applications. Second, a major occurrence which greatly affects the proliferation of CAE software on personal computers is the rapid rise in popularity of programmable logic devices (PLDs).

Many digital designers will be introduced to the personal computer through CAE software which supports PLD design. Three examples of this software, in specific applications, will be shown. All of the programs which will be discussed are available for the IBM PC; some were previously available to run on other personal computers such as the Apple or early versions of CP/M. Software having various pricing levels will be included, from extremely low-cost software priced at a few hundred dollars to more sophisticated logic compilers with prices in the one- to two-thousand dollar range. All programs which are shown run resident on the IBM PC; that is, there is no additional processor or memory hardware which must be purchased and installed in order to run these programs. In contrast to this, some CAE programs which are being moved onto the personal computer by some companies originated in larger virtual machine environments and require a drop-in processor board and often a UNIX operating system when being operated in the PC.

5.2. PLDs PROLIFERATE CAE ON PCs.

In spite of all the acronyms, the above message is actually true. Programmable logic devices are rising in popularity far more quickly than any other digital technology. There are many reasons for this popularity. But the most important are flexibility, ease of use, and the ability to replace many packages of TTL gates. Just as important is the availability of development software for design and analysis at reasonable prices on popular personal computers. In fact, the internal architecture of most PLDs encourages, if not forces, the use of sophisticated development software.

Most PLDs today have an internal architecture consisting primarily of a programmable array which feeds AND gates which in turn feeds OR gates (as shown in Figure 5-1). The diagram in Figure 5-1 describes a portion of the PAL16L8 (originally from Monolithic Memories, to whom the registered trademark PAL belongs), probably today's most popular PLD. Some devices even have internal flip-flops connected to output pins which in turn can feed back into the programming array for the purpose of designing sequential logic. The most popular of these devices are the PAL16R4, PAL16R6 and PAL16R8. The internal architecture of these devices is reflected in Figure 5-2.

When the PAL devices first became available in the late 1970s, their use was greatly enhanced by the availability of a program named PALASM (also from Monolithic Memories), a simple Boolean-expression language which allowed engineers to write logic equations to define the function for a PAL. Since then, a number of advances have been made in software tools for PLDs offering higher levels of design expression. The basic concept remains the same—to provide a package any engineer can use for design and simulation of programmable logic to provide better logic designs in less time while consuming less real estate on a PC board than would comparable TTL logic designs.

The design process for programmable logic is pictured in the diagram of Figure 5-3. This process includes both tools for design (hardware assemblers or compilers) and tools for design analysis (logic simulation). The fact that the

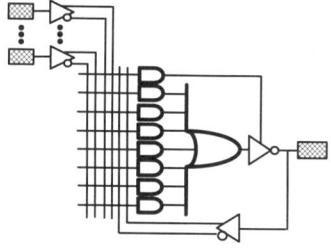

Fig. 5-1. Basic PAL architecture (non-registered).

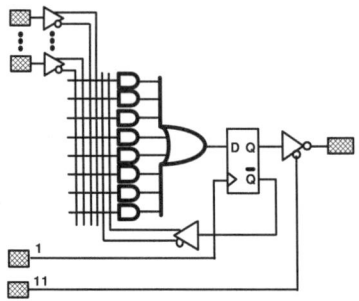

Fig. 5-2. Basic PAL architecture (registered).

most popular PLD tools offer design synthesis capability in addition to analysis is quite significant from a historical standpoint.

Until now, computerized tools for design synthesis including logic reduction or minimization were available only to engineers at large companies having very expensive programs on mainframe computers. Now, for the first time, such tools have become available to all digital design engineers through the popularity of programmable logic. PLDs are everyman's gate array, and the software tools for PLDs are becoming everyman's introduction to CAE.

Many if not most hardware designers today still create their logic designs using pencil and paper. Unless they have been exposed to the microprocessor

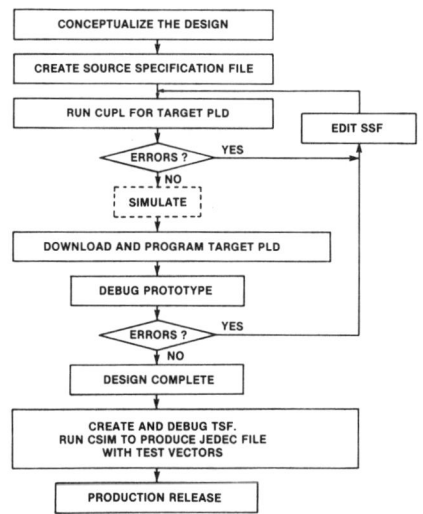

Fig. 5-3. The PLD design process.

and its associated software, they may have never used a computer for their own design work. It therefore follows that for many digital designers, the first introduction to the computer, let alone CAE, is that point at which they first design a PLD into a TTL-based circuit. When exposed in this manner, all will use design synthesis to produce a fuse programming pattern in order to create a programmed PLD.

As shown in Figure 5-3, some will choose to simulate a design before actually programming a part. The simulation (analysis) process will allow a correction of logic design errors before PLDs are actually customized (and sometimes wasted). More important, as in any design process, performing a simulation before actually building hardware eliminates much downstream debugging time. The tradeoff is only the small amount of time spent in the earlier phases of simulation.

Today, however, many engineers newly introduced to PLDs and the computer may not simulate before programming their first parts just because simulation is foreign to them. The customary way to debug logic is with a scope and logic analyzer, not with a computer. As time passes, most will end up performing simulations at some point, either through choice or through an understanding of the value of time spent in simulation. This lesson usually results from extreme pain during system debug, when simulations had not been previously performed.

Even those who have a great deal of luck with designs which work with little debug will eventually be bitten in the case of a complex logic sequencer using PLDs. The complex sequencer or state machine is the most difficult to debug with a scope and logic analyzer but is relatively easy to debug in the earlier design stages with logic simulation.

5.2.1. Three Popular PLD Development Tools

There are many development tools for programmable logic devices on the market today—far more than can be reasonably be covered in detail in this chapter. Most, like PALASM, PLAN, AMAZE, A-PLUS, and HELP, have been developed by various semiconductor manufacturers (Monolithic Memories, National Semiconductor, Signetics, Altera, and Harris) to support only their own devices. Many of these offer only design synthesis support without logic simulation. Of these, we will focus only on PALASM since it includes a logic simulation capability and is the most popular and widespread of these languages. It was also the first successful design and analysis tool for programmable logic. After PALASM, the currently available two programs which are both universal (support devices for all PLD manufacturers) and high level (offer formats of design expression closer to the designer's original thoughts) will be described. These are ABEL from Data I/O and CUPL from P-CAD. In the discussion of these three products, the focus will be on a common applications example which includes both asynchronous and sequential circuit design.

This example involves a fairly typical video subsystem not unlike a simplified version of what one might find in many popular personal computers. The heart of this subsystem is the PLD that performs the primary timing and control functions. The functional block diagram for this video system is shown in Figure 5-4. Notice that both the 8088 microprocessor and 6845 CRT controller are able to address the video screen RAM by way of a multiplexer. This MUX is controlled by a signal from the video controller PLD. The video circuit is designed with what is called "full arbitration," in which the 8088 may read or write the video screen RAM without impacting the information flow between the screen RAM and the video monitor. This eliminates any disturbances which would appear on monitors controlled by lower-performance video circuitry. The video controller PLD controls all timing including the READY signal which places the 8088's CPU in wait states, allowing the microprocessor to access the video RAM transparently.

A diagram showing pin assignments for the video controller PLD is shown in Figure 5-5. Notice this device receives the upper bits of the 8088 address as well as the memory read and write signals. The device performs an address decoding function on these bits and creates the signal VID_ACCESS at the appropriate time. The VID_ACCESS signal in turn feeds back within the device and activates the internal state sequence to provide the necessary arbitration. Here the READY line is driven both active (from having been tri-stated) and low until a transfer has been made. The transfer acknowledge signal (XACK) then becomes active.

The portion of the state sequence in which the 8088 is actually allowed to

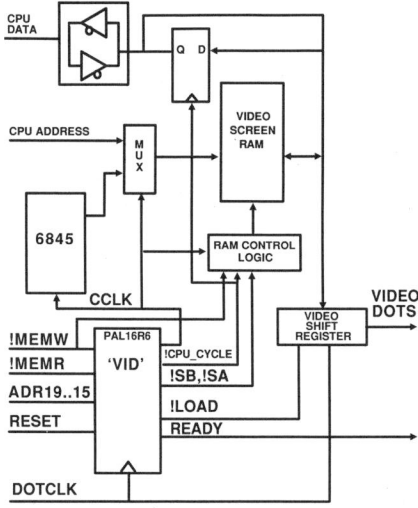

Fig. 5-4. Functional block diagram for video subsystem example.

DIGITAL DESIGN AND ANALYSIS   143

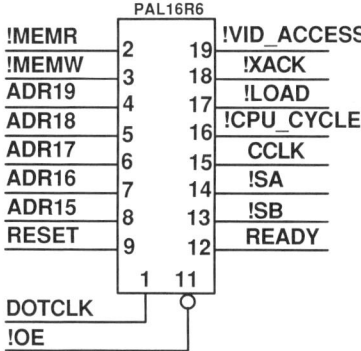

Fig. 5-5. Video controller PLD.

transfer information to and from the video RAM is marked by the signal CPU-CYCLE being active.

The overall dot sequence consists of eight states (there are eight video dots per character displayed on the screen). The video state machine within this PLD keeps track of these eight states according to information expressed in the signals !CCLK, S1, and S0. The character clock (CCLK) serves the purpose as both a state bit and the clock for the 6845 CRT controller chip, and also as the signal which controls the address multiplexer. Further, it provides the clock for the register which stores data just read from the video RAM. Additionally, the signal !LOAD controls the functioning of the video shift register. Both the video shift register and the video controller PLD are clocked by the video dot clock. The device chosen for this example is a PAL16R6, which has six registered outputs, with the internal structure shown in Figure 5-2, and two nonregistered (asynchronous) outputs with internal feedback and programmable tri-state control as shown in Figure 5-1. The signals VID_ACCESS and READY are assigned to the nonregistered output pins while all others will be assigned to registered pins.

The timing diagram shown in Figure 5-6 describes a typical sequence of events including an access of the video circuitry by the microprocessor. The video sequencer continually goes through the eight primary states to display an undisturbed stream of video dots on the screen. Variations in the sequence happen when the video memory is accessed by the microprocessor. Whenever such an access occurs (here starting in state S6), the VID_ACCESS signal will become active causing the READY signal to be actively driven (non-tri-state) going low at the pin. The microprocessor will subsequently enter a wait state which will cause the address and memory control signals to stay at their current values until the READY signal is once again driven active. The next step occurs with the transition from state S2 to state S3 in the presence of an active

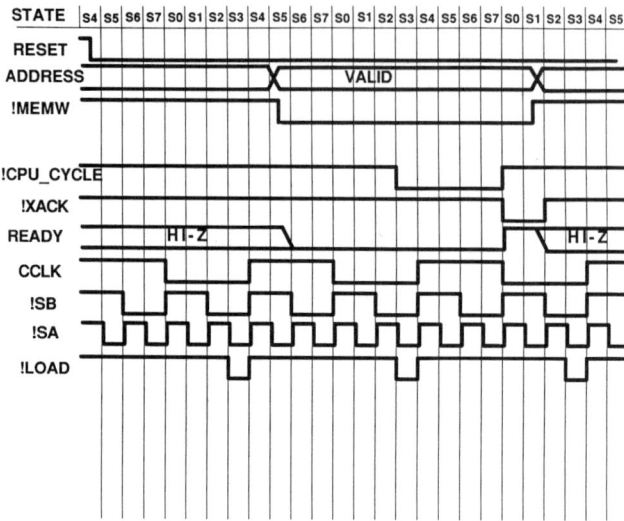

Fig. 5-6. Video controller timing diagram.

VID_ACCESS signal. Here the !CPU_CYCLE signal will become activated and remain as such until the completion of state S7. Following the completion of the !CPU_CYCLE signal and the associated transfer of information between the CPU and the video RAM, the transfer acknowledge signal (!XACK) will go active. This causes READY to go active, both READY and !XACK remaining active until the microprocessor leaves its wait state and deactivates its memory control signals.

5.3. PALASM, THE FIRST PLD DESIGN LANGUAGE

Programmable logic devices actually began in the early 1970s with the appearance of the first FPLA devices from INTERSIL, Fairchild, and Signetics. In spite of this, PLD technology didn't start to accelerate appreciably until the appearance of the PAL device architecture in the late 1970s. This sudden acceleration was not due to any new architectural discoveries. In fact, the PAL architecture is really a subset of the generalized FPLA architecture. The PAL device has fixed connections between the AND gates and OR gates as opposed to programmable connections. If anything, the PAL architecture was simpler and possibly easier to understand for many engineers facing this new technology.

One factor that may have made a direct contribution to the acceleration of PLDs with the introduction of the PAL was the appearance of MMI's PALASM, the first logic design language for any PLD family. At the time it was introduced, PALASM was surprisingly comprehensive. It not only included the ability to deal with symbolic variable names, but also logic simulation capability enabling designs to be verified before devices are actually programmed. The simulation was extremely important in earlier times since the first PAL devices were extremely expensive ($30–$50) and, as is still the case with most PLDs today, were of nonreprogrammable bipolar technology with fusable links for the programming array. In those days, if a wrong pattern was programmed in a PAL, a great deal of money was wasted. But with prices for bipolar PLDs less than $5 each today, wasting devices is not the real loss when PLDs are misprogrammed due to lack of a simulation. Instead, the loss is dollars in engineering time spent finding the problems after the fact.

Figure 5-7 shows the complete PAL design specification for the PAL16R6 video controller according to the PALASM format. The name of the video controller device is VID and therefore the file name containing the PALASM design specification is titled VID.ASM. There are two primary pieces to the PALASM specification: the upper half is the logic design specification, and the lower half is the simulation function table. As we will see in the other language examples, all of the PLD design languages use a function table format for simulation input as well as output. At the top of the PALASM design specification is an area for header information. This allows the user to insert some description of the file in question and the date of the most recent revision. Next is the list of pin assignments which in PALASM are done in the order of the numbering sequence of the pins; in other words, the list begins with DOTCLK because that signal has been assigned to pin 1 and ends with VCC because power has been assigned to pin 20. In PALASM, names which have negative (active-lo) logic assertion levels may be shown in the pin list with this assertion level attached. This is done in PALASM with a slash preceding the variable name.

Next in the PALASM specification are the equations for the output variables. The first four equations are for registered output variables. Notice that the operator for equivalence is :=. This indicates the clocked transfer function associated with a d-type flip-flop. Equivalence for output variables which are not associated with registers (combinatorial outputs) are shown with an = sign alone. Notice that these equations use no parentheses. Also, if variables are not shown directly on the list of pin assignments they do not appear in the equation. PALASM is essentially a lower-level symbolic translator; the logic represented in the equations must correspond exactly to the fuse structure within a target device.

The two nonregistered output pins, VID_ACC and READY, are both assigned to pins which have programmable three-state output control. This is handled in the equations by an IF statement preceding the variable name in the logic equation. In the equation for VID_ACC, the three-state output control is to be

```
p16r6
VID             3-14-86
VIDEO ADDRESS DECODER AND CONTROL SEQUENCER
ASSISTED TECHNOLOGY

DOTCLK /MEMR /MEMW ADR19 ADR18 ADR17 ADR16 ADR15 RESET GND
/OE READY /SB /SA CCLK /CPU_CYC /LOAD /XACK /VID_ACC VCC

XACK        := CPU_CYC * SA * SB * VID_ACC * CCLK * /RESET
             + XACK * VID_ACC * /RESET

CPU_CYC := /XACK * /SA * SB * VID_ACC * /CCLK * /RESET
         + CPU_CYC * SA * SB * VID_ACC * /CCLK * /RESET
         + CPU_CYC * /SB * VID_ACC * CCLK * /RESET
         + CPU_CYC * /SA * VID_ACC * CCLK * /RESET

SA          := /SA * /RESET

SB          := SA * /SB * /RESET
             + /SA * SB * /RESET

IF (VCC) VID_ACC = MEMW * ADR15 * ADR16 * ADR17 * /ADR18 * ADR19
                 + MEMR * ADR15 * ADR16 * ADR17 * /ADR18 * ADR19

IF (VID_ACC) READY = /XACK

CCLK        := SA * SB * CCLK * /RESET
             + /SA * /CCLK * /RESET
             + /SB * /CCLK * /RESET

LOAD        := /SA * SB * /CCLK * /RESET

FUNCTION TABLE

DOTCLK /MEMR /MEMW ADR19 ADR18 ADR17 ADR16 ADR15 RESET /OE
READY CCLK /SB /SA /CPU_CYC /LOAD /XACK /VID_ACC

;                                           /         /
;                                           C         V
;D                                           P         I
;O  /  /  A  A  A  A  R     R               U  /  /  D
;T  M  M  D  D  D  D  E     E  C            _  L  X  _
;C  E  E  R  R  R  R  S  /  A  C  /  /  C  O  A  A
;L  M  M  1  1  1  1  E  O  D  L  S  S  Y  A  C  C
;K  R  W  9  8  7  6  5  T  E  Y  K  B  A  C  D  K  C  COMMENTS
----------------------------------------------------------------
 C  X  X  X  X  X  X  X  H  L  X  H  H  H  H  H  X  POWER ON RESET (S4)
 C  H  H  X  X  X  X  X  L  L  Z  H  H  L  H  H  H  ENTER S5
 C  H  H  X  X  X  X  X  L  L  Z  H  L  H  H  H  H  ENTER S6
 C  H  H  X  X  X  X  X  L  L  Z  H  L  L  H  H  H  ENTER S7
 C  H  H  X  X  X  X  X  L  L  Z  L  H  H  H  H  H  ENTER S0
 C  H  H  X  X  X  X  X  L  L  Z  L  H  L  H  H  H  ENTER S1
 C  H  H  X  X  X  X  X  L  L  Z  L  L  H  H  H  H  ENTER S2
 C  H  H  X  X  X  X  X  L  L  Z  L  L  L  H  L  H  ENTER S3
 C  H  H  X  X  X  X  X  L  L  Z  H  H  H  H  H  H  ENTER S4
 C  H  H  X  X  X  X  X  L  L  Z  H  H  L  H  H  H  ENTER S5
 C  H  L  H  L  H  H  H  L  L  L  H  L  L  H  H  L  ENTER S6, VID ACCESS
 C  H  L  H  L  H  H  H  L  L  L  H  L  L  H  H  L  ENTER S7
 C  H  L  H  L  H  H  H  L  L  L  L  H  H  H  H  L  ENTER S0
 C  H  L  H  L  H  H  H  L  L  L  L  H  L  H  H  L  ENTER S1
 C  H  L  H  L  H  H  H  L  L  L  L  L  H  H  H  L  ENTER S2
 C  H  L  H  L  H  H  H  L  L  L  L  L  L  L  H  L  ENTER S3, STRT CPU_CYC
 C  H  L  H  L  H  H  H  L  L  L  H  H  H  L  H  L  ENTER S4
 C  H  L  H  L  H  H  H  L  L  L  H  H  L  H  H  L  ENTER S5
 C  H  L  H  L  H  H  H  L  L  L  H  L  H  H  H  L  ENTER S6
 C  H  L  H  L  H  H  H  L  L  L  H  L  L  L  H  L  ENTER S7
 C  H  L  H  L  H  H  H  L  L  H  L  H  H  H  L  L  ENTER S0
 C  H  L  H  L  H  H  H  L  L  H  L  H  L  H  L  L  ENTER S1
 C  H  H  X  X  X  X  X  L  L  Z  L  L  H  H  H  H  ENTER S2
----------------------------------------------------------------
```

DESCRIPTION

This device is clocked by the video dot-clock and generates the
clock for the 6845, control for the video shift register and
video memory, as well as arbitration for access to the RAM between
the 8088 and the 6845.

Fig. 5-7. PALASM design specification for video controller (after logic minimization).

always enabled. This is done in PALASM by placing the terms IF (VCC) at the left of its equation. The variable READY has its three-state output controlled by the value of the variable VID_ACC. The equation for the variable READY is therefore preceded by the expression IF (VID_ACC).

At the bottom of the PALASM design specification is the simulation function table. This table indicates the designer's estimation of how the device will behave during a sequence of events. The first element in this section, prior to the function table itself, is an ordering list which contains the variable names in the same order they will appear in the columns of the table. Next comes the function table itself with the labels for each column typed vertically to identify the variables. It is customary that the stimulus (input) variables are shown in the columns on the left side while the response (output) variables are shown on the right side. There is space allowed on the right-hand side of the table for comments which identify what is happening in each row of the table; each row of such a function table is sometimes referred to as a *test vector*. The PALASM function table begins with a dashed line above the table, and at the bottom of the table terminates with a similar dashed line of arbitrary length. Finally, at the bottom of the design specification, following the key word DESCRIPTION, is a verbalization of the function that this particular PLD design performs.

The PALASM design specification shown in Figure 5-7 contains equations which are the result of a logic minimization exercise. Such an exercise must be done manually since the PALASM program does not contain any logic minimization software. In this case, a large number of input variables affect the outputs in question. A manual reduction using a conventional Karnaugh map would be difficult. Figure 5-8 shows the equations originally generated for this example prior to logic minimization. In particular, notice that the equation for the variable SB contains 10 product terms. The registered output to which SB had been assigned has a maximum capability for 8 product terms. It follows that, since the logic as shown in Figure 5-8 would not fit in the PAL16R6 target device, it is necessary to perform some minimization.

```
p16r6
VID             3-14-86
VIDEO ADDRESS DECODER AND CONTROL SEQUENCER
ASSISTED TECHNOLOGY

DOTCLK /MEMR /MEMW ADR19 ADR18 ADR17 ADR16 ADR15 RESET GND
/OE READY /SB /SA CCLK /CPU_CYC /LOAD /XACK /VID_ACC VCC

/CCLK   := /CCLK   * /RESET   * /SB
         +  CCLK   * /RESET   *  SA    *  SB    *  VID_ACC *  XACK
         +  CCLK   *  CPU_CYC * /RESET *  SA    *  SB    *  VID_ACC
         +  CCLK   * /CPU_CYC * /RESET *  SA    *  SB    *  VID_ACC * /XACK
         + /CCLK   * /RESET   * /SA    *  SB
         +  CCLK   * /RESET   *  SA    *  SB    * /VID_ACC
```

Fig. 5-8. PALASM design specification (before minimization).

```
CPU_CYC := /CCLK  *  /RESET   *  /SA  *  SB   *  VID_ACC  *  /XACK
         + /CCLK  *  CPU_CYC  *  /RESET  *  SA   *  SB   *  VID_ACC
         + CCLK   *  CPU_CYC  *  /RESET  *  /SB  *  VID_ACC
         + CCLK   *  CPU_CYC  *  /RESET  *  /SA  *  SB   *  VID_ACC

LOAD := /CCLK  *  /RESET  *  /SA  *  SB

IF (VID_ACC) READY = /XACK

SA := CCLK   *  /RESET   *  /SA  *  VID_ACC  *  XACK
    + CCLK   *  CPU_CYC  *  /RESET  *  /SA  *  VID_ACC
    + CCLK   *  /CPU_CYC *  /RESET  *  /SA  *  VID_ACC  *  /XACK
    + /CCLK  *  /RESET   *  /SA  *  VID_ACC
    + /RESET *  /SA  *  /VID_ACC

SB := CCLK   *  /RESET  *  /SA  *  SB   *  VID_ACC  *  XACK
    + /RESET *  SA  *  /SB  *  /VID_ACC
    + /CCLK  *  /RESET  *  SA  *  /SB  *  VID_ACC
    + /RESET *  /SA  *  SB  *  /VID_ACC
    + /CCLK  *  /RESET  *  /SA  *  SB  *  VID_ACC
    + CCLK   *  /CPU_CYC *  /RESET  *  SA  *  /SB  *  VID_ACC  *  /XACK
    + CCLK   *  CPU_CYC  *  /RESET  *  SA  *  /SB  *  VID_ACC
    + CCLK   *  /RESET   *  SA  *  /SB  *  VID_ACC  *  XACK
    + CCLK   *  /CPU_CYC *  /RESET  *  /SA  *  SB  *  VID_ACC  *  /XACK
    + CCLK   *  CPU_CYC  *  /RESET  *  /SA  *  SB  *  VID_ACC

IF (VCC) VID_ACC = ADR15  *  ADR16  *  ADR17  *  /ADR18  *  ADR19  *  MEMR
                 + ADR15  *  ADR16  *  ADR17  *  /ADR18  *  ADR19  *  MEMW

XACK := CCLK  *  CPU_CYC  *  /RESET  *  SA  *  SB  *  VID_ACC
      + /RESET  *  VID_ACC  *  XACK

FUNCTION TABLE

DOTCLK /MEMR /MEMW ADR19 ADR18 ADR17 ADR16 ADR15 RESET /OE
READY CCLK /SB /SA /CPU_CYC /LOAD /XACK /VID_ACC

;                                             /         /
;                                             C         V
;D                                             P         I
;O  /  /  A  A  A  A  A  R         R          U  /  /   D
;T  M  M  D  D  D  D  D  E         E  C       _  L  X   _
;C  E  E  R  R  R  R  R  S  /      A  C  /  / C  O  A   A
;L  M  M  1  1  1  1  1  E  O      D  L  S  S Y  A  C   C
;K  R  W  9  8  7  6  5  T  E      Y  K  B  A C  D  K   C  COMMENTS
-------------------------------------------------------------------
 C  X  X  X  X  X  X  X  H  L      X  H  H  H H  H  X   POWER ON RESET (S4)
 C  H  H  X  X  X  X  X  L  L      Z  H  H  L  H  H  H  ENTER S5
 C  H  H  X  X  X  X  X  L  L      Z  H  L  H  H  H  H  ENTER S6
 C  H  H  X  X  X  X  X  L  L      Z  H  L  L  H  H  H  ENTER S7
 C  H  H  X  X  X  X  X  L  L      Z  L  H  H  H  H  H  ENTER S0
 C  H  H  X  X  X  X  X  L  L      Z  L  H  L  H  H  H  ENTER S1
 C  H  H  X  X  X  X  X  L  L      Z  L  L  H  H  H  H  ENTER S2
 C  H  H  X  X  X  X  X  L  L      Z  L  L  L  H  H  H  ENTER S3
 C  H  H  X  X  X  X  X  L  L      Z  H  H  H  H  H  H  ENTER S4
 C  H  H  X  X  X  X  X  L  L      Z  H  H  L  H  H  H  ENTER S5
 C  H  L  H  L  H  H  H  L  L      L  H  L  H  H  H  L  ENTER S6, VID_ACCESS
 C  H  L  H  L  H  H  H  L  L      L  H  L  L  H  H  L  ENTER S7
 C  H  L  H  L  H  H  H  L  L      L  L  H  H  H  H  L  ENTER S0
 C  H  L  H  L  H  H  H  L  L      L  L  H  L  H  H  L  ENTER S1
 C  H  L  H  L  H  H  H  L  L      L  L  L  H  H  H  L  ENTER S2
 C  H  L  H  L  H  H  H  L  L      L  L  L  L  L  H  L  ENTER S3, STRT CPU_CYC
 C  H  L  H  L  H  H  H  L  L      L  H  H  L  H  H  L  ENTER S4
 C  H  L  H  L  H  H  H  L  L      L  H  H  L  H  H  L  ENTER S5
 C  H  L  H  L  H  H  H  L  L      L  H  L  H  H  H  L  ENTER S6
 C  H  L  H  L  H  H  H  L  L      L  L  L  L  H  H  L  ENTER S7
 C  H  L  H  L  H  H  H  L  L      H  L  H  H  H  L  L  ENTER S0
 C  H  L  H  L  H  H  H  L  L      H  L  H  L  H  L  L  ENTER S1
 C  H  H  X  X  X  X  X  L  L      Z  L  L  H  H  H  H  ENTER S2
-------------------------------------------------------------------
```

DESCRIPTION

This device is clocked by the video dot-clock and generates the
clock for the 6845, control for the video shift register and
video memory, as well as arbitration for access to the RAM between
the 8088 and the 6845.

Fig. 5-8. (*Continued*)

5.4. CUPL, THE FIRST HIGH-LEVEL UNIVERSAL LANGUAGE FOR PLDs

In the early 1980s it began to be obvious that a number of new players would be entering the group of semiconductor companies making PLDs. New and varied architectures began to appear both from the new companies and from established companies that were evolving their device architectures while creating newer, more innovative, architectures. Over time, more semiconductor companies began to offer logic description languages which supported only the parts they manufactured. It eventually became obvious that universal languages were really needed. These would instead support devices from all manufacturers with a common logic description format. Most of the semiconductor-specific languages tended to be relatively low-level, operating as symbolic-fuse-level translators with some consideration for their specific device architectures. Higher-level language capabilities seemed necessary to enhance the PLD design process further. In May of 1983, the CUPL language first became available, offering both high-level and universal capabilities.

In general, high-level capabilities imply that descriptions can be generated that more closely approximate the original thoughts of the designer. Some of the more elementary capabilities of a high-level language involve the use of parentheses and the distributive property. Also included is expression substitution wherein variables of arbitrary name (intermediate variables) are created and given expressions. These expressions are then substituted for the variable names wherever they are used in other expressions.

Further capabilities of a high-level language involve the ability to group bits together. This is useful for applications which involve address or data buses and refer to those groups of bits in terms of number bases other than binary. Even more sophisticated high-level capabilities involve the modeling of sequential circuits as true state machines with the necessary syntax capability. Also, a high-level language should offer a logic minimization capability such that designers need no longer perform complex and time-consuming Karnaugh map minimization by hand.

Figure 5-9 represents the logic description file for the video controller chip shown in the CUPL language (VID.PLD). For the CUPL language, logic descriptions are kept in files separate from simulator input, function table descriptions. At the top of Figure 5-9 is the header for the logic description file followed by a title block in comments which presents the designer's description of the function for the device. Comment fields for CUPL begin with /* and end with */. Next are the pin declarations, divided into input and output groupings for readability. This is not required since the language will determine which pins are inputs and which are outputs from their usage in the logic description. Note the concept of the indexed variable in which the address bus [ADR19, ADR18, ADR17, ADR16, ADR15] is represented simply as [ADR19. .15].

150 ANALYSIS AND DESIGN OF ELECTRONIC CIRCUITS USING PCs

```
                PARTNO    PLD0000123 ;
                NAME      VID ;
                DATE      4/27/86 ;
                REVISION  02;
                DESIGNER  R.OSANN;
                COMPANY   ASSISTED TECHNOLOGY;
                ASSEMBLY  PC-VIDEO;
                LOCATION  U101;

/**********************************************************************/
/* This device is clocked by the video dot-clock and generates the    */
/* clock for the 6845, control for the video shift register and      */
/* video memory, as well as arbitration for access to the RAM between */
/* the 8088 and the 6845.                                             */
/**********************************************************************/

/*** INPUTS ***/

PIN 1       = DOTCLK      ; /* VIDEO DOT CLOCK                     */
PIN 2       = !MEMR       ; /* 8088 MEMORY READ SIGNAL             */
PIN 3       = !MEMW       ; /* 8088 MEMORY WRITE SIGNAL            */
PIN [4..8]  = [ADR19..15] ; /* 8088 ADDRESS                        */
PIN 9       = RESET       ; /* SYSTEM RESET                        */
PIN 11      = !OE         ; /* OUTPUT ENABLE PIN FOR THIS CHIP     */

/*** OUTPUTS ***/

PIN 12      = READY       ; /* 8088 READY LINE (GEN WAIT STATES)       */
PIN [13..14]= ![SB,SA]    ; /* STATE BITS                              */
PIN 15      = CCLK        ; /* 6845 CHARACTER CLOCK (ALSO STATE BIT)   */
PIN 16      = !CPU_CYCLE  ; /* 8088 ACCESS OF VID RAM IN PROGRESS      */
PIN 17      = !LOAD       ; /* LOAD VIDEO SHIFT REGISTER (OR !SHIFT)   */
PIN 18      = !XACK       ; /* TRANSFER ACKNOWLEDGE AT END OF CPU_CYC  */
PIN 19      = !VID_ACCESS ; /* VIDEO MEMORY ACCESS DECODE              */

$DEFINE S0 4
$DEFINE S1 5
$DEFINE S2 6
$DEFINE S3 7
$DEFINE S4 0
$DEFINE S5 1
$DEFINE S6 2
$DEFINE S7 3

FIELD ADDRESS = [ADR19..15];
FIELD VIDEO_STATE = [!CCLK,SB,SA];
MEMREQ = MEMR # MEMW;


VID_ACCESS = ADDRESS:[B8000..BFFFF] & MEMREQ;
READY      = XACK;
READY.OE   = VID_ACCESS;

SEQUENCE VIDEO_STATE (

PRESENT S0
   IF !RESET & (!VID_ACCESS # VID_ACCESS & !XACK)  NEXT S1 ;
   IF !RESET & VID_ACCESS & XACK                   NEXT S1    OUT XACK ;

PRESENT S1
   IF !RESET & (!VID_ACCESS # VID_ACCESS & !XACK)  NEXT S2 ;
   IF !RESET & VID_ACCESS & XACK                   NEXT S2    OUT XACK ;

PRESENT S2
   IF !RESET & !VID_ACCESS              NEXT S3   OUT LOAD ;
   IF !RESET & VID_ACCESS & !XACK       NEXT S3   OUT [CPU_CYCLE,LOAD] ;
   IF !RESET & VID_ACCESS & XACK        NEXT S3   OUT [XACK,LOAD] ;
```

Fig. 5-9. CUPL logic description for video controller.

```
PRESENT S3
   IF !RESET & (!VID_ACCESS # VID_ACCESS & !(CPU_CYCLE # XACK))
                                           NEXT  S4 ;
   IF !RESET & VID_ACCESS & CPU_CYCLE      NEXT  S4      OUT CPU_CYCLE ;
   IF !RESET & VID_ACCESS & XACK           NEXT  S4      OUT XACK ;

PRESENT S4
   IF !RESET & (!VID_ACCESS # VID_ACCESS & !(CPU_CYCLE # XACK))
                                           NEXT  S5 ;
   IF !RESET & VID_ACCESS & CPU_CYCLE      NEXT  S5      OUT CPU_CYCLE ;
   IF !RESET & VID_ACCESS & XACK           NEXT  S5      OUT XACK ;

PRESENT S5
   IF !RESET & (!VID_ACCESS # VID_ACCESS & !(CPU_CYCLE # XACK))
                                           NEXT  S6 ;
   IF !RESET & VID_ACCESS & CPU_CYCLE      NEXT  S6      OUT CPU_CYCLE ;
   IF !RESET & VID_ACCESS & XACK           NEXT  S6      OUT XACK ;

PRESENT S6
   IF !RESET & (!VID_ACCESS # VID_ACCESS & !(CPU_CYCLE # XACK))
                                           NEXT  S7 ;
   IF !RESET & VID_ACCESS & CPU_CYCLE      NEXT  S7      OUT CPU_CYCLE ;
   IF !RESET & VID_ACCESS & XACK           NEXT  S7      OUT XACK ;

PRESENT S7
   IF !RESET & (!VID_ACCESS # VID_ACCESS & !(CPU_CYCLE # XACK))
                                           NEXT  S0 ;
   IF !RESET & VID_ACCESS & CPU_CYCLE      NEXT  S0      OUT XACK ;
   IF !RESET & VID_ACCESS & XACK           NEXT  S0      OUT XACK ;
```

Fig. 5-9. (*Continued*)

CUPL understands the meaning of the index with regard to both order and significance of bit position. Also note the use of negative signal polarity assertion in the pin list (! preceding signal name). Next in the logic description is the state assignment for the states named S0 through S7, which are each assigned a three-bit hex value. Next are two field assignments in which lists of variables are assigned names. The address bus is assigned the name ADDRESS and the three variables which make up the state bits for the video state machine are assigned the name VIDEO_STATE. The variable MEMREQ is then assigned to be the OR of the memory read and write strobes. An intermediate variable named VID_ACCESS is subsequently assigned an expression indicating that the address bus is within the range of B8000 through BFFFF and the variable MEMREQ is true. (Note the CUPL operators which were borrowed from high level software languages: & = AND, # = OR, ! = NOT). The equation for the variable READY is followed by the equation for READY.OE, the expression controlling the three-state output through which the signal named READY passes. Finally the state machine description begins with the key word SEQUENCE followed by the name of the field of bits (VIDEO_STATE) comprising the state. A { then opens the description block and is followed by a number of present-state/next-state blocks which describe the behavior of the machine. There will be as many present-state blocks in the description as there are states in the state diagram. In this case, since there are eight states in the diagram, there will be eight present state blocks. Within each state a conditional expression is preceded by the key word IF. This is followed by the key word NEXT indicating a next-state transition and, where appropriate, the key word OUT followed by output variables which are asserted at that transition. In the state machine model for the CUPL language, registered outputs always have their assertion associated with the transition from one state to another. This is important for some state machine applications in which output variables may or may not be set in a given state depending on how the machine arrived there. At the end of the state machine description block is another } terminating that particular state machine description. Multiple state machines can be defined within one logic description file by creating another group of state bits and another state machine description block.

The file containing the simulation function table description for the CUPL simulator is shown in Figure 5-10. The file starts with an information header and title block description. These should be the same as the logic description file and are usually copied from that file when the simulation input file is first created. This is followed by the order statement defining the names and sequence of the columns represented in the function table. Also included are directive commands which help space the columns to suit the designer. Following the

```
                PARTNO      PLD0000123 ;
                NAME        VID ;
                DATE        4/27/86 ;
                REVISION    02;
                DESIGNER    R.OSANN;
                COMPANY     ASSISTED TECHNOLOGY;
                ASSEMBLY    PC-VIDEO;
                LOCATION    U101;

/****************************************************************/
/* This device is clocked by the video dot-clock and generates the */
/* clock for the 6845, control for the video shift register and */
/* video memory, as well as arbitration for access to the RAM between */
/* the 8088 and the 6845.                                        */
/****************************************************************/

ORDER:
/*** INPUTS ***/
  DOTCLK,%2,
        !MEMR,%2,
             !MEMW,%2,
                  ADR19,%2,ADR18,%2,ADR17,%2,ADR16,%2,ADR15,%2,
                                                        RESET,%2,
                                                             !OE, %7,
/*** OUTPUTS ***/
  READY,%2,
        CCLK,%2,
             !SB,%2,!SA,%2,
                          !CPU_CYCLE,%2,
                                       !LOAD,%2,
                                              !XACK,%2,
                                                     !VID_ACCESS ;
```

Fig. 5-10. CUPL simulation file for video controller.

```
VECTORS:
$MSG "                                                                        !    ";
$MSG "                                                                   !    V    ";
$MSG "                                                                   C    I    ";
$MSG "                                                                   P    D    ";
$MSG "                                                                   U    _    ";
$MSG "        D                                                          _    A    ";
$MSG "        O  !  !  A  A  A  A  R              R              C  !  !  C    ";
$MSG "        T  M  M  D  D  D  D  E              E  C           Y  L  X  C    ";
$MSG "        C  E  E  R  R  R  R  S  !           A  C  !  !    C  O  A  E    ";
$MSG "        L  M  M  1  1  1  1  E  O           D  L  S  S    L  A  C  S    ";
$MSG "        K  R  W  9  8  7  6  5  T  E        Y  K  B  A    E  D  K  S    ";
$MSG "        ------------------------------      ---------------------------  ";
$MSG "        POWER ON RESET (ENTER S4)                                        ";
        C   X  X  X  X  X  X  X  1  O           X  H  H  H  H  H  H  X
$MSG "                 ENTER S5                                                ";
        C   1  1  X  X  X  X  X  O  O           Z  H  H  L  H  H  H  H
$MSG "                 ENTER S6                                                ";
        C   1  1  X  X  X  X  X  O  O           Z  H  L  H  H  H  H  H
$MSG "                 ENTER S7                                                ";
        C   1  1  X  X  X  X  X  O  O           Z  H  L  L  H  H  H  H
$MSG "                 ENTER SO                                                ";
        C   1  1  X  X  X  X  X  O  O           Z  L  H  H  H  H  H  H
$MSG "                 ENTER S1                                                ";
        C   1  1  X  X  X  X  X  O  O           Z  L  H  L  H  H  H  H
$MSG "                 ENTER S2                                                ";
        C   1  1  X  X  X  X  X  O  O           Z  L  L  H  H  H  H  H
$MSG "                 ENTER S3                                                ";
        C   1  1  X  X  X  X  X  O  O           Z  L  L  L  H  L  H  H
$MSG "                 ENTER S4                                                ";
        C   1  1  X  X  X  X  X  O  O           Z  H  H  H  H  H  H  H
$MSG "                 ENTER S5                                                ";
        C   1  1  X  X  X  X  X  O  O           Z  H  H  L  H  H  H  H
$MSG "                 ENTER S6, START VIDEO RAM ACCESS                        ";
        C   1  O  1  O  1  1  1  O  O           L  H  L  H  H  H  H  L
$MSG "                 ENTER S7                                                ";
        C   1  O  1  O  1  1  1  O  O           L  H  L  L  H  H  H  L
$MSG "                 ENTER SO                                                ";
        C   1  O  1  O  1  1  1  O  O           L  L  H  H  H  H  H  L
$MSG "                 ENTER S1                                                ";
        C   1  O  1  O  1  1  1  O  O           L  L  H  L  H  H  H  L
$MSG "                 ENTER S2                                                ";
        C   1  O  1  O  1  1  1  O  O           L  L  L  H  H  H  H  L
$MSG "                 ENTER S3, START CPU_CYCLE                               ";
        C   1  O  1  O  1  1  1  O  O           L  L  L  L  L  L  H  L
$MSG "                 ENTER S4                                                ";
        C   1  O  1  O  1  1  1  O  O           L  H  H  H  L  H  H  L
$MSG "                 ENTER S5                                                ";
        C   1  O  1  O  1  1  1  O  O           L  H  H  L  L  H  H  L
$MSG "                 ENTER S6                                                ";
        C   1  O  1  O  1  1  1  O  O           L  H  L  H  L  H  H  L
$MSG "                 ENTER S7                                                ";
        C   1  O  1  O  1  1  1  O  O           L  H  L  L  L  H  H  L
$MSG "                 ENTER SO                                                ";
        C   1  O  1  O  1  1  1  O  O           H  L  H  H  H  H  L  L
$MSG "                 ENTER S1                                                ";
        C   1  O  1  O  1  1  1  O  O           H  L  H  L  H  H  L  L
$MSG "                 ENTER S2                                                ";
        C   1  1  X  X  X  X  X  O  O           Z  L  L  H  H  H  H  H
```

Fig. 5-10. (*Continued*)

order statement is the vector function table itself. Here, the labels for the columns have been placed within a message statement so that they will appear in the output file and not be stripped off as comments normally are. This function table is similar to that for PALASM except here the stimulus portion of each vector uses 1s and 0s for logic levels while the output table uses Hs and Ls to represent logic values. Although not shown here, the CUPL simulation table can optionally have an asterisk placed in the position for an output response value. This tells the CUPL simulator that it should create the logic value which would result from the stimulus portion of the vector, rather than compare the stimulus with a correct output value. The message format directive is again used within the function table to add information indicating what each vector in the table represents.

5.5. ABEL, FIRST HIGH-LEVEL PLD LANGUAGE WITH STATE MACHINE CAPABILITY

In mid-1984 the ABEL language first became available, providing many of the high-level capabilities already made popular by CUPL and adding for the first time the concept of a true state machine description capability. The syntax for state machines in the ABEL language is somewhat different from that which later appeared in CUPL as shown in the preceding example. Both are quite capable and highly preferable for complex sequential designs where structure and good documentation are required. ABEL is also a universal language supporting many devices from different manufacturers with a common language format. The logic specification for the video controller shown in the ABEL language is shown in Figure 5-11.

Figure 5-11 starts with a description header and an arbitrary title description shown in comments. Comments in the ABEL language start with a single set of double quotes and extend to the end of a current line. Next come the pin assignments, which are very similar to those in the CUPL language except that information is presented in a reverse order. Notice that all the names shown on the pin declaration are given as the logic true root of the signal in question. The output assertion polarity is arrived at by implication. In this case, the target device, a PAL16R6, has an inverting buffer in each of its outputs, implying a low active insertion for variables attached to these output pins.

After the pin assignments, a number of equations create variable names for different purposes. The first equation assigns the variable name ADDRESS to the list of variables ADR19, ADR18, etc. These address bits are placed in their appropriate positions of significance by adding ''don't-cares'' to the least-significant bit positions in the list. Next, the variable MEMREQ is assigned the OR of the two memory READ and WRITE strobes. Then the variable VID_ACCESS is assigned to the address being within the boundaries of B8000 and BFFFF hex. The video state bits are assigned to be equal to the variable VIDEO_STATE. The state assignment is shown next in binary for the eight states.

The equations for the output variables start with those for READY and the tri-state control for the READY output. This is indicated by the key word ENABLE preceding the variable READY on the left-hand side of its equation.

```
                MODULE    VID
                FLAG
                TITLE   'VIDEO CONTROLLER AND MEMORY ARBITER
                         REV-02    R.OSANN    ASSISTED TECHNOLOGY
                         3/14/86 '
                VID     DEVICE 'P16R6' ;

" This device is clocked by the video dot-clock and generates the
" clock for the 6845, control for the video shift register and
" video memory, as well as arbitration for access to the RAM between
" the 8088 and the 6845.

"ASSIGN PIN AND NODE NAMES

                DOTCLK      PIN 1;    " VIDEO DOT CLOCK
                MEMR        PIN 2;    " 8088 MEMORY READ SIGNAL
                MEMW        PIN 3;    " 8088 MEMORY WRITE SIGNAL
                ADR19       PIN 4;    " 8088 ADDRESS
                ADR18       PIN 5;    " 8088 ADDRESS
                ADR17       PIN 6;    " 8088 ADDRESS
                ADR16       PIN 7;    " 8088 ADDRESS
                ADR15       PIN 8;    " 8088 ADDRESS
                RESET       PIN 9;    " SYSTEM RESET
                OE          PIN 11;   " OUTPUT ENABLE PIN FOR THIS CHIP

                READY       PIN 12 ;  " 8088 READY LINE (GEN WAIT STATES)
                SB          PIN 13 ;  " STATE BITS
                SA          PIN 14 ;  " STATE BITS
                CCLK        PIN 15 ;  " 6845 CHARACTER CLOCK (ALSO STATE BIT)
                CPU_CYCLE   PIN 16 ;  " 8088 ACCESS OF VID RAM IN PROGRESS
                LOAD        PIN 17 ;  " LOAD VIDEO SHIFT REGISTER (OR !SHIFT)
                XACK        PIN 18 ;  " TRANSFER ACKNOWLEDGE AT END OF CPU_CYC
                VID_ACCESS  PIN 19 ;  " VIDEO MEMORY ACCESS DECODE

ADDRESS = [ADR19,ADR18,ADR17,ADR16,ADR15,X,X,X, X,X,X,X, X,X,X,X];
MEMREQ = MEMR # MEMW;
VID_ACCESS = (ADDRESS >= ^hB8000) & (ADDRESS <= ^hBFFFF) & MEMREQ;
L,H,X,Z = 0,1,.X.,.Z.;      "DEFINE FOR SIMULATION

VIDEO_STATE = [CCLK,SB,SA];

S0 = ^b100;
S1 = ^b101;
S2 = ^b110;
S3 = ^b111;
S4 = ^b000;
S5 = ^b001;
S6 = ^b010;
S7 = ^b011;

EQUATIONS

READY        = XACK;
ENABLE READY = VID_ACCESS;

STATE_DIAGRAM VIDEO_STATE

STATE S0:
      IF !RESET          THEN  S1 ;
          XACK := !RESET & VID_ACCESS & XACK;

STATE S1:
      IF !RESET          THEN  S2 ;
          XACK := !RESET & VID_ACCESS & XACK;

STATE S2:
      IF !RESET          THEN  S3 ;
          LOAD := 1;
    CPU_CYCLE := !RESET & VID_ACCESS & !XACK ;
         XACK := !RESET & VID_ACCESS & XACK;
```

Fig. 5-11. ABEL logic specification for video controller.

156 ANALYSIS AND DESIGN OF ELECTRONIC CIRCUITS USING PCs

```
STATE S3:
        IF !RESET              THEN  S4 ;
    CPU_CYCLE := !RESET & VID_ACCESS & CPU_CYCLE ;
         XACK := !RESET & VID_ACCESS & XACK;

STATE S4:
        IF !RESET              THEN  S5 ;
    CPU_CYCLE := !RESET & VID_ACCESS & CPU_CYCLE ;
         XACK := !RESET & VID_ACCESS & XACK;

STATE S5:
        IF !RESET              THEN  S6 ;
    CPU_CYCLE := !RESET & VID_ACCESS & CPU_CYCLE ;
         XACK := !RESET & VID_ACCESS & XACK;

STATE S6:
        IF !RESET              THEN  S7 ;
    CPU_CYCLE := !RESET & VID_ACCESS & CPU_CYCLE ;
         XACK := !RESET & VID_ACCESS & XACK;

STATE S7:
        IF !RESET              THEN  S0 ;
         XACK := !RESET & VID_ACCESS & (XACK # CPU_CYCLE) ;

TEST_VECTORS

  ( [DOTCLK,!MEMR,!MEMW,ADR19,ADR18,ADR17,ADR16,ADR15,RESET,!OE] ->
    [READY,CCLK,!SB,!SA,!CPU_CYCLE,!LOAD,!XACK,!VID_ACCESS] )

[C, X, X, X, X, X, X, X, H, L]->[X, H, H, H, H, H, H, X]; "POWER ON RESET (S4)
[C, H, H, X, X, X, X, X, L, L]->[Z, H, H, L, H, H, H, H]; "ENTER S5
[C, H, H, X, X, X, X, X, L, L]->[Z, H, L, H, H, H, H, H]; "ENTER S6
[C, H, H, X, X, X, X, X, L, L]->[Z, H, L, L, H, H, H, H]; "ENTER S7
[C, H, H, X, X, X, X, X, L, L]->[Z, L, H, H, H, H, H, H]; "ENTER S0
[C, H, H, X, X, X, X, X, L, L]->[Z, L, H, L, H, H, H, H]; "ENTER S1
[C, H, H, X, X, X, X, X, L, L]->[Z, L, H, L, H, H, H, H]; "ENTER S2
[C, H, H, X, X, X, X, X, L, L]->[Z, L, L, L, H, L, H, H]; "ENTER S3
[C, H, H, X, X, X, X, X, L, L]->[Z, H, H, H, H, H, H, H]; "ENTER S4
[C, H, H, X, X, X, X, X, L, L]->[Z, H, L, H, H, H, H, H]; "ENTER S5
[C, H, L, H, L, H, H, H, L, L]->[L, H, L, H, H, H, H, L]; "ENTER S6, VID_ACCESS
[C, H, L, H, L, H, H, H, L, L]->[L, H, L, L, H, H, H, L]; "ENTER S7
[C, H, L, H, L, H, H, H, L, L]->[L, L, H, H, H, H, H, L]; "ENTER S0
[C, H, L, H, L, H, H, H, L, L]->[L, L, H, L, H, H, H, L]; "ENTER S1
[C, H, L, H, L, H, H, H, L, L]->[L, L, L, H, H, H, H, L]; "ENTER S2
[C, H, L, H, L, H, H, H, L, L]->[L, L, L, L, L, L, H, L]; "ENTER S3, CPU_CYCLE
[C, H, L, H, L, H, H, H, L, L]->[L, H, H, H, L, H, H, L]; "ENTER S4
[C, H, L, H, L, H, H, H, L, L]->[L, H, H, L, L, H, H, L]; "ENTER S5
[C, H, L, H, L, H, H, H, L, L]->[L, H, L, H, L, H, H, L]; "ENTER S6
[C, H, L, H, L, H, H, H, L, L]->[L, H, L, L, L, H, H, L]; "ENTER S7
[C, H, L, H, L, H, H, H, L, L]->[H, L, H, H, H, L, L]; "ENTER S0
[C, H, L, H, L, H, H, H, L, L]->[H, L, H, L, H, H, L, L]; "ENTER S1
[C, H, H, X, X, X, X, X, L, L]->[Z, L, L, H, H, H, H, H]; "ENTER S2

END VID
```

Fig. 5-11. (*Continued*)

Finally the state machine description block begins with the key words STATE_DIAGRAM, followed by the variable VIDEO_STATE, representing the state bits. There are then eight present-state blocks beginning with the key word STATE followed by the name of each state and a : . Within each present-state block, the ABEL state machine syntax is basically an IF, THEN, ELSE syntax with IF implying the conditions and the key word THEN preceding a next state. Outputs are handled within each present-state block according to the := operator. This clocked transfer function indicates that output variables will be set on the next clock after the given state is entered.

The ABEL specification file also contains the simulation test vectors. These are shown at the bottom of Figure 5-11 starting with the key word TEST_VECTORS. This key word is followed by an ordering statement which indicates a list of stimulus variables followed by an arrow pointing to a list of response variables. The simulation vector table itself then follows with a straight tabular format similar to that shown in the previous examples. Other capabilities of the ABEL simulator include the representation of groups of bits in hexadecimal. Simulation of state machines can also be done using the symbolic state names. In the table shown, comments are used to label each vector in the simulation input table. At the bottom of the logic specification for the video controller is the key word END followed by the name of the module, VID. This allows multiple PLD descriptions to be included in one file.

5.6. ARRIVAL OF HIGH-LEVEL DESIGN ON PCS

Just as PLDs have become everyone's custom chip, PCs have become everyone's CAE workstation. As PLDs rapidly penetrate the digital design community, software tools which support them will provide the CAE introduction for most designers.

6

The Use of PCs for Industrial/Laboratory Automation

Bruce D. Pollard, Ph.D.
Automation Specialist
Technology and Development Center
ARCO Chemical Company
Newtown Square, Pennsylvania

6.1. INTRODUCTION

Personal computers have found widespread industrial application in data acquisition, computer control and on-line numerical processing. For example, in chemical plants controlled by computers linked to automated measurement systems, measurement systems often consist of instruments connected to personal computers. This chapter describes the use of microcomputers for automating various commercial, industrial, and laboratory systems. The focus is on the proper design and operation of the computer-apparatus *interface*, upon which the overall performance of the system depends.

While an interface can be facilitated in various manners ranging from purchasing "off-the-shelf" hardware and software to preparing an original design utilizing discrete electrical components, all approaches require a reasonable understanding of interfacing basics, which are presented in Section 6.2. Following, in Section 6.3, is a simple example of a typical PC interfacing project. Subsequent sections elaborate on connecting devices with serial interfaces or voltage outputs to personal computers. A later section introduces applications of PC-controlled robotics.

Reference is given to publications and manufacturers' literature throughout the chapter; included with the references are manufacturers' and distributors' addresses. None of these references, however, should be construed as endorsements.

For those with a minimal understanding of the workings of computers, it is recommended that Chapter 2 be reviewed before continuing with this chapter.

6.2. BASICS

6.2.1. Introduction

In some areas of engineering, a little knowledge can be dangerous, but when interfacing a microcomputer to an external environment, any amount is useful. This section touches upon most of the important aspects of interfacing. However, it does not stand alone; research into what has been done in similar applications and reference to the literature cited is mandatory. Introductory texts[1,2,3] should be studied for the concepts and terminology. Personal computer journals[4] and electronics magazines[5,6] have issues and special features devoted to data acquisition and interfacing. In many first-attempt interfacing projects, information-gathering accompanies the all-important interface planning stage. In such cases, a discussion of the decisions with an "old pro" would prove invaluable.

Although the *interface* can be defined as the connection enabling communication between two systems, the physical boundary of a computer interface is not always clear. Figure 6-1 is a schematic showing the interface between the microcomputer system and the application. The two basic aspects of an interface are hardware and software. Depending upon the application, the hardware can be as simple as a cable or as complex as another computer connected between the host computer and the application. Software resides on the host computer and controls the interface so the task of communicating with the application can be accomplished. In its simplest form, this communication might be

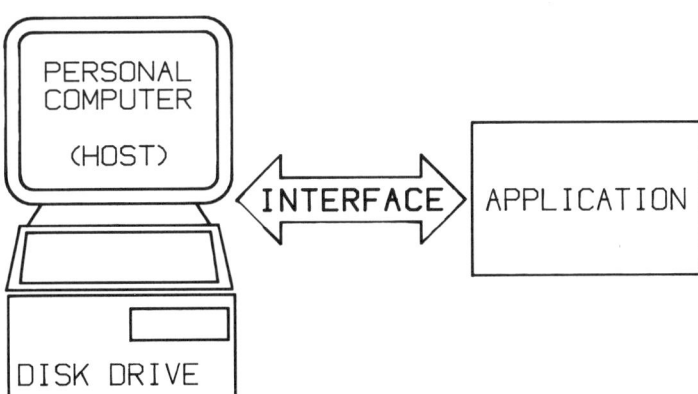

Fig. 6-1. The interface—the hardware and software connection between a personal computer and an external application which may be an instrument, pilot plant, and so forth.

the collection of temperature variations which have been measured by an attached instrument. In fully automated systems which include data acquisition, calculations, and controlled responses, the software is complex and requires special programming techniques. One of these techniques, *real-time programming*, is employed when the interface is required to collect data and respond in a continuous fashion with no time delay between critical steps; process control is a good example of a real-time application.

6.2.2. Choice of Computer

The choice of computer depends on many factors. While cost is an important consideration, it should not dictate a choice which would limit the final functionality of the project. Many times a very expensive system might be great for calculation and multiple users but of limited use for interfacing because of the complex programming required or the inability to operate in a real-time environment. Before a choice is made, simple experiments should be performed to insure the computer can provide the hardware and software facilities as well as the speed to satisfy all the requirements of the particular application in an easy-to-use fashion.

There are several microprocessor architectures[7]. Memory-mapped microprocessors are simpler to program because they address the peripherals as though they are memory locations. The Rockwell Aim 65[8], used in the Section 6.3 example contains a 6502 microprocessor, which is memory-mapped.

The suitability of a microcomputer may depend upon the accessibility and documentation of its input/output lines[9]. Many systems provide a special integrated circuit called a *peripheral interface adapter* (PIA) for such applications. A typical PIA might contain parallel input/output lines, a timer, and a parallel-to-serial conversion register along with control and status lines which can be directly connected to the microprocessor's interrupt system. Many of the less expensive, "toy", systems have custom circuits leading to "game ports"[10]. If proper documentation can be obtained "toy" computers may be a very economical approach. Finally, different technologies used in microelectronics (among them TTL and CMOS) have specific limitations and are not always compatible[11]. The compatibility of a computer and its interface should be verified in the planning stage. One should never expect to drive relays directly from a computer output. Instead, an intermediate driver circuit should be provided[12].

6.2.3. Software

Every computer language has strong and weak points. Certainly for data acquisition and control programs, machine or assembly language provides the most flexibility and speed. However, writing an entire program in assembly language is time-consuming and not practical when a significant portion of the program involves numerical calculations or graphic manipulations which are easy to accomplish with a higher-level language. Most higher-level languages such as BASIC, FORTRAN, or C have provisions for subroutines written in machine language, so there is no reason not to combine the best of both. In this chapter as in the others in this text, BASIC has been chosen as the higher-level language principally because it is almost universally available on personal computers. When BASIC is interpreted line by line, it may run too slow for some real-time applications. Compilers for BASIC offer substantially greater speed, but switching to another faster language such as machine code or C may be in order.

The PEEK and POKE commands available in most versions of BASIC are essential to interface programming. PEEK reads a specific memory location and POKE inserts a value into a given location. POKE is used to specify where a user-written machine language subroutine is located or to code the subroutine from a DATA statement to an open segment in memory. PEEK and POKE can transfer values in and out of the user subroutine which is limited to passing one internal variable. When the microprocessor is memory-mapped and the input/output addresses are known, it is possible that the interface program can be accomplished simply by POKING output data or control codes and PEEKING input or status codes.

Taking the time to lay out the program and consider what must happen step by step is important. If possible, short programs should be written to test each step and to ''benchmark'' the execution time required. These short programs can then be converted to subroutines for use in a master program. If the data are collected in binary or another code different from that in which numbers are stored in the higher-level language used, the data will have to be converted. Many times the addresses of machine-language subroutines stored in ROM and used for data input, output, and conversion in BASIC are provided in the computer documentation. The use of such subroutines saves time preparing and running the program.

There are very simple ways to get data into BASIC variables. The example in Section 6.3 illustrates how the individual digits of a number are input in binary-coded decimal, transferred to BASIC using PEEKS, and combined to a single floating-point value.

6.2.4. Approaches

The several approaches to interfacing include buying a complete ''off-the-shelf'' system; purchasing I/O modules that come with software written in a general form for adaptation or especially for the application; or designing and constructing hardware as well as writing original software. Each approach has advantages and disadvantages. Certainly for one-time or first-time applications, off-the-shelf is the most cost-effective; however, it is not usually adaptable nor a good choice with which to learn interfacing. Off-the-shelf systems for a specific application are most often advertised in the journals or trade periodicals related to that category of application. For example, a system for thermal analysis would be found in a laboratory apparatus magazine. Such systems can be very expensive and are usually designed to do specific tasks with specific hardware. Before a decision to purchase such a system is made, the vendor can be expected to demonstrate that its system will fill the need, that a completely operational system will be delivered, and that software and hardware service will be available over the planned lifetime of the system.

Interface modules that plug into the PC motherboard are now available for most systems. These modules may contain a specific function or a group of functions on one board. Often the vendor supplies a set of subroutines for accessing such a board and in some cases may provide (at additional cost) a *framework* program for preparing a custom application. Because the modules are suitable for a wide range of applications, mass production makes them inexpensive. As framework software availability increases, the module approach should become the most popular.

Some distributors[13,14,15] have catalogs listing a full range of interfacing electronics with application notes. For the IBM PC family and compatibles, Analog Devices[16], National[17] and Data Translation[18] offer a single plug-in board containing many channels of analog collection, digital input/output, and analog output capabilities. Hart Scientific[19] and Interactive Microware[20] offer software frameworks to support these boards. Keithley[21], Hewlett-Packard[22] and Cyborg[23] offer separate units that can be plugged into various computers. The user chooses plug-in boards depending upon the application, for example, an internally compensated thermocouple board for temperature measurements. Each of these companies offers a group of subroutines callable from BASIC as well as framework programs for software development. Such programming eliminates the need for writing machine-language code except in the most difficult of cases, decreases the lead time for developing the system, and usually provides the best performance of the hardware. The framework approach has the additional advantage of providing subroutines which store data in standard file formats, allowing for easy modification and graphic display of the data, the latter of which could take substantial programming if not supplied.

If the automator has a working knowledge of analog and digital electronics, the "home-brew" approach may be a good choice. Sometimes the less design-intensive alternatives do not meet the project's needs and construction is the only choice. Building an interface from the ground up, in addition to being most instructive, also requires the builder to consider the specification of each component and thus determine the final limitations of the interface. Constructing the hardware usually requires writing original software—at least some of which is in machine code. Machine-language I/O routines are not usually too difficult to write, but, if speed or exact timing is necessary, special skills may be required. On the other hand, the versatility of choosing what works best in machine language and what best in a higher level language allows the ultimate approach to interface programming. The preceding is not meant to scare people away from doing a project from the "ground up," but only as a caveat that, for a particular application and set of users, this approach may not be the most cost-effective, particularly if a short time schedule must be met.

6.2.5. Considerations

Each interface application is unique. The proper choice of features requires the designer to consider many factors. These major considerations are discussed briefly in the following paragraphs:

- Data rate and volume
- Codes
- Response times
- Data format
- Control and status signals
- Error detection and correction
- Acquisition Mode
- Graphics

The data-acquisition rate and the amount of data required for a given application are the most important factors and many times dictate design choices. The Nyquist theorem[24] states the minimum acquisition rate required to sample an oscillating signal without losing information is two times the frequency. When a complex waveform is sampled, the acquisition rate should be determined by experiment. Although information theory and the time domain of the experiment establish the acquisition rate and number of samples, practicality is also paramount. It makes no sense to collect 1024 data samples at one microsecond per point, store them in the acquisition module, and then transfer them to the processor in ASCII format at 300 baud (30 characters per second). If each datum consists of three significant figures and a decimal point (four characters), such an approach would require a transfer time in excess of 100 seconds. Therefore one must seek another approach to accomplish the data transfer!

Data can be transferred from the application through the interface to the computer in any of various codes (see Section 2.2). Binary and two's complement binary are the most fundamental but are not suitable for cases in which many decades of dynamic range require an exponent or decimal point. Binary-coded decimal (BCD) supports that need as well as facilitates division of a datum into digits for sequential transfer. ASCII allows a wider range of codes including alphanumerics and is easy to transfer through standard PC ports into variables in higher-level languages. However, ASCII transfer is very inefficient when high speed is of concern.

Some hardware standards have been developed to support the most popular codes. The RS-232 connection standard is based on serial transmission of ASCII. The IEEE-488 bus standard has been developed to connect several devices to one computer interface[25]. The IEEE-488 standard has been implemented by Hewlett-Packard and Tektronix as the HP-IB[26] and HP-GL[27] interconnect respectively. Although not compatible with each other, these products are excellent for connecting personal computers to groups of instruments which have some level of programmability built in.

Careful consideration of response times and resolution must be made before any interface is implemented. Simply put, the *response time* of a system is how long it takes for the output to accurately reflect a change in the input. When the response time of an interface or any subsystem is too long, it will not properly track the change of an input or respond with a required output in a timely fashion. The *resolution* of any system is the smallest change that can be measured. If the resolution is too large, small changes in the input may not be detectable. Practical limits for response time and resolution are determined by the uncertainty (noise and other errors) in the system. While any measurement or control interface introduces some uncertainty, it must be reduced to a level less than that of the system to which it is connected. The introduction of noise should be evaluated during each phase of the interface project by collecting and printing out data-set averages and standard deviations.

The choice of data format is between serial and parallel (see Section 2.2). Both have advantages and disadvantages. In *parallel* transmission, one wire is required for each bit in a data word. It is faster because a whole word can be transmitted in one clock cycle. Most personal computers have parallel data architecture, so no conversion is required. In the *serial* format, data is transmitted on a single cable pair one bit at a time. The advantage here is that transmission over great distances is much less expensive because a multiwire cable is not required. In addition, the RS-232 standard for serial transmission has greatly facilitated the interconnection of devices. Serial transmission is slower, because after receipt by the microcomputer, the bits must be assembled into words before they can be sent, in parallel, to the microprocessor.

In only a few cases can the computer and interface operate without controlling the data flow and checking the status of the data collection device. *Control signals* are outputs from the computer and are used to initiate some action or to modify an action in progress[3]. *Status signals* are inputs to the computer and indicate when data are ready, a process is completed, or some malfunction has occurred. Control and status signals can be combined to form a *handshake* between the computer and a peripheral. With the handshake, a control line can start data conversion and a status line can signal when the data are ready for transfer. Control and status signals are sometimes sent along the same connections as data, but usually they are separate wires in the interface cable. These signals allow the computer to control or track a system in time. Such applications are called *real-time*.

Because most interfaces are expected to operate in environments where electrical noise is present, i.e., commercial or industrial areas, error detection and correction schemes must be applied. The simplest approach is duplicate transmission and comparison. If the data are not identical, the process is repeated until two or more data sets match. While this sounds simple, it is not always possible. The signal level might be changing too fast to allow repetition or the same error could occur twice. More complex schemes have been developed for these cases. Among these are the *parity* system for transmission of single words and *checksums* for blocks of words[28]. None of these are substitutes for careful testing and repeated calibration.

Depending upon the data exchange rate, amount, and control requirements, the interface may use one of several acquisition approaches or modes[29]. The on-the-fly and handshake modes are called *program-controlled* because they require full time attention from the microprocessor. When data are always available, as from a temperature sensor, collecting the data on-the-fly is the simplest approach. Many times this can be accomplished with only a few lines of code. The *handshake* approach requires control and status signals to be passed from the microprocessor through the interface. The handshake insures that data are collected at certain time intervals or after a particular event has occurred, but under program control the processor spends most of its time waiting for the data-ready status.

Most microprocessor architectures provide interrupt lines and instructions so the time between events is not wasted. Data acquisition using this approach is called *interrupt-driven*. Many interface boards contain an interval timer so that once the acquisition rate is set, the microprocessor can spend most of the time reducing data or updating graphics. When the timer runs out, data collection is initiated and the microprocessor is interrupted to collect and process a datum only when the ready status is asserted. A higher-level approach similar to interrupt-driven is the initiation by the microcomputer of a block transfer of data directly from interface to memory. *Direct memory access* (DMA) is particularly useful for high-speed applications. In most cases, microprocessors being so inexpensive, the DMA mode is accomplished by a preprocessor chip on the interface board.

Graphics provide an interface between the user and the computer in any data acquisition and control project; the fundamentals are discussed in Section 2.5.1. Graphics should provide the user with sufficient detail to monitor or interact with the system in a timely and informed manner. Screen redraw time should never limit the response time of the system.

6.3. CONNECTING A DIGITAL pH METER TO A COMPUTER

6.3.1. Introduction

In this example, a digital pH meter (Orion Corp., Model 701A) is interfaced to an AIM 65 microcomputer[8]. The resulting system can be programmed for use as a process monitor, a titration endpoint detector, or as part of an automated pH control apparatus. The interface consists of a cable, connecting the binary coded decimal (BCD) output from the meter to a parallel input on the computer, and, to collect the data, a subroutine written in machine language and callable from BASIC. The considerations of the previous section are illustrated through a description of the interface and how it is achieved.

A good starting point for interfacing an instrument with a computer is to review the principles and operation of that instrument. The measured concentration of hydrogen ions in an aqueous solution expressed on a negative logarithmic scale is called the *solution pH*[30]. The measurement basis is a potential developed between two solutions, one of known pH and the other unknown, across a membrane made of a special kind of glass. The potential of the glass electrode is measured in reference to that of a known electrode using a very-high-impedance voltmeter called a pH meter. Most state of the art pH meters have digital readouts and some sort of electrical digital output. Although some meters may have an RS-232 serial output (discussed with the next application), many have a simpler parallel output consisting of four binary coded lines for each digit of the display. Such an output is most suitable for driving a printer but can also be easily connected to a computer. The response time of a well-built pH measurement system is always limited by that of the glass electrode, which, unless it is especially designed, reaches a stable potential no sooner than 0.5 second. Thus most digital pH meters display a new reading 2 to 10 times a second.

6.3.2. Hardware

Although the choice of computer is important, in this case, because a high measurement rate is not required, cost and instructive value are the determining factors. The AIM 65 is a single-board microcomputer containing 4 K RAM, keyboard, 20-digit readout, and printer. With BASIC and monitor programs in ROM, it sells for about $500 and generally is available from local electronic supply houses. The five-volume set of documentation that comes with the AIM 65 is complete and a good introduction to microcomputers. An interface and connections for a cassette-tape recorder are provided to facilitate permanent storage of short programs.

The AIM 65 is well suited for interface applications. The 8-bit microprocessor uses memory-mapped input/output. This means each input or output device is assigned a group of addresses which can be read or written as though they were memory locations. Thus collecting data from an input port or sending data to a peripheral device utilizes the same programming as reading from and writing to memory positions. A special integrated circuit called the *versatile interface adapter* (VIA) has been included on the AIM 65 board. It has two 8-bit ports which can be assigned as outputs or inputs, two built-in timers and a shift register for converting data between serial and parallel formats. Control and status lines which can be set by the program to support either program or interrupt control are built-in to the VIA. Chapters in the AIM documentation and the specification sheets[31] for the VIA cover complete hardware and software aspects of the VIA, mainly through examples.

Because the VIA is so easy to use, the hardware portion of the interface is simply a cable connecting the outputs from the pH meter to the proper inputs on the microcomputer. According to the wiring diagram, shown in Figure 6-2, each of three binary-coded decimal digits (0.01, 0.1 and 1) has four wires; one each for 1, 2, 4 and 8 times the base value of that digit. The eight wires for the two smallest digits are assigned to input port A, completely filling it, while the four wires from the 1s digit along with a single wire which corresponds to a 1 in the 10s digit are assigned to port B. Thus data ranging between 0.00 and 19.99 can be acquired simply by reading the two ports as though they were memory locations.

In this project, the meter's TTL outputs (transistor-transistor logic, a 0 is indicated by a voltage less than 0.4 V and a 1 is indicated by a voltage greater than 2.4 V) are compatible with the VIA inputs. Even when compatibility is indicated by the documentation, it should be verified by experimentation because, due to design or adjustment, the power supplies of the instrument and computer might not provide the expected voltages. Interface cables may require various kinds of ground or signal-return connections. In this case, a single wire serves the purpose; however when noise immunity is required over long distances or if the computer provides power to run the interface, a digital ground and separate common return or shield may be required. The cable is constructed from lengths of solid telephone wire properly soldered to circuit-board-edge connectors purchased at a local electronic supply house.

One additional wire is required to indicate to the computer when a datum is ready. The pH meter documentation indicates that for the short time the meter converts the voltage to a digital signal the datum on its output is unreliable. An *end of conversion* (EOC) signal goes from TTL high to low when the datum is available. This signal is connected to a control and status line on the microcomputer's VIA. The VIA can be configured to change a bit in one of its registers (assigned memory locations) from 0 to 1 when the high-to-low transition takes place on the EOC wire from the pH meter. Under program-controlled acquisition, the interface software monitors the status register in a tight loop and jumps

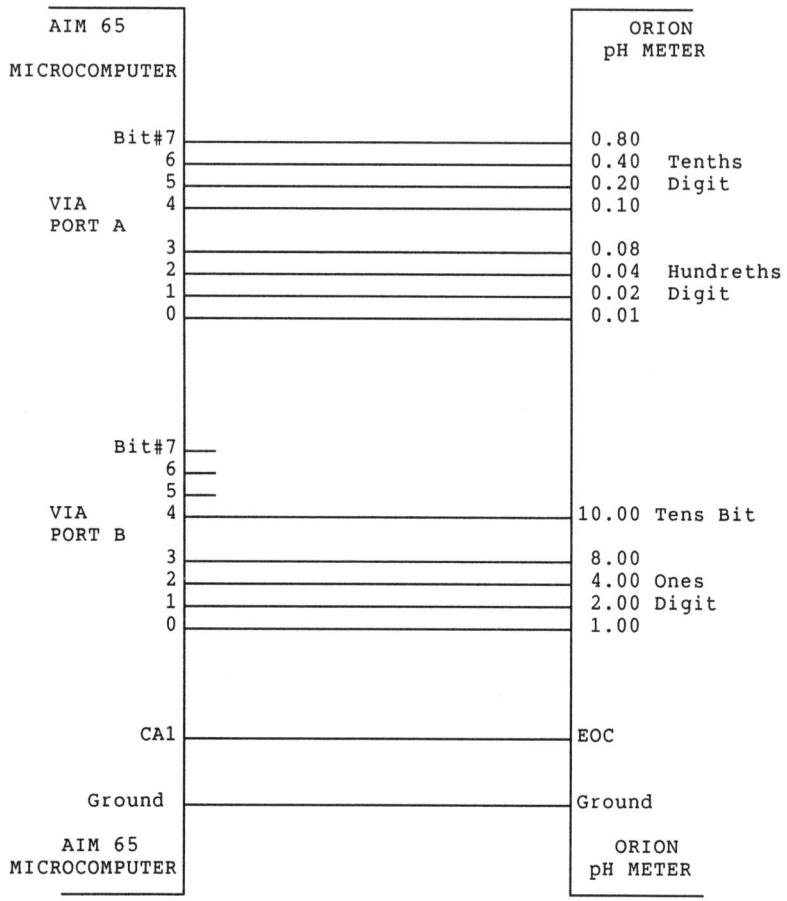

Fig. 6-2. Connection diagram for interfacing an AIM 65 microcomputer to an Orion Model 701 digital pH meter.

out of the loop to collect data from the ports when the value read from the status register (also called a *flag register*) changes.

6.3.3. Software

The software portion of the interface is written in machine language (Table 6-1) as a user subroutine or function which can be called from BASIC with the command containing the code USR. Each time the USR is called, it initializes the VIA by moving 00 into address A002 and A003 (base 16) to set ports A and B to input and into A00C to select a flag to be set when the status line goes

Table 6-1. Machine Language USR Subroutine.

Starting Address	Instruction	Comment
0FB0*	LDA #00	
0FB2	STA A003	SET PORT A TO INPUT
0FB5	STA A002	SET PORT B TO INPUT
0FB8	STA A00C	SET A FLAG WHEN STATUS GOES HIGH TO LOW
0FBB	LDA A00D	LOOK FOR THE FLAG
0FBE	BEQ 0FBB	IF NO FLAG, LOOK AGAIN
0FC0	LDA A001	READ PORT A
0FC3	STA 0FFB	SAVE IT
0FC6	LDA A000	READ PORT B
0FC9	STA 0FFD	SAVE IT
0FCC	AND #0F	BREAK OFF THE 1s DIGIT
0FCE	STA 0FFC	SAVE IT
0FD1	LDA 0FFD	RECALL PORT B
0FD4	ROR A	SWAP
0FD5	ROR A	THE
0FD6	ROR A	DIGITS BY
0FD7	ROR A	ROTATION
0FD8	AND #01	BREAK OFF THE 10 BIT
0FDA	STA 0FFD	SAVE IT
0FDD	LDA 0FFB	RECALL PORT A
0FE0	AND #0F	BREAK OFF THE 0.01s DIGIT
0FE2	STA 0FFA	SAVE IT
0FE5	LDA 0FFB	RECALL PORT B
0FE8	ROR A	SWAP
0FE9	ROR A	THE
0FEA	ROR A	DIGITS BY
0FEB	ROR A	ROTATION
0FEC	ABD #0F	BREAK OFF THE 0.1s DIGIT
0FEE	STA 0FFB	SAVE IT
0FF1	RTS	RETURN TO BASIC
0FFA	Storage location of	0.01's digit
0FFB		0.1's digit
0FFC		1's digit
0FFD		10 bit

*All numbers in base 16.

from high to low. The program loops until the flag indicates a datum is ready and then transfers port A (A001) and port B (A000) into memory. After collection, the data from each port are broken into individual digits which are stored in successive memory locations. Finally the subroutine is terminated and control returns to the BASIC program. Note that both ports are read before the data manipulation steps. In this case, the pH meter's conversion rate is so slow that the entire subroutine runs before the data changes; however, when the conversion rate is fast, delaying collection from the second port could invalidate the data due to contention between consecutive readings. When contention is a possibility, the use of benchmark programs and an oscilloscope to check timing is recommended.

A simple BASIC program (Table 6-2) illustrates collecting a datum and printing it out. BASIC on the AIM 65 requires the 16-bit address of a USR subroutine to be broken in half and POKED into special memory locations before it can be called. In this case, the USR starts at $0FB0_{16}$, so the first byte, 15_{10} ($0F_{16}$) and the second byte 176_{10} ($B0_{16}$) are POKED into locations 05 and 04 respectively. Statement 30 calls the USR. Variables W and U could be used to pass single variables to and from the USR but using them would require calling other machine-language subroutines. Instead, statements 40–70 PEEK each digit out of the location where the USR has stored it. In the final statements, the digits are recombined and printed out. A GOTO command causes the program to repeat the acquisition steps until the program abort button is pushed.

6.3.4. Testing

Validation is a critical step in the interfacing process. Because there is no logical error-checking, e.g., parity or checksums, the ability to pass every possible value from the meter to the computer output must be established as fact. This can be accomplished by running the program and slowly rotating the calibration control on the meter while verifying that the number printed out matches the number on the meter display. An incorrect value might indicate a wrong connection in the interface cable, a programming or typographical error, a malfunction in the meter or computer interface electronics, or a timing problem. While the solution to most of these problems is simple, understanding a timing

Table 6-2. BASIC Program for BCD pH Acquisition.

Statement	Comment
10 POKE 05, 15*	STORE HIGH BYTE OF USR LOCATION
20 POKE 04, 176	STORE LOW BYTE OF USR LOCATION
30 U = USR(W)	CALL THE ACQUISITION SUBROUTINE
40 A = PEEK (4090)	GET THE 0.01s DIGIT
50 B = PEEK (4091)	GET THE 0.1s DIGIT
60 C = PEEK (4092)	GET THE 1s DIGIT
70 D = PEEK (4093)	GET THE 10 BIT
80 P = A/100+B/10+C+D*10	RECOMBINE
90 PRINT "THE PH IS . ."; P	PRINT OUT VALUE
95 GOTO 30	LOOP BACK TO USR
100 STOP	

*All numbers in base 10 unless noted.

problem is difficult, requiring good logical diagnoses through programming changes or electronic troubleshooting procedures using an oscilloscope.

The interface should be evaluated under all possible operating conditions. The program in Table 6-2 can serve as a benchmark for determining that the maximum acquisition rate is acceptable. Tests should be made to see what values result from operational errors (such as leaving the electrode unplugged). If the system is to be installed in a rugged environment, it should be given a field trial. Once the system is ready to go into service, the logical conclusion to the project is preparation of complete documentation including circuit diagrams, test results, and commented program listings along with a good set of operation and maintenance instructions.

6.4. SERIAL INTERFACING (ASYNCHRONOUS RS-232C)

The asynchronous RS-232C serial data interface is perhaps the most common method of attaching peripherals (except printers) to personal computers. Many instruments (e.g., balances, voltmeters, programmable power supplies, and so forth), that is, data acquisition and control devices, have been designed to be interfaced via the "RS-232" port on a PC[28]. Following is enough information for the successful connection and programing of such devices.

More often than not, the serial port on a PC is simply called an "RS-232 port," but such a reference is incomplete. As shown in Figure 6-3 and Table 6-3, the EIA-RS-232C standard[32], established in 1969, specifies electrical characteristics for the connection, defines a set of control signals to be used when a physical "handshake" is required, and assigns pin numbers on a preferred 25-pin connector to each possible signal. The additional information required to describe a serial data connection is the communication mode, speed, format, code, error-checking, and protocol.

Similar electrical characteristics are specified (Figure 6-3) for RS-232 data and control lines; however, there is one difference. While positive voltages on the data lines indicate a "space" or logical 0 and negative voltages indicate a "mark" or logical 1, the positive and negative assignments for control lines are just reversed. The RS-232 voltage levels are not transistor-transistor logic (TTL) compatible so special integrated circuits called receivers and drivers[28] as well as power supplies are required. Manufacturers of RS-232 systems recommend a maximum distance between devices of 50–100 feet, but RS-232 cables have been known to operate over distances exceeding 1000 feet. Due to increased cable capacitance and noise susceptibility, cable runs of such distances may require a substantial decrease in the communication rate and should not be used if errors which are not normally detectable or recoverable occur.

The RS-232 convention defines two kinds of devices: data terminating equip-

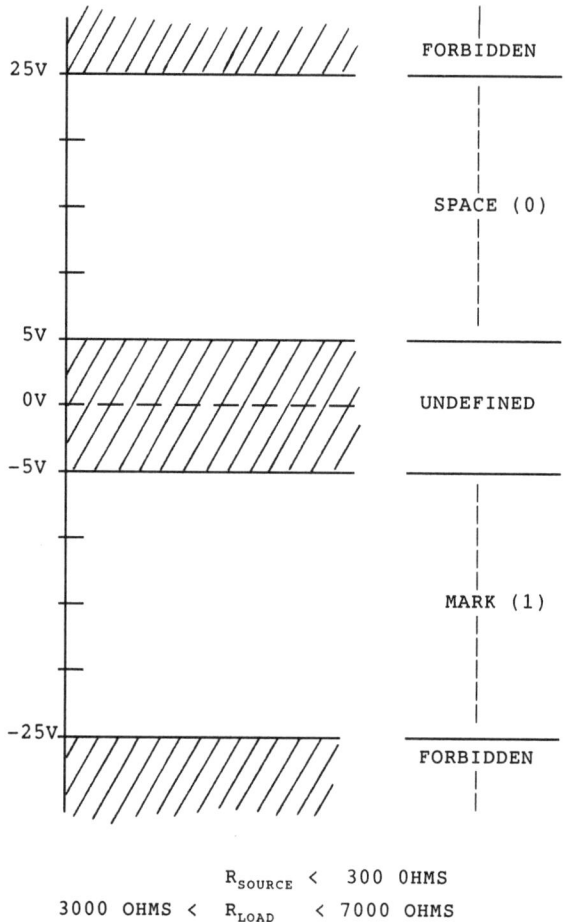

Fig. 6-3. EIA RS-232C electrical specifications.

ment (*DTE*) and data communicating equipment (*DCE*). Which data and control signals originate from a device or are received by it depend upon whether it is DTE or DCE. Table 6-3 shows all the RS-232 signals, highlighting those commonly used in PC connections. Most computers are configured as DTE with a male 25-pin connector and modems (see Section 2.3.3); as DCE, with a matching female 25-pin connector. Instruments and other devices are usually

Table 6-3. RS-232C Serial Conventions

Pin Number[1]	Description	Abbreviation	Source
1	Protective ground[3]	PG	—
2	Transmitted data[2]	TD	DTE
3	Received data[2]	RD	DCE
4	Request to send[3]	RTS	DTE
5	Clear to send[3]	CTS	DCE
6	Data set ready[3]	DSR	DCE
7	Signal ground (common return)[2]	SG	—
8	Line carrier detected[3]	CD	DCE
9	Reserved for testing	—	—
10	Reserved for testing	—	—
11	Unassigned	—	—
12	Secondary line signal detected	SCD	DCE
13	Secondary clear to send	SCTS	DCE
14	Secondary transmitted data	STD	DTE
15	Transmit signal clock	TC	DCE
16	Secondary received data	SRD	DCE
17	Received signal clock	RC	DCE
18	Unassigned	—	—
19	Secondary request to send	SRTS	DCE
20	Data terminal ready[3]	DTR	DTE
21	Signal quality indicator	SQ	DCE
22	Ring indicator	RI	DCE
23	Data rate selector		DTE
24	Auxiliary transmit clock	AUX	DTE
25	Unassigned	—	—

[1] The recommended connector, a 25-pin TRW-Cinch D-Subminiature DB25P and DB25S, is also manufactured by other firms. In general, the male pin connector (DB25P) is installed on the data terminating equipment (DTE) and the female socket connector (DB25S) is mounted on the data communicating equipment (DCE).
[2] Used in minimum three-wire nonstandard, but common asynchronous connection.
[3] Used along with (2) in de facto nine-wire standard asynchronous serial connection between PCs and modems.

DTE but this must be verified by reference to the technical manual or through the use of an inexpensive tester called a "breakout box." Connecting DTE to DCE simply requires a cable with the proper connectors wired from pins 1 to 1, 2 to 2, etc., but connecting DTE to DTE requires the exchange of the send and receive signals on one end along with the exchange of some handshake signals as shown in Figure 6-4. Sometimes the equipment does not make use of all the handshake signals so they need not be exchanged; however, unless the requirements are determined by testing with a breakout box, a wrong assumption could lead to a malfunction.

The possible modes of communication are simplex, half duplex, and full duplex. *Simplex* is one direction only as with television reception. *Half duplex* is bidirectional, but signals are transmitted only one way at a time and the receiver must be notified when its turn to transmit arrives. C.B. radios operate

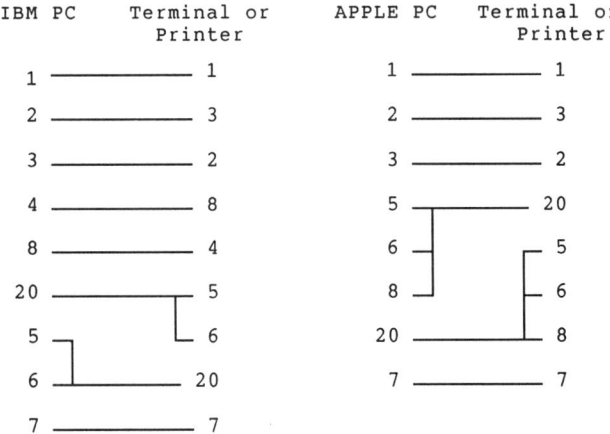

Fig. 6-4. Wiring diagrams for DTE to DTE, "NULL MODEM" connections using female 25-pin "D" connectors.

as half duplex devices. Most PCs communicate in the *full duplex* mode, that is, simultaneously. Full duplex is the most efficient, however, as will be discussed later. When hardware handshakes are not employed, the receiving device must be able to signal the transmitter when it can not handle the data sent to it.

Communications speeds are measured in baud. The *baud* is determined by phase changes on a telephone line between two modems. Usually, in the measurement of signal flow in one direction, one baud equals one bit per second. A good estimate of the unidirectional transmission rate in characters per second is about one-tenth the Baud rate. Typical rates are 1200 and 2400 baud (120 and 240 characters per second) for asynchronous serial data being sent from a computer over phone lines via modems to remote devices. Speeds of 19,200 or 9600 baud are not uncommon for local connections between computers or between a computer and an automation device.

The communications format of serial ports on most PCs is *asynchronous*, meaning that both ends of a link have their own clocks. These clocks must be preset to the same speed. In asynchronous communication, each data word is preceded by one mark or start bit to tell the receiver when a word starts. In addition, usually one mark or stop bit (sometimes two stop bits) are required to complete a data word. In the *synchronous* communications format, the originating device supplies a clock signal and synchronization words when transmission of data commences; there are no start bits for each word. Even though synchronous communication is more efficient, the added complexity and cost of connecting clocks has made it unpopular.

All common PCs use the ASCII code (see Table 2-2) for serial communications. Until recently standard ASCII was a 7-bit code containing all the characters on a standard typewriter keyboard. However, to provide control, alternate, and special characters for graphics and cursor control, an 8-bit version has replaced it. The actual codes used depend upon the computer or peripheral manufacturer. Care must be taken to match the port setup of the PC with the device connected so that if 8-bit words are not recognized by one device, the other will not send them.

Error-checking is a very important part of any communications system. In the case of asynchronous RS-232 connections between devices sending ASCII character strings, the first level of error-checking is the inclusion of a parity bit in each data word[28]. When used, the parity bit of a data word is always set when the total number of marks is even or odd depending on the choice of even or odd parity. Although omission of the parity check is a risky choice, it is not uncommon when the data are inspected on a CRT. The use of an eighth bit to extend the ASCII codes precludes the use of a parity bit.

The communications protocol for serial asynchronous communication consists of the word code and format combined along with a hardware or software handshake convention. In the typical case for a PC, the combined word is 1 start bit + 7 data bits + 1 parity bit + 1 stop bit = 10 bits. The usual hardware handshake is two-part. When it is first switched on, the DTE sets DTR high (positive voltage) and expects to sense a high DSR from the attached DCE, indicating the rest of the communication circuit is on. When it has data to transmit, DTE asserts RTS; if DCE answers with CTS, the data are sent. Sometimes, when the communications link involves two modems, the DTE requires a third signal, CD, be asserted by the DCE to indicate a solid connection between the modems. If the hardware handshake is not used, usually a software handshake is employed. The most common one, XON/XOFF, is used to exchange data at a fast rate, even when the flow must be suspended for the receiving device to perform some other task. The receiving end stores incoming data in a buffer and when the buffer get nearly full sends an ASCII 11_{16} (cntl S on the keyboard, labelled XOFF or DC1 on some charts). When the device on the other end interprets the XOFF, it stops transmitting characters. Later, when the receiving device's input buffer has more space, it sends out an ASCII 13_{16} (cntl Q on the keyboard, labelled XON or DC3 on some charts); and when the device on the other end interprets that, it starts sending again. When the input buffers are big or if both devices can process the data as soon as it is received, XON/XOFF may not be needed. If data flow is principally in one direction and the receiving device's buffer is of moderate size, then XON/XOFF works well. If the buffers are not well balanced, then the transmitting device may not process an XOFF before the receiving device's buffer overflows. In such cases, hardware handshakes or reduced transfer rates are required.

After a connecting cable has been purchased or prepared and attached between the devices, the RS-232 ports must be set to match with respect to the features discussed above. Most PCs are similar to the IBM PC in that the communication parameters are software-selectable from the operating system monitor or through a special configuration program. For example in MS-DOS (or PC-DOS), issuing the command "MODE COM1: 9600,N,7,1,P,X" would set communication port 1 to speed 9600 baud, no parity check, 7-bit ASCII plus 1 stop bit (1 start bit is always assumed), continuous retry on timeout and XON/XOFF handshaking.

The BASIC language interpreters provided with most personal computers contain provisions for sending data out or receiving data from the RS-232 "Com" ports. Usually early in the program the port is initialized with the OPEN command. Special forms of the PRINT and INPUT commands send and receive the data in character-string form. When the character string received is to be treated as a number, it can be converted with the standard string functions. In general, BASIC programs using RS-232 ported devices are limited in speed and may not be suitable for real-time applications. However, acquisition devices which are easily programmable[33,34], having their own memories so they can operate in real time and communicate with the host PC via RS-232, may provide ready answers to simple data acquisition problems. Some older versions of BASIC treat the Com port as a sequential file and require a CLOSE command to send the PRINTed strings out through the port; such versions are to be avoided because they severely limit the speed and utility of the programs they run.

While the RS-232 asynchronous link has many details, it is reliable, available on most PCs, and facilitates a standardized connection to many laboratory and industrial data acquisition and control systems. A system[34] containing 16 analog input channels (12-bit resolution), two digital I/O ports, event counters, and alarms circuits is available for connection to a PC via the RS-232 port. Complete process control systems,[35,36] miniature data loggers,[37] laboratory balances,[38] power controllers[39] and chromatographic instruments[40] are just a few examples. In many cases, special software is available for controlling the device and for processing the data from it, but if the steps required are few and well described in the device's manual, writing a program to accomplish the task should be easy.

6.5. ANALOG SIGNAL ACQUISITION

6.5.1. Introduction

The measurement of a varying input voltage and the storage of this measure is a task often required of industrial and laboratory computer automation systems. Whether the data collection hardware is a board that plugs into the PC or is housed in a separate cabinet, it will most certainly contain an analog-to-digital converter (ADC) to perform this task. A basic knowledge of the ADC is imperative to the success of most automation projects[41]. This section provides an introduction to ADCs, describes the hardware options, and reviews the various types of software available. A short description of how personal computers equipped with data acquisition systems can be used to emulate various kinds of electronic instrumentation is also presented.

As with all electronic subsystems, countless specifications and figures of merit serve to differentiate between the many models of ADCs. The most important of these are conversion rate, resolution, range, code, and noise characteristics. The *conversion rate*, measured in Hertz, is a measure of how many conversions the device can complete in one second. The specified rate could be misleading because it is determined for the device and does not account for the time required for the computer to control and transfer data from it. *Resolution* is how many bits are in the converted word. The typical ADC used in data acquisition has 12 bits, so it can resolve 1 part in $4096(2^{12})$. When combined with the resolution, the input range determines the smallest convertible signal. For example, if a 12-bit ADC has a range of 0 to 10 volts, the smallest voltage change that can be detected is 10 V/4096, or about 2 millivolts. Any of several codes may be used in the digital domain. The most common of these are binary, two's complement binary, and binary-coded decimal (see Section 2.2). The choice of code depends upon the application and programming philosophy. A large group of specifications are required to describe the noise characteristics of a converter. While a discussion of these is beyond the scope of this book (for some laboratory applications, see reference 46), in the jargon of an experienced interface engineer "sensing the 2-millivolt change in the above example over the entire range of the device requires it to be linear, have no missing codes, and have RMS noise levels and conversion errors of 1 bit or less."

As set forth in the basics portions of this chapter, Section 6.2, the required acquisition speed is perhaps the most important consideration at the onset of an automation project. The required speed depends on the application and dictates among other things the choice of computer and amount of memory, the type of ADC, the programming approach, and the cost of the project. It is not advisable to pick the fastest available unit because there are critical trade-offs among sampling rate, data reduction speed, cost, resolution and programming ease. There are indeed no easy ways to calculate the proper sampling rate. The Nyquist theorem, which states that the period of a repetitive waveform can be accurately determined only if the signal is sampled at a frequency at least twice that of the waveform, is a start, but has limited application where the character of a noisy waveform or single event is of interest. A useful rule-of-thumb for integrating Gaussian-shaped peaks and the like is to collect at least 20 points over the span of the event[42]; however, if the information required involves taking the derivative of the signal or if the peaks overlap or are noisy, as many as 200 points per event may be required. A good approach to choosing the proper speed is a two-fold procedure: first, evaluate the input signal, directly with an oscilloscope if possible, paying close attention to the voltage range, time domain, and noise; second, simulate the acquisition and time-limited calculation portions of the project using the chosen computer if one is available. Such a simulation can be as simple as typing in data sets having as many points as would be acquired at various acquisition rates, and running through the calculations; or it can be as complicated as a program to generate the points via a random-number algorithm to superimpose noise over a theoretical waveform. Further, once the simulation process is completed, the entire automation process should be much clearer.

6.5.2. Digital-to-Analog Conversion

Because most ADCs have one as an integral part of their design, it is appropriate to mention the counterpart of the ADC, the *digital-to-analog converter* (DAC). Simply put, the DAC converts a digital number sent to it from a computer or other device to an analog voltage. The defined figures of merit and specifications for the DAC are similar to those of the ADC. The conceptual model of a DAC is a precision voltage source feeding a ladder network of resistors, each increasing in value by a factor of 2 over the last, with switches, each corresponding to 1 bit in the digital word. If the switches are closed according to a binary code, the current flow from the voltage source will correspond in magnitude to the digital value. The current can be converted to voltage using operational amplifiers. A string of inputs to a group of DACS is the basis for the CRT display, some types of plotters, and digital control in general.

6.5.3. Analog-to-Digital Conversion

The three most common types of ADCs will be described according to speed, from fastest to slowest (and not coincidentally, in cost from highest to lowest). The "flash" ADC can operate at video speeds (1 MHz) or faster. It is simply a group of ultrafast operational amplifiers configured as comparators in a binary ladder, one for each bit of resolution. The computer reads their output as fast as it can, or additional circuitry can be added for temporary data storage, called *cache memory*. Flash ADCs are expensive, have no noise immunity, and usually have limited resolution.

Successive approximation ADCs are the most common in computer automation systems requiring conversion rates ranging from 100 kHz to 100 Hz. They consist of at least two op amps, a DAC, and some digital circuitry for clocking and logic. On the command to aquire a datum, the input voltage level is locked into the first op amp configured as a sample and hold. The output from the first stage is input to one side of the second op amp configured as a comparator. The output from a DAC is input to the other side. The DAC is clocked through the possible values in an optimal sequence until the comparator signals equality.

Then the value of the DAC is sent out to the computer. Successive approximation ADCs have limited utility with noisy input signals, but can be built on a single chip to have excellent resolution and internal accuracy.

The *"dual-slope" integrating ADC* is the most inexpensive, has the ability to average out noise on an input signal, but is limited to speeds under 1 kHz. The circuitry usually consists of op amps used as integrators, comparators, and current reference generators and digital circuitry for clocking and logic, all this circuitry on one chip. First, a current proportional to the input signal is integrated into a capacitor for a fixed length of time. Then it is discharged at a known rate until all the charge is gone. The time it takes for the discharge process is counted up into a binary register and output as the digitized value. A 12-bit dual-slope ADC may take 1000 times as long as a successive approximation ADC because the former has no optimal approximation sequence, but the integrating feature offers excellent noise rejection and so they are commonly found in digital voltmeters and similar instruments.

6.5.4. Integrated Acquisition Systems

Although individual ADCs are available for PCs, they are usually purchased as a part of a data acquisition system consisting of a board which plugs into and gets its power from the PC (examples are boards fabricated by Analog Devices[16] and Data Translation[18]), or as a self-contained unit that may communicate with the PC via the serial port or plug into the PC motherboard (examples are units from Keithly[21], and Cyborg[23]). In addition to the ADC, the system might include auxiliary circuitry to support the ADC such as multiplexers, triggers, and signal conditioning; or separate systems such as DAC channels or digital I/O. The multiplexer facilitates the collection of data from several sources in rapid sequence while employing only a single ADC.

Many times much memory and processing time can be saved if rapid data collection is initiated by a signal from the sampled environment instead of the computer. The circuits to perform this task, called *Schmitt triggers*, are located on the acquisition board and may be set to a trigger condition by external controls or sometimes by the acquisition program. Signal conditioning may take the form of special circuitry for temperature measurement with thermocouples[43], but more often special input amplifiers are provided to insure the proper voltage is read from sources with limited current capacities (high impedance) and to protect the entire system from excessive high voltage. A special kind of input amplifier called "programmed-gain" is used in systems where accurate collection from a variety of medium (0–10 V) to low (0–1 mV) sources is required. Such a feature is standard in the systems called "digimeters" sold for collection of data from laboratory instruments by CIS/Beckman[44] and Nelson Analytical[45].

Most data acquisition systems come with software which ranges in complexity from a simple set of BASIC-callable machine-language subroutines to a complete application package. The advantage of purchasing software with the hardware, in addition to saving time, is that the vendor may supply benchmarks or other figures of merit representing the performance of the integrated package. Such information or demonstrations are very useful for choosing the right system. The current trend is for vendors to supply a standalone package called an *acquisition framework*. In addition to flexible acquisition routines, this package usually includes screen graphics, file storage, and an advanced data reduction capability (such as Fourier transforms). A good framework has numerous setup menus and a way to store a set of conditions on disk. It may have a programming language of its own and come with a set of examples for collecting and processing strain, temperature, or pressure data. The examples can usually be modified to do exactly what is required. Frameworks written primarily to be statistical packages but which include an acquisition module have limited use. An important caveat in the evaluation of acquisition frameworks is that some can collect only a limited amount of data. While these may be suitable for some applications, they do not lend themselves to process control or other monitoring systems in which data must be monitored continuously, reduced, stored, and acted upon in real time. A review of some framework programs gives a more detailed comparison of their various features[46]. For readers who are more advanced, one of those reviewed, SALT, is described in a separate article[47] and is available for a nominal fee from the address given in the reference.

6.5.6. Instrument Emulation.

One interesting realization made possible by the development of PCs and data acquisition systems is the emulation of virtually any piece of electronic instrumentation[48]. An example of the emulation process is a digital voltmeter program which collects a datum using the ADC portion of the system, scales it, and prints it out on command. The logical extension which is a common demonstration of framework software is a digital storage oscilloscope. In addition, it is useful to have an accurate clock (sometimes called a *real-time clock*) available on the computer. If the acquisition system has a DAC, a waveform or pulse generator can be emulated simply by programming the PC to output scaled values from a sine or more complex function. Hewlett-Packard[49] has been leading the developments in instrument emulation, making available "smart instruments" based on their series 100 and 300 PCs with special plug-in boards. Moreover, the emulation of quite sophisticated instruments such as phase-sensitive (lock-in) amplifiers[50] and multichannel signal averagers[51] utilizing more generic hardware has been reported. In general, the emulation approach has proven to be very inexpensive and instructive, but limited performance is sometimes a problem.

6.6. PC-BASED ROBOTIC WORKSTATIONS

Robotics is a young, fast-growing interdisciplinary field spanning a variety of goals, skills, and techniques. It piques the interest of industrialists who see robotic automation as the key to increased productivity and academicians who define robotics as the study of the intelligent connection of instrumental perception with mechanical action. This section provides an introduction to the concepts and application of robotics to laboratory and industrial automation as facilitated through personal computers.

The literature on robotics varies in content. Survey texts, such as *Robots*[52], provide a lay description of robotic systems and concepts and are useful in understanding the scope and issues of robotics. Instructional texts, such as Heath[53], teach robotics by developing the concepts around a "teaching robot" system. The most up-to-date and technical information is published in the robotics journals[54] and in proceedings of symposia[55]. Computer journals and magazines such as *BYTE* contain articles about robotics and occasionally devote an entire issue to the subject[56].

Most laboratory and small industrial robotic systems take the form of a workstation which consists of a robotic arm and associated controls. Figure 6-5 is a schematic representation of a typical robotic arm consisting of three parts: the arm, wrist, and hand; and having six degrees of movements—three translational and three rotational. The area of coverage is limited to within the boundary traced out by the rotation of the arm extended, but recently arms have been mounted on tracks to increase mobility and allow the robot to work over extended areas or on larger objects without itself being too large. The inexpensive Armatron robot by Tandy Corporation is an instructive example of a manually controlled robotic arm that can be automated by experimenters. The Hero 1[57] is a freewheeling self-contained instructional unit which can be purchased as a kit. Other much more expensive robotic arms which are designed to be connected to PCs or come with microprocessor controllers are available from Zymark[58], Movemaster[59], Rhino[60], Microbot[61] and others and are intended for actual applications. When a PC is used as a controller for the arm, it is not uncommon for the PC to collect data, control and coordinate the actions of various modules and sensors which are grouped around the arm to make up the workstation.

Regardless of its function, modern robotics has found great success through the careful application of basic control principles[62] to a variety of kinematic elements using feedback data from many types of sensing elements. Most robotic systems utilize discrete control which is exemplified by the control of processes and machinery essentially on-off in their behavior. Gross movement is broken down to a sequence of steps. Continuous feedback as to direction of movement and relative location with respect to each degree of movement is required to signal when each step is completed and to calculate the absolute location.

182 ANALYSIS AND DESIGN OF ELECTRONIC CIRCUITS USING PCs

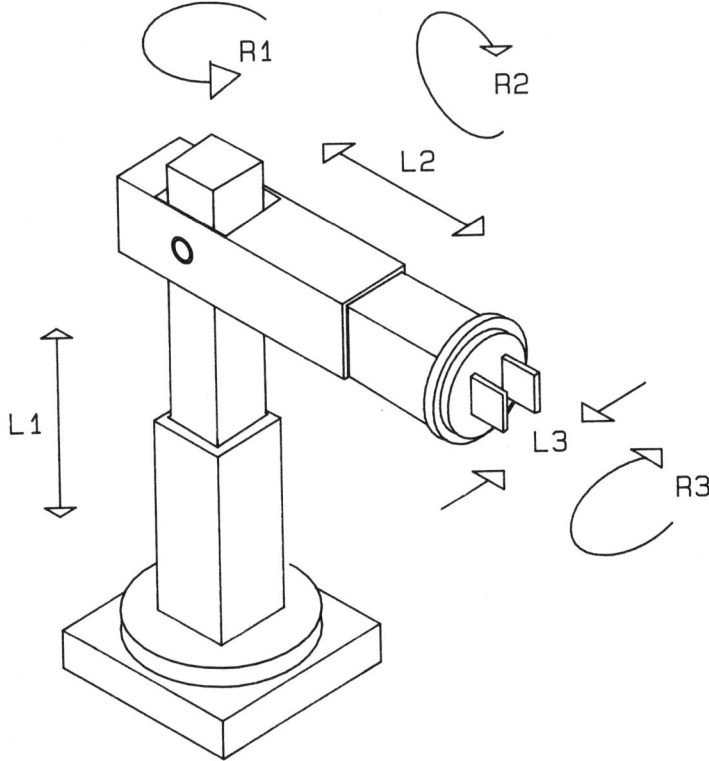

Fig. 6-5. A simple robotic arm having thee degrees of translational (linear) movement and three degrees of rotational movement.

Multilevel control is used to increase the overall speed of the system until proximity sensors or preprogrammed set-points trigger a slower speed for more intricate movements. True proportional control, with the speed or force set to some proportion of the final result based on the level of feedback from a sensor, is not common in preprogrammed robotic movements, but is utilized extensively for control of the activities taking place at each of the workstation modules.

Although pneumatic and hydraulic pistons have been used to actuate robotic movement, electrical motors linked to the moving element through gears or pulleys with cables are the most common approach. In general, DC motors are more suitable for control applications because their speed and torque can be controlled by electrical devices. One special class of DC motor called the *stepping motor* is extremely useful because it can be moved a fraction of a rotation one step at a time. Via a series of pulses sent to the motor through a control circuit, the motor can be made to turn a selected number of revolutions at a specific speed[63]. Provided the gears or cables do not slip or stretch and the motor does not skip a step or two, the final position can be reached in a reproducible and accurate manner. When high precision is required, sensors that confirm each step or indicate absolute position are installed to monitor each degree of movement.

The availability of inexpensive yet sophisticated sensors as a result of the electronic revolution is perhaps the key to the many new developments in robotic automation. Current research is focused on developing senses of sight and touch which are more than primitive[55]. For example, a stereo imaging system has been developed in conjunction with an adaptive computer algorithm to locate and pick up randomly placed geometric solid objects[56]. PC based robotic systems make common use of infrared and visible-light photo-generator/photo-sensor pairs as proximity detectors, as shaft encoders (a binary-coded mask mounted on the shaft of a motor with a photo-impulse counter used to indicate absolute or relative movement), and for other functions. Proximity sensors based on inductive and capacitive changes, such as those in PC keyboards, have not found widespread use in robotics as yet. Usually the sensors are attached directly to the robotic movement controller or are interfaced to the PC through parallel I/O ports. In the latter case, a program running in the PC detects a change in state of the sensors and jumps to a subprogram which causes the robot to perform the required action.

After a robotic workstation has been set up and its basic movements and control functions tested, it must be programmed to perform its function[64]. The movements are programmed in a fashion similar to numerical control procedures used for the machining of complex parts. A robot teaching program breaks down movements from one position to another into a sequence of actions along each degree of movement. Sometimes a hand-held device having push buttons to control movement in each degree (called a *teaching pendant*) is used to facilitate the programming of complex movements. A novel approach is to link a PC-based CAD program to the controller[65]. Each complex movement is given a name and is combined with others to comprise the complete task. The name which represents a subroutine allows the movement to be stored and used repeatedly in the same task or in similar ones. Preprogrammed movement of objects from one location to another, for example from a rack of samples to a balance, requires that all the devices and holders in the workstation be fixed to a baseplate so that the absolute locations remain constant from one run to the next. If permanent location is not possible, the required number of sensors and the sophistication of the movement programming becomes prohibitive to the point of requiring artificial intelligence and a much larger computer. However, for most systems, movement programming, control of devices located around the robot, and data collection and reduction can be handled by a single personal computer. To prevent the robot from running "berserk" and literally "crashing" into the workstation devices, its movements must be observed throughout a test which includes every conceivable combination of movements.

The Zymark laboratory robot[58] is a good example of a complete and versatile workstation. It can be purchased with a set of modules as a turnkey system to perform one of several functions, or the arm and controller can be purchased separately. The turnkey systems contain a large number of check and control sensors enabling it to operate for long periods unattended. For example, on the automated titration system a photo-sensor is placed below the titration cup dispenser so the controller can confirm a cup has been obtained by the hand and a second sensor is located to signal when the cup has been properly placed on the titration module. If the robot fails to grab a cup, it can try again, but if the cup is placed improperly an alarm is sounded if it cannot be righted. Such detail makes the system reliable enough to justify the $20K—$50K expense. Zymark's first robotic controller had a computer link limited to sending data to and from the controller, but the latest-generation controller has been designed to download commands from another computer. This feature allows for the logical integration of a workstation where the main computer is the master and the controller is an intelligent slave.

6.7. SUMMARY AND CONCLUSION

This chapter has been an introduction to the application of PCs to laboratory and industrial data acquisition. It has emphasized the various factors that should be considered before an automation project is undertaken. The example of Seciton 6.3, connecting a pH meter to a PC, illustrates the simplicity of accomplishing such a task. The description of RS-232 asynchronous serial and analog-to-digital interfaces should prove helpful towards the purchase and connection of most data-acquisition devices. Finally, the section on robotics provides an introduction to how control and acquisition are integrated through the PC to accomplish complex tasks.

The PC is the computer of choice for most automation projects. The interface hardware is becoming readily available and there is a large selection of software from which to choose. However, because electronics and programs are required to work in harmony, the factors introduced in this chapter and the tricks of the trade discussed in the many references are of the highest importance. Hopefully the reader will use this chapter as a ticket to ride the automation train on the proper track.

REFERENCES

1. Cassell, D. A. *Microcomputers and Modern Control Engineering*. Reston, VA: Reston, 1983.
2. Carrick, A. *Computers and Instrumentation*. Philadelphia: Hayden Press, 1979.
3. Cooper, J. W. *The Minicomputer in the Laboratory*. New York: Wiley Interscience, 1977.

4. Special Issue: *Real World Interfacing. Byte,* Apr. 1984.
5. Special Issue: "Data Acquisition with PCs." *Electronic Design,* Aug. 15, 1985.
6. "Data Acquisition with PCs". *Electronic Design,* 94–109, October 31, 1984.
7. Dessy, R. E. "Choosing a PC." *Anal. Chem.* 58, 78–91 A, 1986.
8. Rockwell International, Inc., Microelectronics Division, 3310 Miraloma Ave., Anaheim, CA.
9. Leibson, S. "An Input/Output Primer," Parts 1–6. *Byte,* Feb.–Jul. 1982.
10. *VIC-20 Programmer's Guide.* West Chester, PA: Commodore, Inc., 1982.
11. Carr, J. J. *How to Design and Build Electronic Instrumentation.* Blue Ridge Summit, PA: Tab, 1978.
12. *Radio Shack Electronic Designer's Handbook.* Fort Worth: Tandy Corp., 1983.
13. *Industrial IBM PC Source Book.* ICS, Inc., 8601 Aero Dr., San Diego, CA 92123.
14. *PC Enhancement Handbook for Scientists and Engineers.* Cyber Research, Inc., 5 Science Park Center, New Haven, CT 06536.
15. *Data Acquisition and Control Interfaces.* MetraByte, Inc., 254 Tosca Dr. Stoughton, MA 02072.
16. Analog Devices, Inc., Two Technology Way, Norwood, MA 02062.
17. National Instruments, Inc., 12109 Technology Blvd., Austin, TX 78727.
18. Data Translation, Inc., 100 Locke Drive, Marlboro, MA 01752.
19. Hart Scientific, Inc., 177 W. 300 St., Provo, UT 84601.
20. Interactive Microware, Inc., P.O. Box 139, State College, PA 16804.
21. Keithly Instruments, Inc., 28775 Aurora Rd., Cleveland, OH 44193.
22. Hewlett-Packard Co., 3000 Hanover St., Palo Alto, CA 94304.
23. Cyborg, Inc., 55 Chapel St., Newton, MA 02158.
24. Swanson, R., D. J. Thoennes, R. C. Williams, and C. L. Wilkins. "Determination of the Nyquist Frequency." *J. Chem. Ed.* 52(8), 530–33 (1975).
25. Clune, T. R. "Interfacing for Data Acquisition." *Byte,* 269–282, Feb. 1985.
26. "HP-IB Interconnection Guide." Palo Alto, CA: Hewlett-Packard, Inc. (1983).
27. Tektronix, Inc., PO Box 5000, Beaverton, OR 97075.
28. Grofton, P. W. *Mastering Serial Connection.* Berkeley, CA: Sybex, 1986.
29. *Microcomputer Handbook.* Maynard, MA: Digital Equipment Corp., 1980.
30. Bates, R. G. *Determination of pH.* 2nd ed., New York: John Wiley & Sons, 1973.
31. "R6522 VIA Data Sheet." Doc. 29000 D47, Rockwell (ref. 8).
32. Electronics Industries Association, Standards Dept., 2001 I St., Washington, DC 20006.
33. Carlson Instruments, Inc., P.O. Box 245, Warner, NH 03278.
34. TransEra Corp., 3707 North Canyon Rd., Provo, UT 84604.
35. Taylor Instruments Co., 95 Ames St., Rochester, NY 14601.
36. Rosemont, Inc., P.O. Box 35129, Minneapolis, MN 55435.
37. Science Electronics, Inc., P.O. Box 986, Dayton, OH 45401.
38. Mettler Instrument Corp., P.O. Box 71, Hightstown, NJ 08520.
39. ISCO Inc., P.O. Box 5347, Lincoln, NE 68504.
40. Hewlett-Packard, Inc., Analytical Instruments Div., 1820 Embarcadero Rd., Palo Alto, CA 94303.
41. *Analog-Digital Conversion Handbook.* Norwood, MA: Analog Devices, Inc., 1986.
42. Taraszewski, W. J., D. Haworth, and B. Pollard. "Application of Signal-to-Noise Theory to Detection and Integration of Dynamic Signals." *Anal. Chim. Acta* 157(1), 73–82 (1984).
43. *Temperature Measurement Handbook.* Stamford, CT: Omega Engineering, Inc.
44. CIS/Beckman, Inc., 160 Hopper Ave., Waldwick, NJ 07463.
45. Nelson Analytical, Inc., 20370 Town Center Lane, Cupertino, CA 95014.
46. Wirth, P., and L. E. Ford. "Five Laboratory Interfacing Packages." *Byte,* 303–314, Jul. 1986.
47. Fenster, S. D., and L. E. Ford. "SALT." *Byte,* 147–164, Jun. 1985. Software is available from S. Fenster, 4949 S. Woodlawn Ave., Chicago, IL 60615.

48. Special Edition: "PC Based Instruments." *Electronic Design*, Mar. 13, 1986.
49. Rothschild, C. J., R. C. Sismilich, and W. T. Walker. "Low-Cost Automated Instruments for PCs." *Hewlett-Packard J.*, 4–10, May 1986.
50. Oriel, Inc., 10 High St., Stamford, CT 06424.
51. Nithipatikom, K., and B. D. Pollard. "An Instrument for Determining Single and Multiple Room Temperature Phosphorescence Lifetimes." *Appl. Spectrosc.* 39(1), 109–15 (1985).
52. Marsh, P., *Robots*. New York: Crown, 1985.
53. Heath, L., *Fundamentals of Robotics*. Reston, VA: Reston, 1985.
54. *International Journal of Robotics Research*, Cambridge, MA: MIT Press, 1982-1987.
55. Hanafusa, H., and H. Inoue, eds. *Robotics Research: Proceedings of the 2nd International Symposium*. Cambridge MA: MIT Press, 1985.
56. Special Issue: "Robotics." *Byte*, Jan. 1986.
57. Heath/Zenith Systems, Inc., Benton Harbor, MI 49022.
58. Zymark, Inc., Zymark Center, Hopkinton, MA 01748.
59. Mitsubishi (Japan)—dist. by: Hudson Robotics, Inc., 120 Morris Ave., Springfield, NJ 07081.
60. Rhino Robots, Inc., 308 State St., Champaign, IL 61802.
61. Microbot, Inc., 453-H Ravendale Dr., Mountain View, CA 94043.
62. Liptak, B., ed. *Industrial Engineer's Handbook of Process Control*. Radnor, PA: Chilton Press, 1985.
63. *Stepping Motor Guide*. Braintree, MA: Sigma Instruments, Inc., 1984.
64. Singer, A., and P. Rony, "Controlling Robots with Personal Computers." *Machine Design*, 78–82, Sept. 23, 1982.
65. Colonna, R., and K., Pelsue. "CAD/CAM for LASER Working Equipment." *Photonics Spectra*, Mar. 1986.

7
Special Computer Codes

John R. Greenbaum
General Electric Company
Philadelphia, Pennsylvania

7.1 INTRODUCTION

The different types and classes of CAD programs described in earlier chapters can simulate the behavior of most practical engineering circuit designs. However, some designs can be more easily evaluated using less sophisticated codes. This chapter presents especially written computer codes which provide rapid and accurate solutions for different types of circuits. The codes are sufficiently general that they can readily be adapted for specific design applications. Six different types of circuits and methods for their solution are presented.

In addition to an explanation of each of the programs, examples of its use as well as a listing of the entire code are provided. The examples demonstrate both the simplicity of usage and the form and format of input and output data. Most of the programs have been written to be "user-friendly," i.e., they are interactive and/or menu-driven.

The programs in order of presentation follow:

1. Fitting general-order polynomials to data available as either tabulated listings or curves. Very often this information is required as input to one of the standard CAD programs, but must be transformed into equational format for program input.
2. T- and P-pad design and analysis. *Pads* are common-component circuit arrangements used for signal level or strength control. This program eliminates most of the drudgery associated with designing such pads.
3. Network-matching circuits also are tedious to design, especially since they are normally quite frequency-sensitive. A code is presented that permits a simple and straightforward method for designing and analyzing the response of such networks.
4. Bandpass filters containing active components, such as operational amplifiers, are commonly used circuits. The code presented allows rapid design

of such circuits by the entry of only component values. An example is provided demonstrating program use for a worst-case analysis application.
5. RF amplifier design requires that very detailed and exacting analyses be performed while using many independent variables. Some of the variables are input and output impedance variations, bandwidth definitions, and S-parameter variations of transistors. The code presented significantly reduces the time, effort, and possibility of error in designing these circuit types.
6. Feedback systems are commonplace in the world of electronics. The examples, explanations and code provided have applications that range from simple feedback amplifiers, to robotics, to normal servo systems.

Statistics are an integral part of the design engineering process. Tolerances associated with component values and their distribution, as well as the variability of active device characteristics preclude fabricating a "nominal" circuit. Statistical programs permit the development of realistic predictions of performance for production circuits. Special software has been written to provide statistical analyses. A list of available programs for use on the IBM, or compatible PCs, with pertinent memory and cost data is provided in Table 7-1.

7.2. FITTING GENERAL-ORDER POLYNOMIALS

Very often data available in the form of curves are desired for use as input for computer-aided circuit analysis. In fact, most CAD programs require that data be input in either tabulated or equational form. A program to transform these data from either tabulated data or curves to equational representation was written by D. C. and M. C. Lin and published in *Machine Design* (September 1984). The data points are obtained directly from the curves in the form of X, Y point pairs. The program shown in Figure 7-1 is adapted from the published code.

Engineers routinely fit experimental data to a straight line or a curve. Curve-fitting with the right order of polynomial takes time, especially when clear trends are not apparent. This code, which requires no understanding of the mathematics involved, provides the desired values automatically.

The following computer program has been developed to facilitate curve-fitting. The program is based on the least-squares principle and is written in BASIC. The general form of the polynomial is

$$Y = a_1 + a_2 x + a_3 x^2 + \cdots + a_m x^{m-1} + a_{m+1} x^m \qquad (7\text{-}1)$$

where coefficients a_i are found by minimizing the sum of the square of the

Table 7-1. Statistical Process Control Software for the IBM Family of Personal Computers.

Name	Source	RAM (K bytes)	Hardware /Software Requirements	Approximate Acquisition Cost
/SPC SYSTEM	The Crosby Company P.O. Box 2433 Glen Ellyn, IL 60138 (312) 790-1711	96K	IBM PC, XT, Jr MS-DOS 1 disk drive	$850
ANSTAT AUTO COMPII	Quality Measurement Systems, Inc. 2555 Baird Rd. Penfield, NY 14526 (716) 385-3848	128K	IBM PC PC-DOS 2.0 2 disk drives	$3000
ATTRIBUTE DATA CHARTING-SPC2	Pierson Company 8392 Stoney Creek Court Davison, MI 48423 (313) 653-7065	128K	IBM PC MS-DOS 2 disk drives	$175
CHART-PAK	Quality Resources 840 McKinley Plymouth, MI 48170 (313) 453-4616	32K	IBM PC MS-DOS 2 disk drives	$100
CUSTOM/QC	Stochos, Inc. 14 North College St. Schenectady, NY 12305 (518) 372-5426 (800) 426-4014	256K	IBM PC, XT, AT and compatibles MS-DOS, PC-DOS 2.0+	Basic $500 Statistics $250 Process Eval. $200 Sampling Plans $15 Freq. Curves $100 All Modules $995
GAGING ANALYST	Perry Johnson Seminars, Inc. 3000 Town Seminar Suite 1075 Southfield, MI 48075 (313) 356-4410	128K	IBM PC, XT, AT and compatibles hard disk/dual disk drive	$1195

Table 7-1. (Continued)

Name	Source	RAM (K bytes)	Hardware /Software Requirements	Approximate Acquisition Cost
MOVING AVERAGE & RANGE CHARTING- SPC3	Pierson Company 8392 Stoney Creek Court Davison, MI 48423 (313) 653-7065	128K	IBM PC MS-DOS 2 disk drives	$175
NWA QUALITY ANALYST	Northwest Analytical, Inc. 520 N.W. Davis St. Portland, OR 97209 (503) 224-7727	128K	IBM PC/XT/AT PC-DOS	$495
PROCESS CONTROL PACKAGE	Q-Consultants Box 22183 Tucson, AZ 85734 (602) 295-6000	64K	IBM PC MS-DOS 1 disk drive	$125
Q STATS 1.0	Applied Concepts Company 5430 South 12th Ave. Suite C Tuscon, AZ 85706 (602) 294-1188	16K	IBM PC 1 disk drive	$99, $199 with CHARTPAK-64 (product of Abacus Software, Inc.)
QC-PRO ATTRIBUTE DATA ANALYSIS	GTG Microstat Ltd. 2419 Drew Rd. Mississauga, Ontario, Canada L5S 1A1 (416) 678-1001	128K	IBM PC 2 disk drives PC-DOS	$238
QC-PRO STATISTICAL PROCESS CONTROL	GTG Microstat Ltd. 2419 Drew Rd. Mississauga, Ontario, Canada L5S 1A1 (416) 678-1001	128K	IBM PC 2 disk drives PC-DOS	$460

SPECIAL COMPUTER CODES 191

Name	Address	Memory	System	Price
QCPAC	Marcel Dekker 270 Madison Ave. New York, NY (212) 696-9000	64K	IBM PC PC-DOS 1 disk drive	$165
QCPACK	Applied Concepts Inc. 5430 South Ave. Suite C Tuscon, AZ 85706 (602) 294-1188	128K	IBM PC/XT/Jr PC-DOS 1.1+ 1 2-sided-disk drive 80 column monitor	$100
QUALISTAT	Advanced Technology Inspection 839 N. Rochester Rd. Clawson, MI 48017 (313) 589-3213	256K	IBM PC DOS 3.0+	$995
QUALITY ALERT	Penton Software, Inc. 420 Lexington Ave. Suite 2846 New York, NY 10017 (800) 221-3414	64K	IBM PC, XT, AT, PCjr PC-DOS 2.0 1 or 2 disk drives 192K+ graphics card required color/composite monitor IBM PC, XT, AT, PCjr PC-DOS 2.0 1 or 2 disk drives 128K+ graphics card required color/composite monitor	$395 (Version 1.5) $795 (Version 2.0)
SPC ANALYST	Perry Johnson Seminars, Inc. 3000 Town Center Suite 1075 Southfield, MI 48075 (313) 356-4410	128K	IBM PC, XT, AT 1 dual-disk drive/ hard disk IBM color graphics board HP/IBM plotter support	$1495 $500

Table 7-1. (Continued)

Name	Source	RAM (K bytes)	Hardware /Software Requirements	Approximate Acquisition Cost
SQCpack	PQ Systems 270 Regency Ridge Suite 214 P.O. Box 633 Dayton, Ohio 45459 (800) 547-1565	192K	IBM PC, XT, AT and compatibles 2 disk drives graphics adaptor	$495
SQE	Plan-Test Associates 3443 North Central Ave. Suite 609 Phoenix, AZ 85012 (602) 956-0180	128K	IBM PC MS-DOS 1 disk drive	$350
STATGRAPHICS	Statistical Graphics Corporation Research Park 2 Wall Street Princeton, NJ 08540 (609) 924-9374	320K	IBM PC MS-DOS 2 disk drives/ 1 disk and hard disk 8087 supported	$895
STATISTICAL PROCESS CONTROL Version 3	Pierson Company 8392 Stoney Creek Court Davison, MI 48423 (313) 653-7065	128K	IBM PC MS-DOS 2 disk drives	$400
STAT-PAK	Quality Resources 840 McKinley Plymouth, MI 48170 (313) 453-4616	16K	IBM PC MS-DOS 2 disk drives	$100

Author: footnote unclear

```
10 CLEAR 5000
20 DIM A(25,26),C(25,26),W(25),X(25),Y(25),Z(25)
30 INPUT"NUMBER OF POINTS";N
35 LPRINT" The number of input points is ";N
40 INPUT"ORDER OF POLYNOMIAL";L
45 LPRINT" The order of the polynomial is";L
50 M=L+1
60 R=L+2
70 FOR I=1 TO N
80 PRINT"INPUT X(";I;"),Y(";I;")"
90 INPUT X(I),Y(I)
120 NEXT I
130 FOR J=1 TO N
140 C(J,1)=1
150 NEXT J
160 FOR I=2 TO M
170 FOR J=1 TO N
175 IF X(I)=0 THE X(I)=.000001
178 CLS : PRINT @540, "WORKING"
180 C(J,I)=C(J,I-1)*X(J)
190 NEXT J
200 NEXT I
210 FOR I=1 TO M
220 FOR J=1 TO M
230 A(I,J)=0
240 FOR K=1 TO N
250 A(I,J)=A(I,J)+C(K,I)*C(K,J)
260 NEXT K
270 NEXT J
280 NEXT I
290 FOR I= 1 TO M
300 A(I,R)=0
310 FOR K=1 TO N
320 A(I,R)=A(I,R)+C(K,I)*Y(K)
330 NEXT K
340 NEXT I
350 GOSUB 500
360 FOR I=1TO N
370 W(I)=0
380 FOR J=1 TO M
390 W(I)=W(I)+A(J,R)*X(I)[(J-1)
400 NEXT J
410 Z(I)=ABS(W(I)-Y(I))/Y(I)*100
420 NEXT I
430 FOR I=1 TO M : LPRINT
431 IF I=1 THEN LPRINT"CONSTANT=";A(I,R) ELSE GOTO 432
432 IF I=2 THEN LPRINT"X=";A(I,R) ELSE GOTO 433
433 IF I=3 THEN LPRINT"X SQUARED =";A(I,R) ELSE GOTO 434
434 IF I=4 THEN LPRINT"X CUBED =";A(I,R)
435 IF I=5 THEN LPRINT"X FORTH =";A(I,R)
440 LPRINT
450 NEXT I
460 FOR I=1 TO N
470 LPRINT"X(";I;")=";X(I),"Y(";I;")=";Y(I),"W(";I;")=";W(I);" Z (";I;")=";Z(I)
480 NEXT I
490 LPRINT
500 FOR K=1 TO M
510 S=A(K,K)
520 FOR I=K TO R
530 A(K,I)=A(K,I)/S
534 CLS : PRINT@540, "WORKING"
540 NEXT I
550 FOR J=1 TO M
560 IF J=K THEN 610
570 S=A(J,K)
580 FOR I=K TO R
590 A(J,I)=A(J,I)-A(K,I)*S
600 NEXT I
610 NEXT J
620 NEXT K
630 RETURN
```

Fig. 7-1. Program of xxing general-order polynomials.

difference between the given data and the predictions. Combining Equation 7-1 with the least-squares equations for a_i gives

$$S = \sum_{i=1}^{n} \left(Y_i - \sum_{j=1}^{m+1} a_j X_i^{j-1} \right)^2 \qquad (7\text{-}2)$$

where n is the number of data sets and S is the minimized sum.

Differentiating Equation 7-2 with respect to a_i and setting the resulting equations equal to zero yields the matrix equation

$$AG = B \qquad (7\text{-}3)$$

where:

$$[A] = \begin{matrix} n & \Sigma X_i & \Sigma X_i^2 & \cdots & \Sigma X_i^m \\ \Sigma X_i & \Sigma X_i^2 & \Sigma X_i^3 & \cdots & \Sigma X_i^{m+1} \\ \vdots & & & & \\ \Sigma X_i^m & \Sigma X_i^{m+1} & \Sigma X_i^{m+2} & \cdots & \Sigma X_i^{2m} \end{matrix}$$

$$G = \begin{matrix} a_1 \\ a_2 \\ \vdots \\ a_{m+1} \end{matrix}$$

$$[B] = \begin{matrix} \Sigma Y_i \\ \Sigma X_i Y_i \\ \Sigma X_i^2 Y_i \\ \vdots \\ \Sigma X_i^m Y_i \end{matrix}$$

Given n sets of experimental data, $[A]$ and $[B]$ can be easily calculated. Solving Equation 7-3 by matrix inversion yields the coefficients needed for Equation 7-1.

An augmented matrix $[A]$ is created for the matrix inversion by attaching $[B]$ to the last column of $[A]$. The Gaussian elimination method reduces the original $[A]$ to a unit matrix and leaves the last column of the augmented matrix as the solution. Besides calculating a_i, the program calculates the predicted Y_i from Equation 7-2 and the percentage deviation of the prediction from the original data for each point. This shows how well the equation fits.

Program operation is straightforward. The order of the polynomial to be fitted

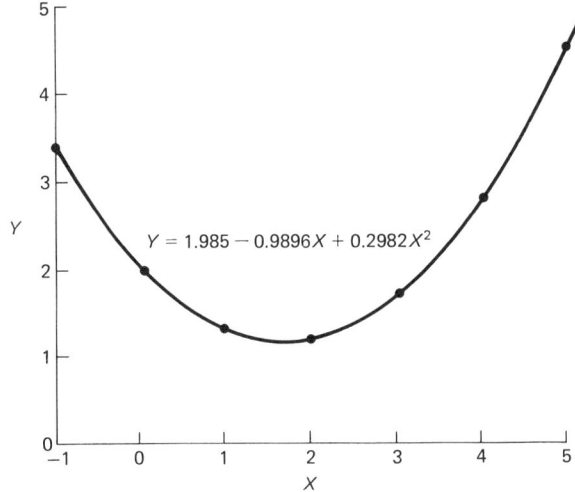

Fig. 7-2. Plot of second-order polynomials.

and the number of data sets are entered in response to program prompts. The data are then entered in X, Y pairs, again in response to program prompting. The results of two sample runs using the Figure 7-2 data points are shown in Figures 7-3 and 7-4. The first is for seven data point pairs fitted to a second-order polynomial. The second is fitting the same data to a fourth-order polynomial.

The results indicate the number of point pairs entered, the polynomial order, and the calculated values for the equation constant term, for X, for X squared, X cubed, and X to the fourth power, as the order requires. The first column lists the input data, $X(I)$; the second column, the corresponding $Y(I)$ value. The

```
The number of input points is  7
The order of the polynomial is 2

CONSTANT= 1.985


X=-.989643


X SQUARED = .298214
X( 1 )=-1        Y( 1 )= 3.29     W( 1 )= 3.27286     Z( 1 )= .521064
X( 2 )= 0        Y( 2 )= 1.95     W( 2 )= 1.985       Z( 2 )= 1.79486
X( 3 )= 1        Y( 3 )= 1.31     W( 3 )= 1.29357     Z( 3 )= 1.25411
X( 4 )= 2        Y( 4 )= 1.19     W( 4 )= 1.19857     Z( 4 )= .720275
X( 5 )= 3        Y( 5 )= 1.72     W( 5 )= 1.7         Z( 5 )= 1.16278
X( 6 )= 4        Y( 6 )= 2.79     W( 6 )= 2.79786     Z( 6 )= .281625
X( 7 )= 5        Y( 7 )= 4.49     W( 7 )= 4.49215     Z( 7 )= .0477581
```

Fig. 7-3. Example of second-order polynomial.

```
The number of input points is   7
The order of the polynomial is  4

CONSTANT=  1.96168


X=-.988733


X SQUARED =  .324573


X CUBED =-.0113837


X FORTH =  1.24937E-03

X( 1 )=-1      Y( 1 )= 3.29    W( 1 )= 3.28761    Z( 1 )= .0725111
X( 2 )= 0      Y( 2 )= 1.95    W( 2 )= 1.96168    Z( 2 )= .598724
X( 3 )= 1      Y( 3 )= 1.31    W( 3 )= 1.28738    Z( 3 )= 1.72665
X( 4 )= 2      Y( 4 )= 1.19    W( 4 )= 1.21142    Z( 4 )= 1.80016
X( 5 )= 3      Y( 5 )= 1.72    W( 5 )= 1.71047    Z( 5 )= .553838
X( 6 )= 4      Y( 6 )= 2.79    W( 6 )= 2.7912     Z( 6 )= .0428812
X( 7 )= 5      Y( 7 )= 4.49    W( 7 )= 4.49024    Z( 7 )= 5.25689E-03
```

Fig. 7-4. Example of fourth-order polynomial.

third column, $W(I)$, lists the calculated value for $Y(I)$ and the fourth column lists the percentage deviation of the calculated value from the original data.

The equivalent representation of Figure 7-3 is

$$Y = 1.9858 - 0.9896X + 0.2982X^2$$

The equational representation of Figure 7-4 is

$$Y = 1.9617 - 0.9887X + 0.3246X^2 - 0.0114X^3 + 0.00125X^4$$

7.3. DESIGNING T- AND Pi-PADS

Audio and RF circuits and systems generally require attenuators to control the signal level. Although the process of designing such circuits is relatively straightforward, it is also both tedious and time-consuming. A program written by Philip Arnold, published in *Microwaves & RF* (May 1985), eases the process of designing attenuator pads.

Two programs, both written in BASIC, allow the design and analysis of T- or Pi-pads. They can be used to design attenuators that operate up to about 1 GHz. The 1-GHz frequency limit is set when real-world chip resistors are considered. If RCR type resistors are to be used, then frequencies of about 150

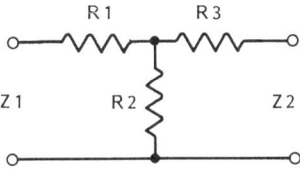

Fig. 7-5. General T-pad.

MHz should probably be considered as an upper limit since these resistors are about 0.050 inches long and this length creates significant parasitics. Although no new theory is developed, the simple programs remove the drudgery by eliminating the need for reference books and many calculations. The programs are extensively menu-driven, making them quite easy to use.

When using the programs, it is important to remember that Z1 and Z2 refer to transmission-line impedances, not to complex loads. The program TPAD calculates the following values (Figure 7-5):

- Given Z1 = Z2, the program provides the values for R1, R2, and R3 for any desired level of attenuation.
- Given Z1 ≠ Z2, the program calculates the theoretical values of R1, R2, and R3 for any level of attenuation. If the values of Z1 and Z2 are different enough and the desired attenuation is sufficiently small that that R1 or R3 are negative numbers, the program indicates that the results are invalid and

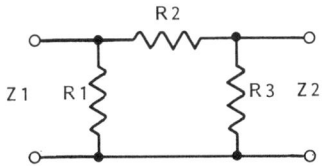

Fig. 7-6. General TT-pad.

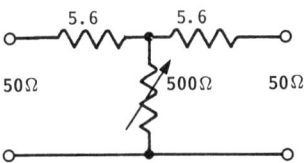

Fig. 7-7. Example of T-pad results.

198 ANALYSIS AND DESIGN OF ELECTRONIC CIRCUITS USING PCs

Table 7-2. Sample TPAD Run.

R1 = R3 (Ω)	Attenuation (dB)	R2 (Ω) Computed	R2 (Ω) Measured	VSWR into 50 Ω Computed	VSWR into 50 Ω Measured
5.6	1.46	430	470	1.1	1.12
5.6	1.95	220	225	1.0	1.06
5.6	2.48	140	150	1.1	1.11

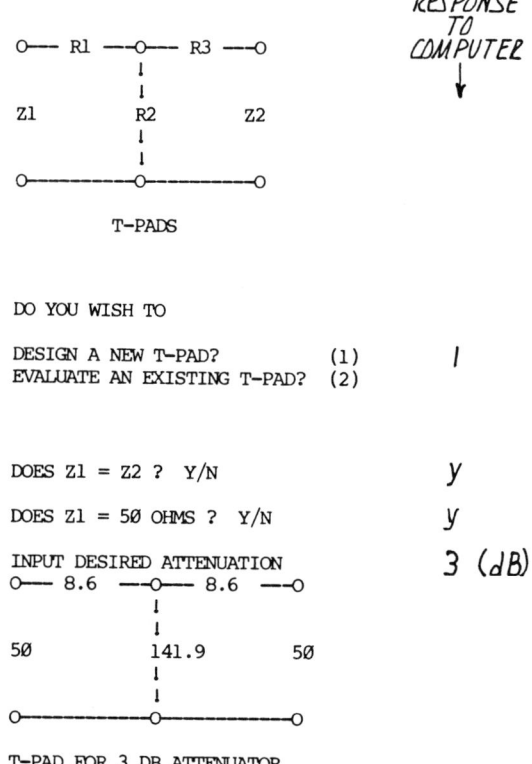

Fig. 7-8. Example of T-pad, 3-dB attenuation.

SPECIAL COMPUTER CODES

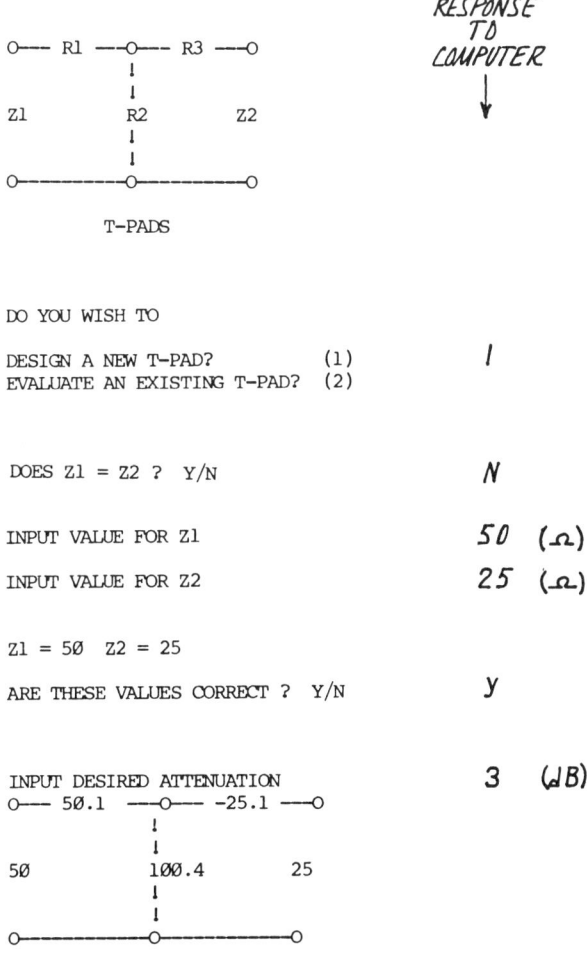

SAMPLE RUN
UNEQUAL TERMINATIONS

T-PADS

RESPONSE TO COMPUTER

DO YOU WISH TO

DESIGN A NEW T-PAD?   (1)   *1*
EVALUATE AN EXISTING T-PAD? (2)

DOES Z1 = Z2 ? Y/N   *N*

INPUT VALUE FOR Z1   *50* (Ω)

INPUT VALUE FOR Z2   *25* (Ω)

Z1 = 50   Z2 = 25

ARE THESE VALUES CORRECT ? Y/N   *y*

INPUT DESIRED ATTENUATION   *3* (dB)

```
O--- 50.1 ---O--- -25.1 ---O
          |
          |
50      100.4        25
          |
          |
O------------O------------O
```

T-PAD FOR A 3 DB ATTENUATOR

INVALID T-PAD , NEGATIVE VALUE RESISTOR
LOWEST VALUE OF ATTEN. = 7.66 DB

Fig. 7-8. (*Continued*)

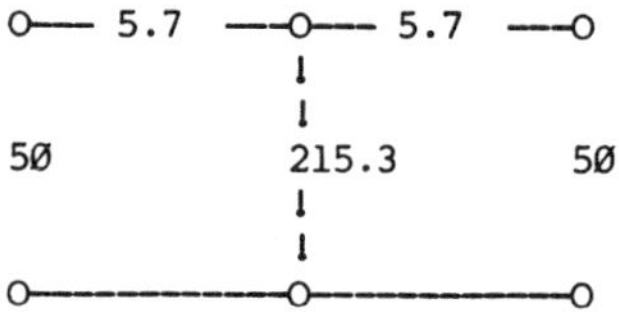

```
O── 5.7 ──O── 5.7 ──O
            │
            │                          EXACT DESIGN FOR A
50        215.3        50              2 dB T-PAD
            │
            │
O──────────O──────────O

T-PAD FOR 2 DB ATTENUATOR


O── 5.6 ──O── 5.6 ──O
            │
            │                          PRACTICAL DESIGN FOR A
50         220         50              2 dB T-PAD
            │
            │
O──────────O──────────O

ATTEN(DB)   VSWR   REF COEF   REF PWR(DB)
  -1.95      1        0           0


O── 5.6 ──O── 5.6 ──O
            │
            │                          COMPUTED DESIGN FOR ONE END
50         430         50              OF A 1.5→2.5 dB ADJUSTABLE
            │                          T-PAD
            │
O──────────O──────────O

ATTEN(DB)   VSWR   REF COEF   REF PWR(DB)
  -1.46     1.1     .046          .01


O── 5.6 ──O── 5.6 ──O
            │
            │                          COMPUTED DESIGN FOR THE OTHER
50         140         50              END OF A 1.5→2.5 dB
            │                          ADJUSTABLE T-PAD
            │
O──────────O──────────O

ATTEN(DB)   VSWR   REF COEF   REF PWR(DB)
  -2.48     1.1     .048          .01
```

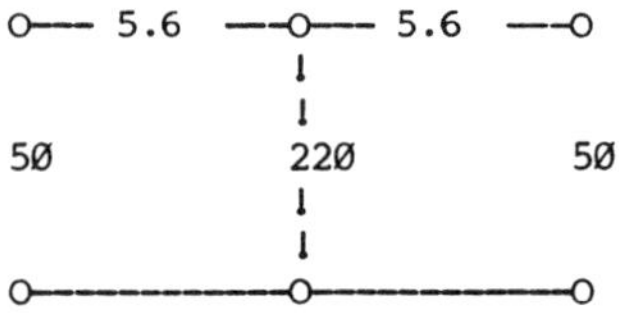

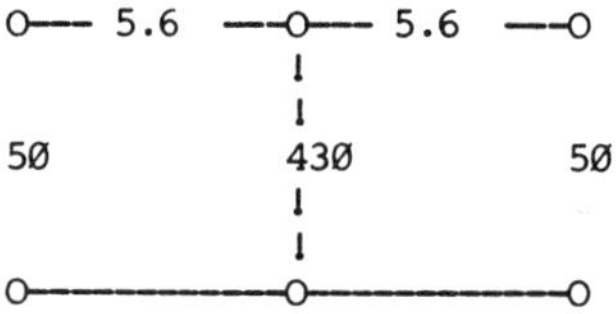

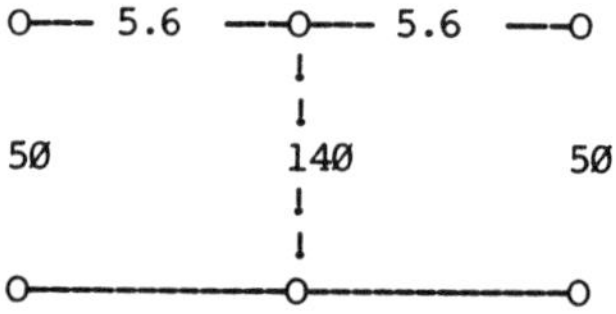

Fig. 7-9. Example of T-pad 2 dB attenuation.

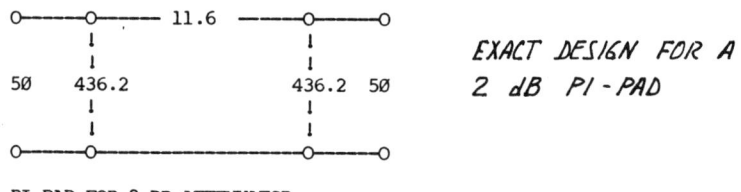

```
O────O──── 11.6 ────O────O
     │              │
     │              │                  EXACT DESIGN FOR A
50  436.2         436.2  50            2 dB PI-PAD
     │              │
     │              │
O────O──────────────O────O
```

PI-PAD FOR 2 DB ATTENUATOR

Fig. 7-10. Example of T-pad, 2 dB exact designs.

automatically calculates the lowest attenuation possible for a matched condition.
- Given the values of Z1, Z2, R1, R2, and R3, the program calculates attenuation (in decibels), VSWR, reflection coefficient, and the amount of reflected power (in decibels).

The program PIPAD is identical to TPAD in all respects, except that it uses the Pi topology (Figure 7-6).

As an example, the TPAD program is used to design a variable attenuator pad (Figure 7-7). Nominal attenuation is 2.0 dB and is variable from 1.5 to 2.5

```
10   REM   PROGRAM NAME IS "TPAD"
20   REM   P. ARNOLD   10/11/83
30   REM      CALCULATES THE RESISTOR VALUES FOR   'T-PADS'
40   REM      INPUT & OUTPUT TERMINATIONS ARE ARBITRARY
50   REM      ALSO CALCULATES THE LOSS AND VSWR FOR ANY ARBITRARY
60   REM      RESISTOR T-PAD WITH ARBITRARY TERMINATONS
95   C1 = 2.3025851
100  GOSUB 2100
110  VTAB (15): HTAB (10)
120  INVERSE : PRINT "T-PADS": NORMAL
130  PRINT : PRINT
140  PRINT "DO YOU WISH TO"
150  PRINT
160  PRINT "DESIGN A NEW T-PAD?          (1)"
170  PRINT "EVALUATE AN EXISTING T-PAD?  (2)"
180  GET A
190  IF A = 1 THEN 220
200  IF A = 2 THEN 1000
210  GOTO 100
220  GOSUB 2100
230  PRINT "DOES Z1 = Z2 ? Y/N  "
240  GET A$
250  IF A$ = "N" THEN 2500
255  REM    EQUAL IMPEDANCES
260  PRINT
270  PRINT "DOES Z1 = 50 OHMS ? Y/N  "
280  GET A$
290  IF A$ = "N" THEN 320
300  Z1 = 50
310  GOTO 325
320  INPUT "INPUT VALUE FOR Z1 ";Z1
325  PRINT
330  INPUT "INPUT DESIRED ATTENUATION ";P
340  K = EXP (C1 * P / 20)
350  R1 = (K - 1) / (K + 1) * Z1
360  R2 = 2 * Z1 * (K / (K ^ 2 - 1))
370  GOSUB 2200
380  PRINT "T-PAD FOR ";P;" DB ATTENUATOR"
390  PRINT : PRINT
400  PRINT "DO YOU WANT A HARD COPY? Y/N  "
410  GET A$
420  IF A$ = "N" THEN 1500
```

Fig. 7-11. TPAD and PIPAD programs.

```
430   PR# 1
440   PRINT  CHR$ (9);"80N"
450   PRINT : PRINT : PRINT
460   GOSUB 2210
470   PRINT
480   PRINT "T-PAD FOR ";P;" DB ATTENUATOR"
490   PRINT : PRINT : PRINT
500   PR# 0
510   GOTO 1500
1000  REM    EVALUATE AN EXISTING T-PAD
1005  GOSUB 2100
1010  PRINT
1015  PRINT "DOES Z1 = Z2 ?   Y/N"
1020  GET A$
1025  IF A$ = "N" THEN 1450
1030  PRINT : PRINT "DOES Z1 = 50 OHMS ?   Y/N"
1035  GET A$
1040  IF A$ = "N" THEN 1054
1045  Z1 = 50:Z2 = 50
1050  GOTO 1059
1054  PRINT
1055  INPUT "INPUT VALUE FOR Z1 ";Z1:Z2 = Z1
1059  PRINT
1060  INPUT "INPUT VALUES FOR R1,R2 & R3 ";R1,R2,R3
1065  GOSUB 2300
1070  PRINT
1075  PRINT "ARE THESE VALUES CORRECT ?   Y/N"
1080  GET A$
1085  IF A$ = "N" THEN 1305
1090  REM    CALCULATE ATTEN, VSWR, REFL COEF, REFL POWER
1095  RS = 1 / ((1 / R2) + (1 / (R3 + Z2)))
1100  ZN = R1 + RS
1105  DP = RS ^ 2 * Z2 / (ZN * ((R3 + Z2) ^ 2))
1106  DP = 10 / C1 *  LOG (DP)
1110  RO = (ZN - Z1) / (ZN + Z1)
1111  RO =  ABS (RO)
1115  VS = (1 + RO) / (1 - RO)
1120  RP = 10 / C1 *  LOG (1 / (1 - RO ^ 2))
1125  PRINT
1130  DP =  INT ((DP + .006) * 100) / 100
1135  RO =  INT ((RO + .0006) * 1000) / 1000
1140  VS =  INT ((VS + .006) * 100) / 100
1145  RP =  INT ((RP + .006) * 100) / 100
1150  GOSUB 2300
1152  GOSUB 1155
1153  GOTO 1170
1155  PRINT "ATTEN(DB)    VSWR    REF COEF   REF PWR(DB)"
1160  PRINT  TAB( 3);DP;
1161  POKE 36,12: PRINT VS;
1162  POKE 36,21: PRINT RO;
1163  POKE 36,32: PRINT RP
1165  RETURN
1170  PRINT : PRINT
1175  PRINT "DO YOU WANT A HARD COPY ?   Y/N"
1180  GET A$
1185  IF A$ = "N" THEN 1207
1190  PR# 1
1195  PRINT  CHR$ (9);"80N"
1200  GOSUB 2315
```

Fig. 7-11. (*Continued*)

SPECIAL COMPUTER CODES

```
1205    GOSUB 1155
1206    PR# 0
1207    PRINT : PRINT
1210    PRINT "DO YOU WANT TO EVALUATE ANOTHER T-PAD ?  Y/N"
1215    GET A$
1220    IF A$ = "N" THEN 1500
1225    GOTO 1005
1300    REM    ERROR, CHANGE VALUES
1305    GOSUB 2315
1310    PRINT
1315    PRINT "WHAT NEEDS TO BE CHANGED ?"
1320    PRINT
1325    PRINT   TAB( 10);"R1    (1)"
1330    PRINT   TAB( 10);"R2    (2)"
1335    PRINT   TAB( 10);"R3    (3)"
1340    PRINT   TAB( 10);"Z1    (4)"
1345    PRINT   TAB( 10);"Z2    (5)"
1350    GET A
1352    PRINT
1355    IF A = 1 THEN 1395
1360    IF A = 2 THEN 1405
1365    IF A = 3 THEN 1415
1370    IF A = 4 THEN 1425
1375    IF A = 5 THEN 1435
1380    GOTO 1305
1395    PRINT "R1 = ";R1;: INPUT "   NEW VALUE =   ";R1
1400    GOTO 1065
1405    PRINT "R2 = ";R2;: INPUT "   NEW VALUE =   ";R2
1410    GOTO 1065
1415    PRINT "R3 = ";R3;: INPUT "   NEW VALUE =   ";R3
1420    GOTO 1065
1425    PRINT "Z1 = ";Z1;: INPUT "   NEW VALUE =   ";Z1
1430    GOTO 1065
1435    PRINT "Z2 = ";Z2;: INPUT "   NEW VALUE =   ";Z2
1440    GOTO 1065
1450    REM    Z1 UNEQUAL TO Z2
1455    PRINT
1460    INPUT "INPUT VALUE FOR Z1   ";Z1
1465    PRINT
1470    INPUT "INPUT VALUE FOR Z2   ";Z2
1475    GOTO 1059
1500    PRINT
1505    PRINT "RUN AGAIN ?  Y/N"
1510    GET A$
1515    IF A$ = "N" THEN 1525
1520    GOTO 100
1525    PRINT  CHR$ (4);"CATALOG"
1527    PRINT  CHR$ (4);"CATALOG"
1530    END
2000    REM    Z1 EQUAL TO Z2
2005    HOME
2010    PRINT "O---- R1 ----O---- R1 ----O"
2015    PRINT "              !"
2020    PRINT "              !"
2025    PRINT "Z1           R2           Z1"
2030    PRINT "              !"
2035    PRINT "              !"
2040    PRINT "O---------------O---------------O"
2045    PRINT
```

Fig. 7-11. (*Continued*)

```
2050    RETURN
2100    REM     Z1 UNEQUAL TO  Z2
2105    HOME
2110    PRINT "O---- R1 ----O---- R3 ----O"
2115    PRINT "              !"
2120    PRINT "              !"
2125    PRINT "Z1        R2         Z2"
2130    PRINT "              !"
2135    PRINT "              !"
2140    PRINT "O---------------O---------------O"
2145    PRINT
2150    RETURN
2200    REM     PRINT OUT FOR  Z1  EQUAL TO  Z2
2202    R1 =  INT ((R1 + .06) * 10) / 10
2203    R2 =  INT ((R2 + .06) * 10) / 10
2204    Z1 =  INT ((Z1 + .06) * 10) / 10
2205    HOME
2210    PRINT "O---- ";R1;
2212    POKE 36,9: PRINT " ----O---- ";R1;
2213    POKE 36,22: PRINT " ----O"
2215    PRINT  TAB( 14);"!"
2220    PRINT  TAB( 14);"!"
2225    PRINT Z1;
2226    POKE 36,13: PRINT R2;
2227    POKE 36,26: PRINT Z1
2230    PRINT  TAB( 14);"!"
2235    PRINT  TAB( 14);"!"
2240    PRINT "O-----------------O-----------------O"
2245    PRINT
2250    RETURN
2300    REM     PRINT OUT FOR  Z1 UNEQUAL TO  Z2
2302    R1 =  INT ((R1 + .05) * 10) / 10
2304    R2 =  INT ((R2 + .05) * 10) / 10
2306    R3 =  INT ((R3 + .05) * 10) / 10
2308    Z1 =  INT ((Z1 + .05) * 10) / 10
2310    Z2 =  INT ((Z2 + .05) * 10) / 10
2315    HOME
2320    PRINT "O---- ";R1;
2325    POKE 36,9: PRINT " ----O---- ";R3;
2330    POKE 36,22: PRINT " ----O"
2335    PRINT  TAB( 14);"!"
2340    PRINT  TAB( 14);"!"
2345    PRINT Z1;
2350    POKE 36,13: PRINT R2;
2355    POKE 36,26: PRINT Z2
2360    PRINT  TAB( 14);"!"
2365    PRINT  TAB( 14);"!"
2370    PRINT "O-----------------O-----------------O"
2375    PRINT : RETURN
2500.   REM     Z1 UNEQUAL TO  Z2
2502    GOSUB 2105
2505    PRINT
2510    INPUT "INPUT VALUE FOR Z1 ";Z1
2515    PRINT
2520    INPUT "INPUT VALUE FOR Z2 ";Z2
2525    PRINT
2530    GOSUB 2105
2535    PRINT
2540    PRINT "Z1 = ";Z1;"   Z2 = ";Z2
```

Fig. 7-11. (*Continued*)

```
2545    PRINT
2550    PRINT "ARE THESE VALUES CORRECT ?   Y/N"
2555    GET A$
2560    IF A$ = "N" THEN 2502
2565    GOTO 3000
2600    REM    PRINT OUT FOR Z1 UNEQUAL TO Z2
2605 R1 =   INT ((R1 + .06) * 10) / 10
2610 R2 =   INT ((R2 + .06) * 10) / 10
2615 R3 =   INT ((R3 + .06) * 10) / 10
2620 Z1 =   INT ((Z1 + .06) * 10) / 10
2625 Z2 =   INT ((Z2 + .06) * 10) / 10
2630    HOME
2635    PRINT "O--- ";R1;
2640    POKE 36,10: PRINT " ---O--- ";R3;
2645    POKE 36,22: PRINT " ---O"
2650    PRINT  TAB( 14);"!"
2655    PRINT  TAB( 14);"!"
2660    PRINT Z1;
2665    POKE 36,13: PRINT R2;
2670    POKE 36,26: PRINT Z2
2675    PRINT  TAB( 14);"!"
2680    PRINT  TAB( 14);"!"
2685    PRINT "O--------------O--------------O"
2690    PRINT : RETURN
3000    REM    Z1 UNEQUAL TO Z2
3005    PRINT
3010    INPUT "INPUT DESIRED ATTENUATION ";P
3015 K =   EXP (C1 * P / 20)
3020 C2 = K / (K ^ 2 - 1)
3025 C3 =   SQR (Z1 * Z2)
3030 C4 = (K ^ 2 + 1) / (K ^ 2 - 1)
3035 R1 = Z1 * C4 - 2 * C3 * C2
3040 R2 = 2 * C3 * C2
3045 R3 = Z2 * C4 - 2 * C3 * C2
3050    GOSUB 2600
3055    PRINT
3060    PRINT "T-PAD FOR A ";P;" DB ATTENUATOR"
3063    GOSUB 3500
3065    PRINT
3070    PRINT "DO YOU WANT A HARD COPY ?   Y/N"
3075    GET A$
3080    IF A$ = "N" THEN 1500
3085    PR# 1
3090    PRINT  CHR$ (9);"80N"
3095    PRINT : PRINT : PRINT
3100    GOSUB 2635
3105    PRINT
3110    PRINT "T-PAD FOR A ";P;" DB ATTENUATOR"
3112    GOSUB 3500
3115    PRINT : PRINT : PRINT
3120    PR# 0
3125    GOTO 1500
3500    REM    CHECK FOR INVALID T-PADS
3505    IF R1 < 0 THEN 3550
3515    IF R3 < 0 THEN 3550
3520    GOTO 3635
3550    PRINT
3555    PRINT "INVALID T-PAD , NEGATIVE VALUE RESISTOR"
3560    IF Z1 > Z2 GOTO 3580
```

Fig. 7-11. (*Continued*)

```
3565 E1 = Z2
3570 E2 = Z1
3575   GOTO 3590
3580 E1 = Z1
3585 E2 = Z2
3590 E3 = - 2 * SQR (E1 * E2) / E2
3595 E4 =   SQR (E3 ^ 2 - 4) / 2
3600 E5 =   ABS (E3 / 2 + E4)
3605 E6 = 20 / C1 * LOG (E5)
3615 E6 =   INT (E6 * 100) / 100
3620 E6 =   ABS (E6)
3625   PRINT "LOWEST VALUE OF ATTEN. = ";E6;" DB"
3630   PRINT
3635   RETURN

10   REM    PROGRAM NAME IS "PIPAD"
20   REM    P. ARNOLD   10/11/83
30   REM       CALCULATES THE RESISTOR VALUES FOR  'PI-PADS'
40   REM       INPUT & OUTPUT TERMINATIONS ARE ARBITRARY
50   REM       ALSO CALCULATES THE LOSS AND VSWR FOR ANY ARBITRARY
60   REM       RESISTOR PI-PAD WITH ARBITRARY TERMINATIONS
95 C1 = 2.3025851
100   GOSUB 2100
110   VTAB (15): HTAB (10)
120   INVERSE : PRINT "PI-PADS": NORMAL
130   PRINT : PRINT
140   PRINT "DO YOU WISH TO"
150   PRINT
160   PRINT "DESIGN A NEW PI-PAD ?          (1)"
170   PRINT "EVALUATE AN EXISTING PI-PAD ? (2)"
180   GET A
190   IF A = 1 THEN 220
200   IF A = 2 THEN 1000
210   GOTO 100
220   GOSUB 2100
230   PRINT "DOES Z1 = Z2 ?   Y/N  "
240   GET A$
250   IF A$ = "N" THEN 2500
255   REM    EQUAL IMPEDANCES
260   PRINT
270   PRINT "DOES Z1 = 50 OHMS ?   Y/N  "
280   GET A$
290   IF A$ = "N" THEN 320
300 Z1 = 50
310   GOTO 325
320   INPUT "INPUT VALUE FOR Z1 ";Z1
325   PRINT
330   INPUT "INPUT DESIRED ATTENUATION (DB) ";P
340 K = EXP (C1 * P / 20)
350 R2 = (K ^ 2 - 1) * Z1 / (2 * K)
360 R1 = (K + 1) / (K - 1) * Z1
370   GOSUB 2200
380   PRINT "PI-PAD FOR ";P;" DB ATTENUATOR"
390   PRINT : PRINT
400   PRINT "DO YOU WANT A HARD COPY?  Y/N  "
410   GET A$
420   IF A$ = "N" THEN 1500
430   PR# 1
440   PRINT CHR$ (9);"80N"
```

Fig. 7-11. (*Continued*)

SPECIAL COMPUTER CODES

```
450   PRINT : PRINT : PRINT
460   GOSUB 2210
470   PRINT
480   PRINT "PI-PAD FOR ";P;" DB ATTENUATOR"
490   PRINT : PRINT : PRINT
500   PR# 0
510   GOTO 1500
1000  REM    EVALUATE AN EXISTING T-PAD
1005  GOSUB 2100
1010  PRINT
1015  PRINT "DOES Z1 = Z2 ?  Y/N"
1020  GET A$
1025  IF A$ = "N" THEN 1450
1030  PRINT : PRINT "DOES Z1 = 50 OHMS ?  Y/N"
1035  GET A$
1040  IF A$ = "N" THEN 1054
1045  Z1 = 50:Z2 = 50
1050  GOTO 1059
1054  PRINT
1055  INPUT "INPUT VALUE FOR Z1 ";Z1:Z2 = Z1
1059  PRINT
1060  INPUT "INPUT VALUES FOR R1,R2 & R3 ";R1,R2,R3
1065  GOSUB 2600
1070  PRINT
1075  PRINT "ARE THESE VALUES CORRECT ?  Y/N"
1080  GET A$
1085  IF A$ = "N" THEN 1305
1090  REM    CALCULATE ATTEN, VSWR, REFL COEF, REFL POWER
1095  RB = R2 + 1 / (1 / R3 + 1 / Z2)
1100  ZN = 1 / (1 / R1 + 1 / RB)
1105  DP = (((1 - RB / (RB + R1)) * R3 / (R3 + Z2)) ^ 2) * Z2 / ZN
1106  DP = 10 / C1 *  LOG (DP)
1110  RO = (ZN - Z1) / (ZN + Z1)
1111  RO =  ABS (RO)
1115  VS = (1 + RO) / (1 - RO)
1120  RP = 10 / C1 *  LOG (1 / (1 - RO ^ 2))
1125  PRINT
1130  DP =  INT ((DP + .006) * 100) / 100
1135  RO =  INT ((RO + .0006) * 1000) / 1000
1140  VS =  INT ((VS + .006) * 100) / 100
1145  RP =  INT ((RP + .006) * 100) / 100
1150  GOSUB 2600
1152  GOSUB 1155
1153  GOTO 1170
1155  PRINT "ATTEN(DB)   VSWR    REF COEF   REF PWR(DB)"
1160  PRINT  TAB( 3);DP;
1161  POKE 36,12: PRINT VS;
1162  POKE 36,21: PRINT RO;
1163  POKE 36,32: PRINT RP
1165  RETURN
1170  PRINT : PRINT
1175  PRINT "DO YOU WANT A HARD COPY ?  Y/N"
1180  GET A$
1185  IF A$ = "N" THEN 1207
1190  PR# 1
1195  PRINT  CHR$ (9);"80N"
1200  GOSUB 2630
1205  GOSUB 1155
1206  PRINT : PRINT : PRINT : PRINT : PR# 0
1207  PRINT : PRINT
1210  PRINT "DO YOU WANT TO EVALUATE ANOTHER T-PAD ?  Y/N"
1215  GET A$
```

Fig. 7-11. (*Continued*)

```
1220   IF A$ = "N" THEN 1500
1225   GOTO 1005
1300   REM    ERROR,  CHANGE VALUES
1305   GOSUB 2630
1310   PRINT
1315   PRINT "WHAT NEEDS TO BE CHANGED ?"
1320   PRINT
1325   PRINT   TAB( 10);"R1    (1)"
1330   PRINT   TAB( 10);"R2    (2)"
1335   PRINT   TAB( 10);"R3    (3)"
1340   PRINT   TAB( 10);"Z1    (4)"
1345   PRINT   TAB( 10);"Z2    (5)"
1350   GET A
1352   PRINT
1355   IF A = 1 THEN 1395
1360   IF A = 2 THEN 1405
1365   IF A = 3 THEN 1415
1370   IF A = 4 THEN 1425
1375   IF A = 5 THEN 1435
1380   GOTO 1305
1395   PRINT "R1 = ";R1; SPC( 5);"NEW VALUE = ";: INPUT R1
1400   GOTO 1065
1405   PRINT "R2 = ";R2; SPC( 5);"NEW VALUE = ";: INPUT R2
1410   GOTO 1065
1415   PRINT "R3 = ";R3; SPC( 5);"NEW VALUE = ";: INPUT R3
1420   GOTO 1065
1425   PRINT "Z1 = ";Z1; SPC( 5);"NEW VALUE = ";: INPUT Z1
1430   GOTO 1065
1435   PRINT "Z2 = ";Z2; SPC( 5);"NEW VALUE = ";: INPUT Z2
1440   GOTO 1065
1450   REM    Z1 UNEQUAL TO Z2
1455   PRINT
1460   INPUT "INPUT VALUE FOR Z1   ";Z1
1465   PRINT
1470   INPUT "INPUT VALUE FOR Z2   ";Z2
1475   GOTO 1059
1500   PRINT
1505   PRINT "RUN AGAIN ?  Y/N"
1510   GET A$
1515   IF A$ = "N" THEN 1525
1520   GOTO 100
1525   PRINT   CHR$ (4);"CATALOG"
1527   PRINT   CHR$ (4);"CATALOG"
1530   END
2000   REM    Z1  EQUAL TO Z2
2005   HOME
2010   PRINT "O--------O-------- R2 --------O--------O"
2015   PRINT "         !                    !"
2020   PRINT "         !                    !"
2025   PRINT "Z1       R1                   R3       Z1"
2030   PRINT "         !                    !"
2035   PRINT "         !                    !"
2040   PRINT "O--------O--------------------O--------O"
2045   PRINT
2050   RETURN
2100   REM    Z1  UNEQUAL TO  Z2
2105   HOME
2110   PRINT "O--------O-------- R2 --------O--------O"
2115   PRINT "         !                    !"
2120   PRINT "         !                    !"
2125   PRINT "Z1       R1                   R3       Z2"
2130   PRINT "         !                    !"
2135   PRINT "         !                    !"
2140   PRINT "O--------O--------------------O--------O"
```

Fig. 7-11. (*Continued*)

```
2145    PRINT
2150    RETURN
2200    REM    PRINT OUT FOR  Z1   EQUAL TO   Z2
2202 R1 =   INT ((R1 + .06) * 10) / 10
2203 R2 =   INT ((R2 + .06) * 10) / 10
2204 Z1 =   INT ((Z1 + .06) * 10) / 10
2205    HOME
2210    PRINT "O--------O-------- ";R2;
2212    POKE 36,20: PRINT " -------O--------O"
2213    PRINT "           !              !"
2214    PRINT "           !              !"
2215    PRINT Z1;: POKE 36,6: PRINT R1;: POKE 36,26: PRINT R1;: POKE 36,33
        : PRINT Z1
2220    PRINT "           !              !"
2225    PRINT "           !              !"
2230    PRINT "O--------O--------------------O--------O"
2245    PRINT
2250    RETURN
2500    REM    Z1 UNEQUAL TO Z2
2502    GOSUB 2105
2505    PRINT
2510    INPUT "INPUT VALUE FOR Z1 ";Z1
2515    PRINT
2520    INPUT "INPUT VALUE FOR Z2 ";Z2
2525    PRINT
2530    GOSUB 2105
2535    PRINT
2540    PRINT "Z1 = ";Z1;"  Z2 = ";Z2
2545    PRINT
2550    PRINT "ARE THESE VALUES CORRECT ?   Y/N"
2555    GET A$
2560    IF A$ = "N" THEN 2502
2565    GOTO 3000
2600    REM     PRINT OUT FOR Z1 UNEQUAL TO Z2
2605 R1 =   INT ((R1 + .06) * 10) / 10
2610 R2 =   INT ((R2 + .06) * 10) / 10
2615 R3 =   INT ((R3 + .06) * 10) / 10
2620 Z1 =   INT ((Z1 + .06) * 10) / 10
2625 Z2 =   INT ((Z2 + .06) * 10) / 10
2630    HOME
2635    PRINT "O--------O-------- ";R2;
2640    POKE 36,20: PRINT " -------O--------O"
2650    POKE 36,7: PRINT "!";: POKE 36,27: PRINT "!"
2655    POKE 36,7: PRINT "!";: POKE 36,27: PRINT "!"
2660    PRINT Z1;: POKE 36,6: PRINT R1;: POKE 36,26: PRINT R3;: POKE 36,33
        : PRINT Z2
2675    POKE 36,7: PRINT "!";: POKE 36,27: PRINT "!"
2680    POKE 36,7: PRINT "!";: POKE 36,27: PRINT "!"
2685    PRINT "O--------O--------------------O--------O"
2690    PRINT : RETURN
3000    REM    Z1 UNEQUAL TO Z2
3005    PRINT
3010    INPUT "INPUT DESIRED ATTENUATION ";P
3015 K =   EXP (C1 * P / 20)
3020 C2 = K ^ 2 - 1
3025 C3 =   SQR (Z1 / Z2)
3030 C4 = K ^ 2 - 2 * K * C3 + 1
3035 R1 = Z1 * C2 / C4
3040 R2 =   SQR (Z1 * Z2) * C2 / 2 / K
3045 R3 = Z2 * C2 / (K ^ 2 - 2 * K / C3 + 1)
3050    GOSUB 2600
3055    PRINT
3060    PRINT "PI-PAD FOR A ";P;" DB ATTENUATOR"
3063    GOSUB 3500
3065    PRINT
3070    PRINT "DO YOU WANT A HARD COPY ?   Y/N"
```

Fig. 7-11. (*Continued*)

```
3075  GET A$
3080  IF A$ = "N" THEN 1500
3085  PR# 1
3090  PRINT   CHR$ (9);"80N"
3095  PRINT : PRINT : PRINT : PRINT
3100  GOSUB 2635
3105  PRINT
3110  PRINT "T-PAD FOR A ";P;" DB ATTENUATOR"
3112  GOSUB 3500
3115  PRINT : PRINT : PRINT : PRINT
3120  PR# 0
3125  GOTO 1500
3500  REM   CHECK FOR INVALID T-PADS
3505  IF R1 < 0 THEN 3550
3515  IF R3 < 0 THEN 3550
3520  RETURN
3550  PRINT
3555  PRINT "INVALID PI-PAD , NEGATIVE VALUE RESISTOR"
3560  IF Z1 > Z2 GOTO 3580
3565  E1 = Z2
3570  E2 = Z1
3575  GOTO 3590
3580  E1 = Z1
3585  E2 = Z2
3590  E3 =  - 2 * SQR (E1 * E2) / E2
3595  E4 = SQR (E3 ^ 2 - 4) / 2
3600  E5 = ABS (E3 / 2 + E4)
3605  E6 = 20 / C1 * LOG (E5)
3615  E6 = INT (E6 * 100) / 100
3620  E6 = ABS (E6)
3625  PRINT "LOWEST VALUE OF ATTEN. = ";E6;" DB"
3630  PRINT
3635  RETURN
```

Fig. 7-11. (*Continued*)

dB. The results of such an analysis, with comparison measurements made on a tested unit, are shown in Table 7-2. Step-by-step examples of the use of TPAD and PIPAD are shown in Figures 7-8 through 7-10.

A listing of the codes for both TPAD and PIPAD are provided in Figure 7-11.

7.4. CALCULATING "L"-NETWORKS

The program in Figure 7-12, written by Dick Wright, Electrospace Systems, Inc., 1601 N. Plano Road, Richardson, Texas, and published in *Microwaves and RF* (May 1985), fills one need of all circuit designers. That need is to match two unequal impedances to insure maximum information transfer.

The "L" network is one of the simplest and most widely used circuits for matching two unequal impedances over a relatively narrow range of frequencies. It consists of only two components; a coil and a capacitor which can be used as single or double pairs. These components may be used in eight possible

```
100 CLS:KEY OFF:WIDTH 80
120 LOCATE 10,30:PRINT"IMPEDANCE MATCHING PROGRAM"
140 LOCATE 12,30:PRINT"   Using L-Type Networks"
160 LOCATE 23,1:PRINT". . .hit any key to begin"
180 IF INKEY$ = "" THEN 180
200 CLEAR:CLS
220        '
240        'L-TYPE IMPEDANCE MATCHING NETWORKS
260        'Dick Wright
280        'Electrospace Systems, Inc.
300        'May, 1984
320        'Written for IBM PC
340        'BASICA version 2.0
360        '
380'
400'*************************** initialization *****************
420'variable declaration
440'TWOPI=6.28
460'DR=deg/radian conversion
480'R=load resistance, ohms
500'X=load reactance, ohms
520'G=load conductance, mhos
540'B=load susceptance, mhos
560'Z=load impedance, ohms
580'Y=load admittance, mhos
600'-------------------------------------------------------------
620'
640       CLS:KEY OFF
660       'DEFINT I,J,K,N
680       Z$ = "#####.##":Y$ = "#.######":X1$ = "#.#"
700R$ = "####.#":X$ = "#####.##":IN$ = "###.##"
720TWOPI = 8 * ATN(1)
740DR = 4 * ATN(1)/180'degree/radian conversion
760       IND$ = "Inductance":CAP$ = "Capacitance"
780       EL1$ = "Shunt Capacitance":EL2$ = "Shunt Inductance"
800       EL3$ = "Series Capacitance":EL4$ = "Series Inductance"
820       LN$ = STRING$(72,45):C$ = "pf":I$ = "uh"
840DEF FNCAP(X) = ABS(1/(TWOPI * FREQ * X)) * 1000000!  'cap function
860DEF FNIND(X) = ABS(X/(TWOPI * FREQ))'ind function
880'
900PRINT:PRINT
920       PRINT"Type the load impedance in rectangular form"
940       PRINT:INPUT"    First, the resistance in ohms: ",R
960RHOLD = R
980       PRINT:INPUT"    OK! Now the reactance in ohms (observe polarity): ",X
1000IF X < 0 THEN SIGNX$ = "-j" ELSE SIGNX$ = "+j"
1020XHOLD = X
1040      PRINT:INPUT"Type the source resistance level in ohms";Z0
1060PRINT:INPUT"    Type the frequency, in MHz: ",FREQ
1080Z1 = Z0 + (Z0 * .01):Z2 = Z0 - (Z0 * .01)
1100Y1 = 1/Z1:Y2 = 1/Z2
1120      CLS
1140      PRINT;"Load Impedance: ";TAB(50);"Frequency: ";FREQ;" MHz"
1160      PRINT;TAB(10);"R = ";
1180      PRINT USING R$;R;
1200      PRINT;TAB(50);"Date: ";DATE$
1220      PRINT;TAB(10);"X = " + SIGNX$;
1240      PRINT USING X$;ABS(X)
1260      PRINT;"Source Impedance: ";Z0;" ohms"
1280      PRINT
1300'
1320'********************* mainline ***************************
1340'
1360      PRINT
1380GOSUB 8560'validate input data
1400GOSUB 8200'convert impedance to admittance
1420GHOLD = G:BHOLD = B
1440GOSUB 1560'determine type of network
```

Fig. 7-12. T-program.

212 ANALYSIS AND DESIGN OF ELECTRONIC CIRCUITS USING PCs

```
1460'
1480       CLS:PRINT"To run program again, hit <ENTER>"
1500       PRINT:PRINT". . .any other key will end program"
1520       X$ = INPUT$(1):IF X$ = CHR$(13) THEN 200 ELSE 9640
1540'
1560       '------------------------ types subroutine -------------------
1580       '
1600IF R>=Z2 AND R<=Z1 THEN NTWK$="NETWORK X":GOSUB 1760
1620       IF G>=Y1 AND G<=Y2 THEN NTWK$="NETWORK Y":GOSUB 2500
1640IF R>Z1 THEN NTWK$="NETWORKS A,B":GOSUB 3120:GOSUB 3740
1660IF G>Y2 THEN NTWK$="NETWORKS C,D":GOSUB 4360:GOSUB 4980
1680           IF X>=0 AND (R<Z2 OR R>Z1) AND (G>Y2 OR G<Y1) THEN NTWK$="NETWORKS
               A,E,G,D":GOSUB 3120:GOSUB 5600:GOSUB 4980:GOSUB 6900
1700           IF X<0 AND (R<Z2 OR R>Z1) AND (G>Y2 OR G<Y1) THEN NTWK$="NETWORKS
               B,F,H,C":GOSUB 3740:GOSUB 6260:GOSUB 4360:GOSUB 7560
1720       RETURN
1740'
1760'------------------- NETWORK X subroutine --------------------
1780       '
1800       GOSUB 8660
1820TITLE$ = "NETWORK X"
1840       IF XHOLD < 0 THEN 2140 ELSE 1860
1860       'calculate series capacitance
1880       FL$ = EL3$
1900       NOTE1$ = FL$ + " between input and load"
1920C = FNCAP(-XHOLD)
1940       SERIESEL = C
1960       R = RHOLD
1980S2$ = C$
2000REACT = -X
2020RESISTANCE = R
2040GOSUB 8840
2060       PRINT:PRINT:PRINT". . .hit any key to continue"
2080       IF INKEY$ = "" THEN 2080
2100       GOSUB 9700
2120       RETURN
2140       '           calculate series inductance
2160FL$ = EL4$
2180NOTE1$ = FL$ + " between input and load"
2200L = FNIND(-XHOLD)
2220       SERIESEL = L
2240       R = RHOLD
2260S2$ = I$
2280REACT = -X
2300RESISTANCE = R
2320       GOSUB 8840
2340       PRINT:PRINT:PRINT". . .hit any key to continue"
2360       IF INKEY$ = "" THEN 2360
2380       GOSUB 9700
2400RETURN
2420       '
2440       '-------------------- NETWORK Y subroutine --------------------
2460       '
2480       GOSUB 8660
2500       TITLE$ = "NETWORK Y"
2520       IF XHOLD < 0 THEN 2860 ELSE 2560
2540       '            calculate shunt capacitance
2560       FL$ = EL1$
2580       NOTE1$ = FL$ + " across load"
2600       C = FNCAP(1/BHOLD)
2620       SERIESEL = C
2640       R = RHOLD
2660        S2$ = C$
2680       REACT = -1/B
2700       RESISTANCE = 1/G
2720       GOSUB 8840
2740       PRINT:PRINT:PRINT". . .hit any key to continue"
2760       IF INKEY$ = "" THEN 2760
```

Fig. 7-12. (*Continued*)

```
2780        GOSUB 9700
2800        RETURN
2860    '                 calculate shunt inductance
2880        FL$ = EL2$
2900        NOTE1$ = FL$ + " across load"
2920        L = FNIND(1/BHOLD)
2940        SERIESEL = L
2960        R = RHOLD
2980        S2$ = I$
3000        RESISTANCE = 1/G
3020        GOSUB 8840
3040        PRINT:PRINT:PRINT". . .hit any key to continue"
3060        IF INKEY$ = "" THEN 3060
3080        GOSUB 9700
3090        RETURN
3100    '
3120'------------------- NETWORK A subroutine --------------------
3140'mainline
3160GOSUB 8660
3180EL$ = EL1$:FL$ = EL4$
3200TITLE$ = "NETWORK A"
3220NOTE1$ = EL$ + " across load"
3240NOTE2$ = FL$ + " between input and " + EL$
3260S1$ = C$:S2$ = I$
3280WHILE R < Z2 OR R > Z1
3300        I = I + 1
3320            IF R < Z0 AND I <> 1 THEN N = -1 ELSE N = 1
3340        JB = 1/(2^(I-1) * Z0 * N)
3360        JBTOTAL = JBTOTAL + JB
3380            B = B + JB
3400        GOSUB 8420
3420            IF I > 20 THEN 3640'test for non-realizable component
3440    WEND
3460'--------------finalization
3480JX = -1/JBTOTAL
3500IF X < .0001 AND X > -.0001 THEN I = 21:GOTO 3640
3520C = FNCAP(JX)
3540L = FNIND(X)
3560SHUNTEL = C:SERIESEL = L
3580RESISTANCE = R
3600REACTANCE = X - X
3620REACT = -X
3640GOSUB 8840
3660    PRINT:PRINT:PRINT". . .hit any key to continue"
3680    IF INKEY$ = "" THEN 3680
3700    GOSUB 9700
3720RETURN
3740'------------------- NETWORK B subroutine -------------------
3760GOSUB 8660
3780'mainline
3800EL$ = EL2$:FL$ = EL3$
3820TITLE$ = "NETWORK B"
3840    NOTE1$ = EL$ + " across load"
3860NOTE2$ = FL$ + " between input and " + EL$
3880S1$ = I$:S2$ = C$
3900WHILE R < Z2 OR R > Z1
3920        I = I + 1
3940            IF R < Z0 AND I <> 1 THEN N = 1 ELSE N = -1
3960        JB = 1/(2^(I-1) * Z0 * N)
3980        JBTOTAL = JBTOTAL + JB
4000            B = B + JB
4020        GOSUB 8420
4040            IF I > 20 THEN 4260'test for non-realizable component
4060    WEND
4080'--------------finalization
4100JX = -1/JBTOTAL
4120IF X < .0001 AND X > -.0001 THEN I = 21:GOTO 4260
4140C = FNCAP(X)
```

Fig. 7-12. (*Continued*)

214 ANALYSIS AND DESIGN OF ELECTRONIC CIRCUITS USING PCs

```
4160L = FNIND(JX)
4180SHUNTEL = L:SERIESEL = C
4200RESISTANCE = R
4220REACTANCE = X - X
4240REACT = -X
4260GOSUB 8840
4280    PRINT:PRINT:PRINT". . .hit any key to continue"
4300    IF INKEY$ = "" THEN 4300
4320    GOSUB 9700
4340RETURN
4360'------------------- NETWORK C subroutine --------------------
4380'mainline
4400GOSUB 8660
4420EL$ = EL4$:FL$ = EL1$
4440TITLE$ = "NETWORK C"
4460NOTE1$ = FL$ + " across input"
4480NOTE2$ = EL$ + " between load and " + FL$
4500S1$ = I$:S2$ = C$
4520X = 0:GOSUB 8200
4540WHILE G < Y1 OR G > Y2
4560    I = I + 1
4580    IF G > Y2  THEN N = 1 ELSE N = -1
4600    JX = (25 * N)/2^(I-1)
4620    JXTOTAL = JXTOTAL + JX
4640    X = X + JX
4660    GOSUB 8200
4680        IF I > 20 THEN 4880'test for non-realizable component
4700WEND
4720XC = 1/B:JX = JXTOTAL - XHOLD
4740RESISTANCE = 1/G
4760IF XC < .0001 AND XC > -.0001 THEN I = 21:GOTO 4880
4780C = FNCAP(XC)
4800L = FNIND(JX)
4820SHUNTEL = L:SERIESEL = C
4840REACTANCE = X - X
4860REACT = XC
4880GOSUB 8840
4900    PRINT:PRINT:PRINT". . .hit any key to continue"
4920    IF INKEY$ = "" THEN 4920
4940    GOSUB 9700
4960RETURN
4980'------------------- NETWORK D subroutine --------------------
5000'mainline
5020GOSUB 8660
5040EL$ = EL3$:FL$ = EL2$
5060TITLE$ = "NETWORK D"
5080NOTE1$ = FL$ + " across input"
5100NOTE2$ = EL$ + " between load and " + FL$
5120S1$ = C$:S2$ = I$
5140X = 0:GOSUB 8200
5160WHILE G < Y1 OR G > Y2
5180    I = I + 1
5200    IF G > Y2  THEN N = -1 ELSE N = 1
5220    JX = (25 * N)/2^(I-1)
5240    JXTOTAL = JXTOTAL + JX
5260    X = X + JX
5280        IF I > 20 THEN 5500'test for non-realizable component
5300    GOSUB 8200
5320WEND
5340XL = 1/B:JX = JXTOTAL - XHOLD
5360RESISTANCE = 1/G
5380IF JX < .0001 AND JX > -.0001 THEN I = 21:GOTO 5500
5400C = FNCAP(JX)
5420L = FNIND(XL)
5440SHUNTEL = C:SERIESEL = L
5460REACTANCE = X - X
5480REACT = XL
5500GOSUB 8840
5520    PRINT:PRINT:PRINT". . .hit any key to continue"
5540    IF INKEY$ = "" THEN 5540
```

Fig. 7-12. (*Continued*)

SPECIAL COMPUTER CODES 215

```
5560    GOSUB 9700
5580RETURN
5600'------------------- NETWORK E subroutine --------------------
5620'mainline
5640GOSUB 8660
5660EL$ = EL1$:FL$ = EL3$
5680TITLE$ = "NETWORK E"
5700NOTE1$ = EL$ + " across load"
5720NOTE2$ = FL$ + " between input and " + EL$
5740S1$ = C$:S2$ = C$
5760M = -1/B
5780WHILE R < Z2 OR R > Z1
5800    I = I + 1
5820        IF R > Z0 AND I <> 1 THEN N = -1 ELSE N = 1
5840    JB = 1/(2^(I-1) * M * N)
5860    JBTOTAL = JBTOTAL + JB
5880        B = B + JB
5900    GOSUB 8420
5920    IF I > 20 THEN 6160
5940    WEND
5960'--------------finalization
5980JX = -1/JBTOTAL
6000IF X < .0001 AND X > -.0001 THEN I = 21:GOTO 6160
6020IF JX < .0001 AND JX > -.0001 THEN I = 21:GOTO 6160
6040C1 = FNCAP(JX)
6060C2 = FNCAP(X)
6080SHUNTEL = C1:SERIESEL = C2
6100RESISTANCE = R
6120REACTANCE = X - X
6140REACT = -X
6160GOSUB 8840
6180    PRINT:PRINT:PRINT". . .hit any key to continue"
6200    IF INKEY$ = "" THEN 6200
6220    GOSUB 9700
6240RETURN
6260'------------------- NETWORK F subroutine --------------------
6280'mainline
6300GOSUB 8660
6320EL$ = EL2$:FL$ = EL4$
6340TITLE$ = "NETWORK F"
6360NOTE1$ = EL$ + " across load"
6380NOTE2$ = FL$ + " between input and " + EL$
6400S1$ = I$:S2$ = I$
6420    M = 1/B
6440WHILE R < Z2 OR R > Z1
6460    I = I + 1
6480        IF R > Z0 AND I <> 1 THEN N = 1 ELSE N = -1
6500    JB = 1/(2^(I-1) * M * N)
6520    JBTOTAL = JBTOTAL + JB
6540        B = B + JB
6560    GOSUB 8420
6580        IF I > 20 THEN 6780'test for non-realizable component
6600    WEND
6620'--------------finalization
6640JX = -1/JBTOTAL
6660L1 = FNIND(JX)
6680L2 = FNIND(X)
6700SHUNTEL = L1:SERIESEL = L2
6720RESISTANCE = R
6740REACTANCE = X - X
6760REACT = -X
6780GOSUB 8840
6800    PRINT:PRINT:PRINT". . .hit any key to continue"
6820    IF INKEY$ = "" THEN 6820
6840    GOSUB 9700
6860RETURN
6880    '------------------- NETWORK G subroutine --------------------
6900'
6920'mainline
6940GOSUB 8660
```

Fig. 7-12. (*Continued*)

```
6960EL$ = EL3$:FL$ = EL1$
6980TITLE$ = "NETWORK G"
7000NOTE1$ = FL$ + " across input"
7020NOTE2$ = EL$ + " between load and " + FL$
7040S1$ = C$:S2$ = C$
7060      M = X/2
7080GOSUB 8200
7100IF G >= Y1 OR G <= Y2 THEN G = Y1 - .001
7120WHILE G < Y1 OR G > Y2
7140      I = I + 1
7160      IF G > Y2  THEN N = 1 ELSE N = -1
7180      JX = (M * N)/2^(I-1)
7200      JXTOTAL = JXTOTAL + JX
7220      X = X + JX
7240      GOSUB 8200
7260          IF I > 20 THEN 7440'test for non-realizable component
7280WEND
7300XC = -1/B:JX = JXTOTAL
7320RESISTANCE = 1/G
7340C1 = FNCAP(JX)
7360C2 = FNCAP(XC)
7380SHUNTEL = C1:SERIESEL = C2
7400REACTANCE = X - X
7420REACT = -XC
7440GOSUB 8840
7460     PRINT:PRINT:PRINT". . .hit any key to continue"
7480     IF INKEY$ = "" THEN 7480
7500     GOSUB 9700
7520RETURN
7540     '---------------------- NETWORK H subroutine ------------------------
7560     '
7580'mainline
7600GOSUB 8660
7620EL$ = EL4$:FL$ = EL2$
7640TITLE$ = "NETWORK H"
7660NOTE1$ = FL$ + " across input"
7680NOTE2$ = EL$ + " between load and " + FL$
7700S1$ = I$:S2$ = I$
7720      M = X/2
7740GOSUB 8200
7760WHILE G < Y1 OR G > Y2
7780      I = I + 1
7800      IF G > Y2  THEN N = 1 ELSE N = -1
7820      JX = (M * N)/2^(I-1)
7840      JXTOTAL = JXTOTAL + JX
7860      X = X + JX
7880      GOSUB 8200
7900          IF I > 20 THEN 8080'test for non-realizable component
7920WEND
7940XL = -1/B:JX = JXTOTAL
7960RESISTANCE = 1/G
7980L1 = FNIND(JX)
8000L2 = FNIND(XL)
8020SHUNTEL = L1:SERIESEL = L2
8040REACTANCE = X - X
8060REACT = -XL
8080GOSUB 8840
8100     PRINT:PRINT:PRINT". . .hit any key to continue"
8120     IF INKEY$ = "" THEN 8120
8140     GOSUB 9700
8160RETURN
8180'
8200     '-------------- convert Z to Y subroutine ---------------------
8220     '
8240     THETA = ATN(X/R)
8260     Z = R/COS(THETA)
8280     Y = 1/Z
8300     G = Y * COS(-THETA)
```

Fig. 7-12. (*Continued*)

```
8320    B = Y * SIN(-THETA)
8340    RETURN
8360'
8380'-------------- convert Y to Z subroutine ----------------------
8400'
8420    THETAY = ATN(B/G)
8440    Y = G/COS(THETAY)
8460    Z = 1/Y
8480    R = Z * COS(-THETAY)
8500    X = Z * SIN(-THETAY)
8520RETURN
8540    '
8560'-------------------- validate subroutine ---------------------
8580'
8600IF R < .1 THEN PRINT"Value of resistance is too small"
8620RETURN
8640'
8660'-------------------- initialize variables sub -----------------
8680'
8700JBTOTAL = 0:JXTOTAL = 0
8720I = 0:JB = 0:JX = 0
8740R = RHOLD:X = XHOLD
8760G = GHOLD:B = BHOLD
8780    NOTE1$ = " ":NOTE2$ = " "
8800RETURN
8820'
8840'********************* finalization *************************
8860'
8880 LOCATE 7,33:PRINT TITLE$
8900 LOCATE 8,33:PRINT STRING$(9,45)
8920 LOCATE 9,1:PRINT NOTE1$
8940 LOCATE 10,1:PRINT NOTE2$
8960 LOCATE 11,1:PRINT LN$
8980 LOCATE 12,11:PRINT EL$:LOCATE 12,36:PRINT"!":LOCATE 12,47:PRINT FL$
9000 LOCATE 13,1:PRINT LN$
9020 LOCATE 14,1:PRINT"Reactance":LOCATE 14,12:PRINT"Value"
9040 LOCATE 14,22:PRINT"R":LOCATE 14,30:PRINT"jX":LOCATE 14,36:PRINT"!"
9060 LOCATE 14,37:PRINT"Reactance":LOCATE 14,49:PRINT"Value"
9080 LOCATE 14,59:PRINT"R":LOCATE 14,68:PRINT"jX"
9100 LOCATE 15,1:PRINT LN$
9120    PRINT;TAB(36);"!"
9140IF I>=20 THEN PRINT;"                        OUT OF RANGE":GOTO 9500
9160    PRINT USING X$;JX;
9180PRINT;TAB(9);
9200    PRINT USING X$;SHUNTEL;
9220    PRINT;S1$;TAB(19);
9240    PRINT USING R$;R;
9260PRINT;TAB(26);
9280    PRINT USING X$;X;
9300PRINT;TAB(36);"!";
9320    PRINT;TAB(37);
9340    PRINT USING X$;REACT;
9360    PRINT;TAB(46);
9380PRINT USING X$;SERIESEL;
9400    PRINT;S2$;
9420    PRINT;TAB(57);
9440    PRINT USING R$;RESISTANCE;
9460    PRINT;TAB(63);
9480    PRINT USING X$;REACTANCE
9500PRINT;TAB(36);"!":PRINT;LN$
9520    PRINT"Zin = ";
9540    PRINT USING R$;RESISTANCE;
9560    PRINT" + j";
9580    PRINT USING X1$;REACTANCE;
9600    PRINT"                    . . .Shift PrtSc for hardcopy"
9620RETURN
9640    CLS:PRINT". . .program terminated."
9660    PRINT:PRINT"Have a nice day!    ";CHR$(2)
```

Fig. 7-12. (*Continued*)

```
9670    PRINT"Maranatha!"
9680    END
9700 '************************ clear screen sub *****************************
9720 '
9740 FOR I% = 7 TO 23
9760    LOCATE I%,1:PRINT STRING$(72,0)
9780 NEXT I%
9800 RETURN
```

Fig. 7-12. (*Continued*)

combinations for matching any real impedance to any other real impedance. The eight combinations are shown in Figure 7-13. As a minimum, two combinations will provide a solution to any impedance transformation problem. The maximum combination is four, and is a function of absolute levels of impedance. The final choice is determined by actual component values needed. Of course, some may not be possible.

This program, written in BASIC for use on PCs, will calculate all combinations of "L" matching networks for any given load impedance to any given real input impedance. The following user information is input via the keyboard:

- Load impedance in rectangular form.
- Input impedance.
- Frequency.

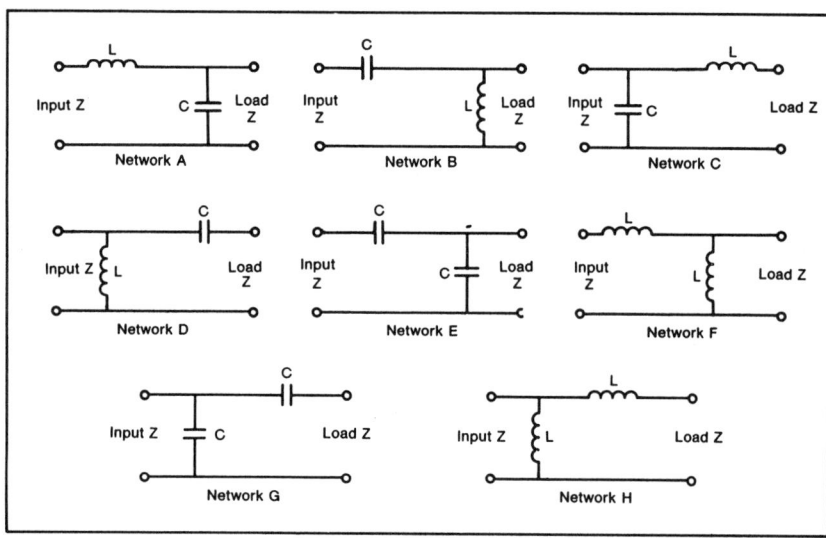

Fig. 7-13. Eight LC combinations.

The program determines which of the eight possible combinations will successfully provide the required impedance match. It then calculates the component reactances and values based on the input frequency.

The program presents the mathematical solutions to the "L" matching circuits in Phillip Smith's book *Electronic Applications of the Smith Chart* (New York: McGraw-Hill, 1969). When viewed on a Smith chart (for an example, see Figure 7-26), the eight combinations of networks can be represented by eight circular boundaries, each of which can be transformed into a pure resistance. For any given load impedance and input resistance, the program determines all possible combinations of boundaries applicable to the solution. It then calculates the reactance values necessary for each network, and computes the inductance and capacitance values required as a function of frequency. Losses in an "L" network are usually small for RF applications and will not seriously limit the range of impedance values that can be transformed. For this reason, losses are not considered in the calculations.

Program control is straightforward, as shown in Figure 7-14. The program listing is just over 400 lines and is 10 Kbytes in length. If it is run under a BASIC interpreter as opposed to a compiler, removing all of the REM statements will decrease the execution time. A source-code listing for the IBM PC is provided at the end of the chapter.

The program requires the user to respond only to prompts. A request appears first for the load impedance in rectangular form—the resistive component followed by a carriage return, and then the reactive component and a carriage return. A prompt for the source impedance appears next. This is resistive only and must be entered in ohms. Finally the program asks for the frequency, which must be in megahertz. The screen then displays the input data, date and the solution to the first matching network. On the IBM PC, hitting Shft PrScr will cause the screen to dump to the printer. Depressing any other key will cause the first matching network to be erased and the solution to the second matching network will appear. This continues until all combinations of matching have been displayed.

If a value of load impedance happens to lie very near a constant susceptance line equal to $1/Z_0$, an "L" network is not necessary for matching. A simple shunt element is all that is required. A network called "Network Y" is displayed showing the parameters of the shunt element. The case of a value of load impedance that lies near a constant resistance line equal to Z_0 has been taken into consideration by an additional network element called "Network X." This is merely a series component used to tune out the reactance of the load impedance.

A typical output screen is shown in Table 7-3. The input data are shown at the top. In this example, a load impedance of $35 + j150$ is connected to a resistive 50-ohm source impedance. Four "L" networks provide the necessary

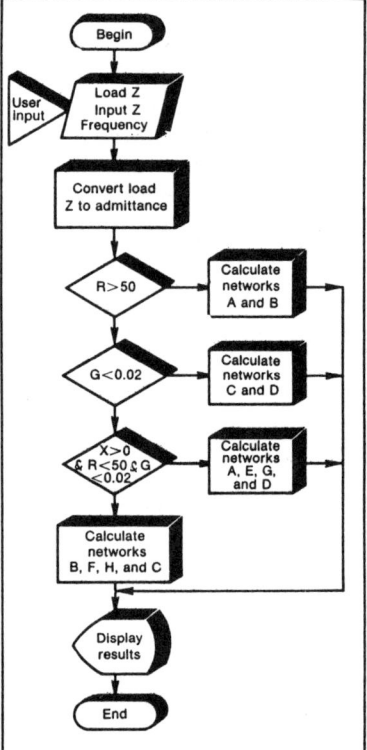

Fig. 7-14. Program control flow chart.

transformation at a frequency of 15 MHz; these are the Networks A, E, D, and G of Figure 7-13.

The listing of the calculated results for each network also provides information describing the proper connection for the network, i.e., Network A indicates that a shunt capacitance is connected across the load and the series inductance is connected between the input (or source) and the shunt capacitance. Shown next is the reactance value and the capacitance of the shunt capacitor ($-j86.49$ ohms and 122.68 pF) along with the impedance for this combination of shunt capacitance and load impedance. In this case the combined impedance is $49.8 - j176.82$ ohms. Finally, the series inductance required to complete the transformation is shown (a reactance of $+j176.82$ ohms which is an inductance of 1.88 μH at this frequency). The impedance of this combination is $49.8 + j0$. Networks E, D, and G show the same information for those network types.

Table 7-3. IMPDNCE.OUT Output Data File.

NETWORK A

R = 35.0   Source impedance: 50 Ω
X = +j 150.00

Shunt capacitance across load
Series inductance between input and shunt capacitance

	Shunt Capacitance				Series Inductance		
Reactance	Value	R	jX	Reactance	Value	R	jX
−86.49	122.68 pF	49.8	−176.82	176.82	188 μH	49.8	0.00

NETWORK E

Shunt capacitance across load
Series inductance between input and shunt capacitance

	Shunt Capacitance				Series Capacitance		
Reactance	Value	R	jX	Reactance	Value	R	jX
−920.24	11.53 pF	49.9	−176.95	−176.95	59.96 pF	49.9	0.00

NETWORK D

Shunt inductance across input
Series capacitance between load and shunt inductance

	Series Capacitance				Shunt Inductance		
Reactance	Value	R	jX	Reactance	Value	R	jX
−172.66	61.45 pF	35.0	−22.66	76.73	0.81 μH	49.7	0.00

NETWORK G

Shunt capacitance across input
Series capacitance between load and shunt capacitance

	Series Capacitance				Shunt Inductance		
Reactance	Value	R	jX	Reactance	Value	R	jX
−127.15	83.45 pF	35.0	−22.85	−76.46	138.77 pF	49.9	0.00

7.5. BANDPASS FILTERS

The design of bandpass filters is a tedious task. Additionally, if it is desired to determine the circuit response on the basis of component tolerance and end of life-value variability, temperature excursions, and so forth, considerable time and effort are required. The computer code presented in Figure 7-15 provides

```
7 REM WRITTEN BY DESIGN & EVALUATION ASSOCIATES.
8 REM EQUATIONS TO PROVIDE FREQUENCY RESPONSE FOR A
9 REM  BAND PASS FILTER. EXAMPLE IS FOR A 10 KHZ CIRCUIT
10 C1=9.999999E-10
20 C2=9.999999E-10
30 R1=22100
40 R2=22100
50 R3=22100
60 R4=40100
70 R5=10000
80 R6=1210
90 RA=R1*R3/(R1+R3)
100 A=1/RA/C1
110 B=1/R2/C2
120 D=1/RA/C2
130 F=1/R1/C2
140 G=1/R3/C2
150 K=(R5+R6)/(R4+R5+R6)
160 A1=F/K
170 B1=A+B+D-G/K
180 W0=A*B
190 F(1)=1:F(2)=1.5:F(3)=2:F(4)=3:F(5)=5:F(6)=7
195 LPRINT "    FREQ";"       GAIN   "   ; "     ANGLE"
200 FOR N1=1 TO 5
210 FOR N2=1 TO 6
220 FR=F(N2)*10[N1
230 W=FR*2*3.1415926
240 DE=(B1[2)*W0+(W0-W[2)*(W0-W[2)
245 M=A1*W/(DE[.5)
250 P1=ATN((W0-W[2)/B1/W)
255 P1=P1*57.29578
260 LPRINT USING "######   ";FR;
265 LPRINT USING"###.#####   ";M,P1
270 NEXT N2
275 NEXT N1
280 FOR N1=0 TO 40 STEP 4
285 FR=(8+N1/10)*1000
290 W=FR*2*3.1415926
295 DE=(B1[2)*W0+(W0-W[2)*(W0-W[2)
300 M=A1*W/(DE[.5)
305 P1=ATN((W02-W[2)/B1/W)
310 P1=P1*57.29578
315 LPRINT USING "######   ";FR;
320 LPRINT USING "###.#####   ";M,P1
325 NEXT N1
330 END
```

Fig. 7-15. Program for design of bandpass filter.

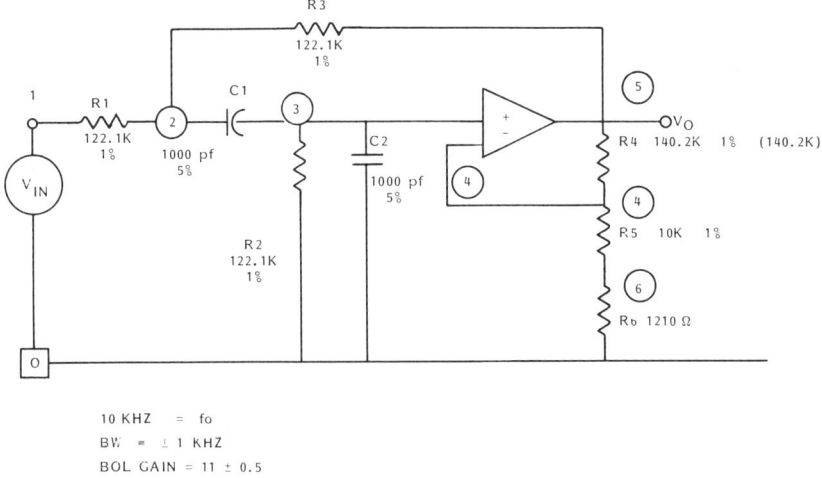

10 KHZ = fo
BW = ± 1 KHZ
BOL GAIN = 11 ± 0.5

Fig. 7-16. Active band pad filter circuit.

an accurate and rapid means of eliminating the tedium of the design and analysis process. This code was developed by Design and Evaluation Associates.*

An example is used to demonstrate the simplicity of use of the code. The circuit to be designed, a 10-kHz active bandpass filter, is shown in Figure 7-16. It is a two-pole design with gain greater than 1. Resistors R1, R2, and R3, and capacitors C1 and C2 determine the frequency response of the filter. Resistors R4, R5, and R6 set the gain; R6 has been selected for a gain of 11 ± 0.5. The operational amplifier, U1, has a rather high gain at the relative low frequency of interest; this allows it to be modeled as an ideal amplifier. The acceptable bandwidth of the circuit is to be ±1 kHz.

The values for the passive components are entered in lines 10 through 80 in the code. To obtain more detailed information describing circuit response close to the resonant frequency, lines 280 through 325 repeat the analysis in 400-Hz increments. The results of the analysis are shown on Figure 7-17.

The circuit models from which the desired equations are developed for entry into the computer code are shown on Figures 7-18, 7-19, and 7-20, which are seen to be successive simplifications of Figure 7-16. The equations that describe the different impedances are shown with the appropriate model. Equations (5) and (6) are a progression of development to provide the gain and frequency equations (7) and (9), respectively.

---

*Glendate Executive Campus
1000 White Horse Road
Voorhees, NJ 08043

FREQ	GAIN	ANGLE
10	0.00304	89.98320
15	0.00457	89.97480
20	0.00609	89.96640
30	0.00913	89.94960
50	0.01522	89.91590
70	0.02131	89.88230
100	0.03045	89.83180
150	0.04568	89.74770
200	0.06092	89.66350
300	0.09141	89.49500
500	0.15257	89.15700
700	0.21406	88.81720
1000	0.30719	88.30210
1500	0.46598	87.42270
2000	0.63125	86.50530
3000	0.99213	84.49150
5000	1.94784	79.05420
7000	3.66823	68.71910
10000	10.55280	6.97518
15000	3.94995	-69.36180
20000	2.21307	-78.38420
30000	1.24095	-83.45540
50000	0.68774	-86.36440
70000	0.48107	-87.45530
100000	0.33308	-88.23760
150000	0.22077	-88.83160
200000	0.16524	-89.12550
300000	0.11000	-89.41780
500000	0.06595	-89.65090
700000	0.04710	-89.75070
8000	5.23243	58.48240
8400	6.09795	52.35900
8800	7.13519	44.44810
9200	8.32779	34.26520
9600	9.56040	21.58540
10000	10.55280	6.97518
10400	10.94260	-7.97113
10800	10.59690	-21.43930
11200	9.75150	-32.48440
11600	8.74325	-41.11860
12000	7.78167	-47.78250

Fig. 7-17. Listing of BP filter analysis results—nominal parameter valve.

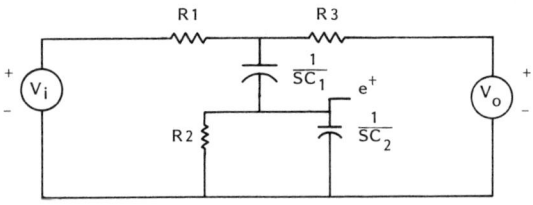

Fig. 7-18. Simplified circuit model 1.

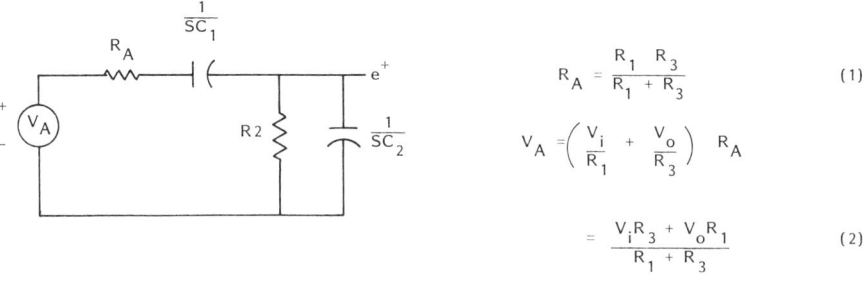

$$R_A = \frac{R_1 R_3}{R_1 + R_3} \tag{1}$$

$$V_A = \left(\frac{V_i}{R_1} + \frac{V_o}{R_3} \right) R_A$$

$$= \frac{V_i R_3 + V_o R_1}{R_1 + R_3} \tag{2}$$

Fig. 7-19. Simplified circuit model 2.

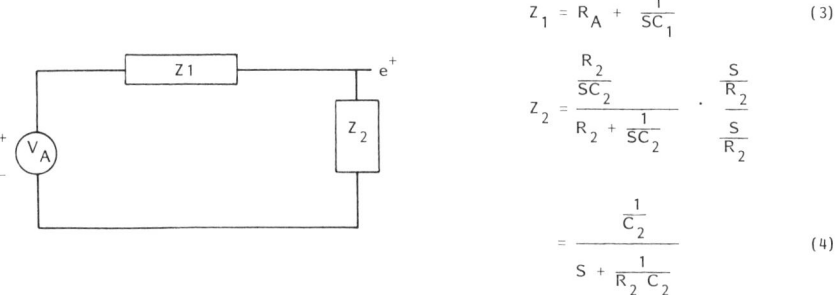

$$Z_1 = R_A + \frac{1}{SC_1} \tag{3}$$

$$Z_2 = \frac{\frac{R_2}{SC_2}}{R_2 + \frac{1}{SC_2}} \cdot \frac{\frac{S}{R_2}}{\frac{S}{R_2}}$$

$$= \frac{\frac{1}{C_2}}{S + \frac{1}{R_2 C_2}} \tag{4}$$

$$e^+ = \frac{\frac{V_i R_3 + V_o R_1}{R_1 \cdot R_3 \cdot C_2} \cdot S}{(S + \frac{1}{R_A C_1}) \cdot (S + \frac{1}{R_2 C_2}) + \frac{S}{R_A C_2}} \tag{5}$$

$$\frac{V_o (R_5 + R_6)}{R_4 + R_5 + R_6} = \frac{\frac{V_i R_3 + V_o R_1}{R_1 \cdot R_3 \cdot C_2} \cdot S}{(S + \frac{1}{R_A C_1}) \cdot (S + \frac{1}{R_2 C_2}) + \frac{S}{R_A C_2}} \tag{6}$$

$$\therefore V_o = \left[\frac{\frac{V_i R_3 + V_o R_1}{R_1 \cdot R_3 \cdot C_2} \cdot S}{(S + \frac{1}{R_A C_1})(S + \frac{1}{R_2 C_2}) + \frac{S}{R_A C_2}} \right] \cdot \left[\frac{R_4 + R_5 + R_6}{R_5 + R_6} \right] \tag{7}$$

$$W_o^2 = \frac{1}{R_A C_1} \cdot \frac{1}{R_2 C_2} \tag{8}$$

$$f_o = \frac{1}{2\pi} \left(\frac{R_1 + R_3}{R_1 \cdot R_2 \cdot R_3 \cdot C_1 \cdot C_2} \right)^{1/2} \tag{9}$$

Fig. 7-20. Simplified circuit model 3.

$$A_{fo} = \frac{\frac{1}{R_1C_2}}{\frac{R_5+R_6}{R_4+R_5+R_6}\left[\frac{R_1+R_3}{R_1R_3}\left(\frac{1}{C_1}+\frac{1}{C_2}\right)+\frac{1}{R_2C_2}\right]-\frac{1}{R_3C_2}} \quad (10)$$

COMPONENT	C1	C2	R1	R2	R3	R4	R5	R6	RESULTING GAIN	DIRECTION FOR
NOM. VALUE	1000E-12	1000E-12	22100	22100	22100	40200	10000	1210	11.07974	MAX. MIN.
MAX. VALUE	1022E-12	1022E-12	22333	22333	22333	40624	10106	1223		
MIN. VALUE	977E-12	977E-12	21961	21961	21961	39948	9937	1202		
CONDITION #1	1022	1000	22100	22100	22100	40200	10000	1210	12.396	↑
2	977								9.948	↓
3	1000	1022							10.015	↓
4		977							12.465	↑
5		1000	22333						11.546	↑
6			21961						10.819	↓
7			22100	22333					11.366	↑
8				21961					10.913	↓
9				22100	22333				10.402	↓
10					21961				11.536	↑
11					22100	40624			12.296	↑
12						39948			10.457	↓
13						40200	10106		10.173	↓
14							9937		10.702	↑
15							10000	1223	10.960	↓
16								1202	11.155	↑
17								1210	11.07974	CHK OK
VALUE FOR:										
A_{fo} MAX	1022	977	22333	22333	21961	40624	9937	1202	21.186	
A_{fo} MIN	977	1022	21961	21961	22333	39948	10106	1223	7.453	

Fig. 7-21. Hand-calculation results.

Directional sensitivity effects of the various passive components on Afo min and max can be obtained manually (using a hand calculator) if necessary, by solving equation 10 which is the expanded equivalent of Equation 7.4-9. The results of a hand calculated solution are shown on Figure 7-21. Entering the appropriate values in lines 10 through 80 in the code provide the results shown on Figures 7-22 and 7-23 which list the maximum and minimum results, respectively. The change in circuit performance resulting from component value variation is determined by comparing these results with each other as well as with the nominal circuit response values shown on Figure 7-17. The maximum and minimum values resulting from the computer analyses are 20.66 and 7.59; the manually calculated values are 21.19 and 7.45 respectively.

7.6. CAD CUTS MATH IN RF AMPLIFIER DESIGN

The program in Figure 7-24 was written by M. Richharia and A. D. Dave of the Space Applications Center, Ahmedabad, India; it was originally published in *Microwaves and RF* (July 1985).

FREQ	GAIN	ANGLE
10	0.00326	89.99110
15	0.00490	89.98660
20	0.00653	89.98210
30	0.00979	89.97320
50	0.01632	89.95530
70	0.02285	89.93750
100	0.03264	89.91070
150	0.04897	89.86590
200	0.06530	89.82120
300	0.09799	89.73170
500	0.16357	89.55210
700	0.22953	89.37150
1000	0.32950	89.09770
1500	0.50025	88.62980
2000	0.67853	88.14100
3000	1.07064	87.06400
5000	2.13825	84.11320
7000	4.23825	78.19510
10000	20.65510	9.18541
15000	4.11761	-78.90760
20000	2.27603	-83.85900
30000	1.27497	-86.55530
50000	0.70692	-88.08880
70000	0.49458	-88.66270
100000	0.34248	-89.07390
150000	0.22701	-89.38610
200000	0.16992	-89.54050
300000	0.11312	-89.69410
500000	0.06782	-89.81660
700000	0.04843	-89.86900
8000	6.47924	71.64890
8400	7.93254	67.23540
8800	9.97065	60.75580
9200	12.89640	50.64250
9600	16.88640	34.17130
10000	20.65510	9.18541
10400	20.57100	-18.51630
10800	17.06690	-39.11960
11200	13.55110	-51.89850
11600	10.96930	-59.86560
12000	9.14661	-65.12680

Fig. 7-22. Listing of BP filter analysis results—maximum parameter values.

RF amplifiers have been designed using the S-parameter technique for some time[1-3]. Of major importance in this technique is the design of suitable networks to match the input and output impedances of the device over the operating bandwidth. A graphical approach using a Smith chart is commonly employed for this task, but cumbersome calculations are required before the chart can be used.

The FORTRAN IV program described here eliminates much of the drudgery involved in the use of a Smith chart for impedance matching. The program is made up of several subroutines that can be used individually or together, as the need arises. The program is interactive, which makes the design process simple and flexible. Table 7-4 lists the equations that were used to write the program.

FREQ	GAIN	ANGLE
10	0.00283	89.97640
15	0.00425	89.96450
20	0.00567	89.95270
30	0.00850	89.92910
50	0.01417	89.88180
70	0.01984	89.83450
100	0.02835	89.76350
150	0.04253	89.64520
200	0.05672	89.52680
300	0.08511	89.28990
500	0.14203	88.81490
700	0.19924	88.33720
1000	0.28581	87.61350
1500	0.43315	86.37920
2000	0.58600	85.09350
3000	0.91717	82.28230
5000	1.76929	74.81430
7000	3.17880	61.41680
10000	7.26230	5.73533
15000	3.74833	-61.84160
20000	2.14739	-73.76040
30000	1.20829	-80.77800
50000	0.66976	-84.86440
70000	0.46847	-86.40340
100000	0.32434	-87.50830
150000	0.21497	-88.34800
200000	0.16090	-88.76340
300000	0.10711	-89.17670
500000	0.06422	-89.50640
700000	0.04586	-89.64750
8000	4.30304	49.49310
8400	4.86223	43.02900
8800	5.47911	35.35470
9200	6.12781	26.42210
9600	6.75193	16.39770
10000	7.26229	5.73533
10400	7.56053	-4.89458
10800	7.58923	-14.83150
11200	7.36795	-23.64400
11600	6.97566	-31.18610
12000	6.50222	-37.51690

Fig. 7-23. Listing of BP filter minimum values.

```
100   C                CAD: RF AMPLIFIER DESIGN
200   C                ------------------------
300   C     PROGRAM FOR COMPUTER AIDED DESIGN OF RF AMPLIFIERS USING
400   C     S-PARAMETERS.
500
600   C     INPUTS TO PROGRAM:
700
800   C
900   C
1000  C          N = NUMBER OF FREQUENCIES FOR WHICH DATA ARE AVAILABLE
1100  C          FREQ(I) = FREQUENCY OF THE ITH POINT FOR WHICH S-PARAMETERS ARE
1200  C                AVAILABLE
1300  C          RH1(I),THT1(I) = MAGNITUDE AND ANGLE OF S11 - ITH FREQUENCY
1400  C          RH2(I),THT2(I) = MAGNITUDE AND ANGLE OF S12 - ITH FREQUENCY
1500  C          RH3(I),THT3(I) = MAGNITUDE AND ANGLE OF S21 - ITH FREQUENCY
1600  C          RH4(I),THT4(I) = MAGNITUDE AND ANGLE OF S22 - ITH FREQUENCY
1700  C             DBG = DESIRED AMPLIFIER GAIN (DB)
1800  C             FMIN = MINIMUM FREQUENCY OF PASS-BAND (MHZ)
1900  C             FCENT = CENTRE FREQUENCY FO PASS-BAND (MHZ)
```

Fig. 7-24. Program for design of RF program.

```
2000    C           FMAX = MAXIMUM FREQUENCY OF PASS-BAND (MHZ)
2100    C              IT = OPTIMUM NARROW-BAND DESIGN FLAG,SET TO 1,WHEN
2200    C                   CONJUGATE MATCHING REQUIRED AT INPUT AND OUTPUT
2300    C              IA = FLAG ,SET TO 1, WHEN MATCHING REQUIRED FOR INPUT
2400    C              IB = FLAG ,SET TO 1, WHEN MATCHING REQUIRED FOR OUTPUT
2500    C
2600    C         OUTPUTS GENERATED:
2700    C
2800    C         INPUT MISMATCH CIRCLE - CENTRE AND RADIUS
2900    C         OUTPUT MISMATCH CIRCLES-CENTRE AND RADIUS
3000    C         INPUT PLANE STABILITY CIRCLE - CENTRE AND RADIUS
3100    C         OUTPUT PLANE STABILITY CIRCLE - CENTRE AND RADIUS
3200    C         MAXIMUM INCREASE IN GAIN POSSIBLE BY INPUT MATCHING (G1MAX)
3300    C         MAXIMUM INCREASE IN GAIN POSSIBLE BY OUTPUT MATCHING (G2MAX)
3400    C         MAXIMUM AVAILABLE GAIN WHEN SOURCE AND LOAD CAN BE CONJUGATELY
3500    C         MATCHED TO A DEVICE
3600    C         NORMALIZED VALUE OF SOURCE AND LOAD FOR SUCH A CONJUGATE MATCH
3700    C
3800              DATA N,FREQ(1)/1,200./
3900              DATA RH1(1),THT1(1),RH2(1),THT2(1),RH3(1),THT3(1),RH4(1),THT4(1)
4000            1 /.316,-72.633,.066,68.055,5.207,100.063,.675,-20.83/
4100              DATA FCENT,DBG,IT,IA,IB/200.,15.,0,0,1/
4200
4300
4400              COMPLEX S11(20),S12(20),S21(20),S22(20),TMS,TML
4500              COMMON/BLOCK1/S11,S12,S21,S22,FREQ(20)
4600              DIMENSION G0(20),XMIS(20),G1MAX(20),G2MAX(20),R1(20),R2(20)
4700            1 ,RH1(20),RH2(20),RH3(20),RH4(20),THT1(20),THT2(20),THT3(20)
4800            1 ,THT4(20),RHO1(20),RHO2(20)
4900              WRITE(6,11)
5000    11        FORMAT(1H1,35X,'***** RF AMPLIFIER DESIGN *****'////)
5100    C
5200    C     WRITE THE GIVEN INPUT PARAMETERS
5300    C
5400              WRITE(6,21)
5500    21        FORMAT(///1H0,35X,'S-PARAMETERS (MAG,ANGLE)'/36X,
5600            1 '------------------------'/1H0,2X,'FREQ(MHZ)',8X,
5700            1 'S11',16X,'S12',15X,'S21',15X,'S22'/)
5800              DO 10 I=1,N
5900              WRITE(6,31)FREQ(I),RH1(I),THT1(I),RH2(I),THT2(I),RH3(I),THT3(I),
6000            1 RH4(I),THT4(I)
6100    31        FORMAT(1H0,3X,F6.1,6X,F5.3,',',F6.1,6X,F5.3,',',F6.1,7X,F5.3,',',
6200            1 ,F6.1,7X,F5.3,',',F6.1)
6300    10        CONTINUE
6400              IF(N.EQ.1)WRITE(6,41)DBG,FCENT
6500    41        FORMAT(///5X,'AMPLIFIER GAIN =',F6.1,' DB',10X,'CENTRE FREQ =',F6.1,
6600            1 ' MHZ',2X,'(NARROW BAND DESIGN)')
6700
6800              IF(N.NE.1)WRITE(6,51)DBG,FMIN,FCENT,FMAX
6900    51        FORMAT(/////5X,'AMPLIFIER GAIN =',F6.1,' DB'//5X,'MIN FREQ =',F6.1
7000            1 ,5X,'CENTRE FREQ =',F6.1,' MHZ',5X,'MAX FREQ =',F6.1,' MHZ')
7100    C
7200    C     CONVERT S-PARAMETERS TO REAL AND IMAGINARY PARTS
7300    C
7400              DO 20 I=1,N
7500              CALL IMAG(RH1(I),THT1(I),S11(I))
7600              CALL IMAG(RH2(I),THT2(I),S12(I))
7700              CALL IMAG(RH3(I),THT3(I),S21(I))
7800              CALL IMAG(RH4(I),THT4(I),S22(I))
7900    20        CONTINUE
8000
8100    C
8200    C     COMPUTE STABILITY CIRCLES
8300    C
8400              CALL STABLE(N,XK)
8500              IF(N.EQ.1.AND.XK.GT.1..AND.IT.EQ.1)CALL NAROWB(XK,TMS,TML)
8600              IF(N.EQ.1.AND.XK.GT.1..AND.IT.EQ.1)GO TO 90
8700    C
8800    C     COMPUTE TRANSDUCER FORWARD GAIN OF DEVICE IN DB
8900    C
9000              WRITE(6,61)
9100    61        FORMAT(//35X,'GAIN REQUIRED THROUGH MATCHING'/35X,'--------------'
9200            1 '----------------')
9300              DO 30 I=1,N
```

Fig. 7-24. (*Continued*)

```
?400              CALL GS21(S21(I),GO(I))
?500       30     CONTINUE
?600    C
?700    C   COMPUTE MISMATCH REQUIRED AT EACH FREQUENCY
?800    C
?900              DO 40 I=1,N
0000              XMIS(I)=DBG-GO(I)
0100       40     CONTINUE
0200    C
0300    C   COMPUTE MAXIMUM INCREASE IN GAIN POSSIBLE BY MATCHING :
0400    C   (a)SOURCE  (b)LOAD
0500    C   COMPUTE REQUIRED GAIN IMPROVEMENT THROUGH MATCHING: A MINUS SIGN IMPLIES
0600    C   INSERTION LOSS
0700    C
0800              WRITE(6,71)
0900       71     FORMAT(1H0,7X,'S. NO. ',5X,'FREQ(MHZ)',5X,'G1MAX(DB)',5X,'G2MAX',
1000            1 '(DB)',5X,'REQD GAIN IMPROVEMENT (DB)'/)
1100              DO 50 I=1,N
1200              CALL GMAX(S11(I),G1MAX(I))
1300              CALL GMAX(S22(I),G2MAX(I))
1400              WRITE(6,81)I,FREQ(I),G1MAX(I),G2MAX(I),XMIS(I)
1500       81     FORMAT(1H0,7X,I2,8X,F6.1,9X,F5.2,10X,F5.2,13X,F5.2)
1600       50     CONTINUE
1700
1800              WRITE(6,91)
1900       91     FORMAT(//35X,'GAIN CIRCLES'/35X,'------------')
2000
2100    C
2200    C   BRANCH WHEN MISMATCH CIRCLES ARE REQUIRED FOR OUTPUT
2300    C
2400
2500              IF(IB.EQ.1)GO TO 70
2600    C
2700    C   COMPUTE MISMATCH CIRCLE FOR INPUT
2800    C
2900              WRITE(6,101)
3000      101     FORMAT(1H0,15X,'MATCHING REQUIRED FOR SOURCE ONLY'//1H0,15X,
3100            1 'FREQ',5X,'CENTRE(MAG,ANGLE)',5X,'RADIUS'/16X,'----',
3200            1 5X,'-----------------',5X,'------')
3300              DO 60 I=1,N
3400              CALL CIRCLE(S11(I),XMIS(I),R1(I),RHO1(I))
3500              A=-THT1(I)
3600              WRITE(6,111)FREQ(I),R1(I),A,RHO1(I)
3700      111     FORMAT(1H0,13X,F6.1,6X,F5.3,',',F6.1,8X,F5.3)
3800       60     CONTINUE
3900              GO TO 90
4000    C
4100    C   COMPUTE MISMATCH CIRCLE FOR OUTPUT
4200    C
4300       70     WRITE(6,121)
4400      121     FORMAT(1H0,15X,'MATCHING REQUIRED FOR LOAD ONLY'//1H0,15X,'FREQ'
4500            1 ,5X,'CENTRE(MAG,ANGLE)',5X,'RADIUS'/16X,'----',
4600            1 5X,'-----------------',5X,'------')
4700              DO 80 I=1,N
4800              CALL CIRCLE(S22(I),XMIS(I),R2(I),RHO2(I))
4900              B=-THT4(I)
5000              WRITE(6,111)FREQ(I),R2(I),B,RHO2(I)
5100       80     CONTINUE
5200       90     STOP
5300              END
5400
5500
5600
5700              SUBROUTINE CIRCLE(S,XM,R,RHO)
5800    C   ----------------------------
5900    C
6000    C   SUBROUTINE TO COMPUTE CONSTANT GAIN CIRCLES.
6100    C
6200    C   INPUTS TO SUBROUTINE:
6300    C
6400    C   S = S11 OR S22 PARAMETER IN REAL,IMAGINARY FORM
6500    C   XM = GAIN CIRCLE (DB) REQUIRED
6600    C
6700    C   OUTPUTS GENERATED:
6800    C
```

Fig. 7-24. (*Continued*)

```
16900    C    R = DISTANCE OF CENTRE OF CIRCLE FROM CENTRE OF SMITH CHART
17000    C    RHO = RADIUS OF CIRCLE
17100         COMPLEX S
17200         SS=CABS(S)
17300         XMS=10.**(XM/10.)
17400         G=XMS*(1.-(SS)**2)
17500         R=(G*SS)/(1.-SS**2*(1.-G))
17600         RHO=(SQRT(1.-G))*(1.-SS**2)/(1.-SS**2*(1.-G))
17700         RETURN
17800         END
17900
18000
18100         SUBROUTINE STABLE(N,XK)
18200    C    ----------------------
18300    C
18400    C    SUBROUTINE TO CHECK STABILITY
18500    C
18600    C    INPUTS TO SUBROUTINE:
18700    C
18800    C    N = NUMBER OF DATA POINTS
18900    C    OTHER PARAMETERS ARE PASSED THROUGH THE COMMON BLOCK -
19000    C    BLOCK1
19100    C
19200    C    OUTPUTS GENERATED:
19300    C
19400    C    XK = STABILITY FACTOR K
19500    C    Z1,T1 = CENTRE OF INPUT STABILITY CIRCLE ON A SMITH CHART
19600    C    (MAGNITUDE, ANGLE)
19700    C    RS1 = RADIUS OF INPUT STABILITY CIRCLE
19800    C    Z2,T2 = CENTRE OF OUTPUT STABILITY CIRCLE ON A SMITH CHART
19900    C    (MAGNITUDE, ANGLE)
20000    C    RS2 =RADIUS OF OUTPUT STABILITY CIRCLE
20100         COMPLEX XD,XM,XN,C1,C2,CS1,CS2
20200         COMMON/BLOCK1/S11(20),S12(20),S21(20),S22(20),FREQ(20)
20300         COMPLEX S11,S12,S21,S22
20400
20500         WRITE(6,1)
20600    1    FORMAT(//35X,'STABILITY PARAMETERS'/35X,'--------------------')
20700
20800    C    STABILITY PARAMETERS ARE COMPUTED FOR EACH FREQUENCY
20900         DO 30 I=1,N
21000         WRITE(6,11)FREQ(I)
21100    11   FORMAT(1H0,5X,'FREQUENCY =',F6.1,' MHZ')
21200         ICHEK1=0
21300         ICHEK2=0
21400         CHECK1=CABS(S11(I))
21500         CHECK2=CABS(S22(I))
21600         XD=S11(I)*S22(I)-S12(I)*S21(I)
21700         XM=S11(I)-XD*CONJG(S22(I))
21800         XN=S22(I)-XD*(CONJG(S11(I)))
21900         D1=CABS(S12(I)*S21(I))
22000         XM=CONJG(XM)
22100         XMAB=CABS(XM)
22200         XN=CONJG(XN)
22300         XNAB=CABS(XN)
22400         XDA=CABS(XD)
22500    C    CHECK IS MADE FOR UNCONDITIONAL STABILITY
22600         CHECK3=ABS((D1-XMA)/((CABS(S11(I)))**2-XDA**2))
22700         CHECK4=ABS((D1-XNA)/((CABS(S22(I)))**2-XDA**2))
22800         IF(CHECK1.LT.1..AND.CHECK2.LT.1.)ICHEK1=1
22900         IF(CHECK3.GT.1..AND.CHECK4.GT.1.)ICHEK2=1
23000         IF(ICHEK1.EQ.1.AND.ICHEK2.EQ.1)GO TO 20
23100    10   WRITE(6,21)
23200    21   FORMAT(1H0,5X,'DEVICE CONDITIONALLY STABLE AT THIS FREQUENCY')
23300    20   XK=(1.+XDA**2-CABS(S11(I))**2-CABS(S22(I))**2)/
23400       1 (2.*D1)
23500    C
23600    C    CENTRE AND RADIUS OF STABILITY CIRCLE ARE COMPUTED ON INPUT PLANE
23700    C
23800
23900         C1=S11(I)-XD*(CONJG(S22(I)))
24000         C2=S22(I)-XD*(CONJG(S11(I)))
24100         CS1=CONJG(C1)/(CHECK1**2-XDA**2)
24200         RS1=(D1/(CHECK1**2-XDA**2))
24300         RS1=ABS(RS1)
```

Fig. 7-24. (*Continued*)

```
24400      C     CENTRE AND RADIUS OF STABILITY CIRCLE ARE COMPUTED ON OUTPUT PLANE
24500                  CS2=CONJG(C2)/(CHECK2**2-XDA**2)
24600                  RS2=(D1/((CHECK2**2-XDA**2)))
24700                  RS2=ABS(RS2)
24800                  CALL MAGTHT(CS1,Z1,T1)
24900                  CALL MAGTHT(CS2,Z2,T2)
25000                  WRITE(6,31)XK,Z1,T1,RS1,Z2,T2,RS2
25100      31     FORMAT(1H0,5X,'STABILITY FACTOR (K) =',F7.2/1H0,5X,
25200           1 'CENTRE OF INPUT STABILITY CIRCLE (MAG,ANGLE)=',F7.2,',',F7.2,5X,
25300           1 'RADIUS =',F7.2/1H0,5X,'CENTRE OF OUTPUT STABILITY CIRCLE',
25400           1 ' (MAG,ANGLE)=',F7.2,',',F7.2,5X,'RADIUS =',F7.2)
25500      30     CONTINUE
25600
25700             RETURN
25800             END
25900
26000
26100
26200             SUBROUTINE IMAG(A,B,C)
26300      C     ----------------------
26400      C
26500      C     SUBROUTINE TO CONVERT A COMPLEX NUMBER GIVEN IN MAGNITUDE,
26600      C     ANGLE FORM TO REAL, IMAGINARY FORM
26700      C
26800      C     INPUTS TO SUBROUTINE:
26900      C
27000      C     A = MAGNITUDE OF THE VECTOR
27100      C     B = ANGLE OF THE VECTOR
27200
27300      C     OUTPUTS GENERATED:
27400      C
27500      C     C = COMPLEX NUMBER (X + jY)
27600             COMPLEX C
27700             RAD=.0174533
27800             X=A*COS(B*RAD)
27900             Y=A*SIN(B*RAD)
28000             C=CMPLX(X,Y)
28100             RETURN
28200             END
28300
28400
28500
28600
28700             SUBROUTINE MAGTHT(X,Y,Z)
28800      C     ------------------------
28900      C
29000      C     SUBROUTINE TO CONVERT A COMPLEX NUMBER GIVEN IN REAL, IMAGINARY
29100      C     FORM TO MAGNITUDE, ANGLE FORM
29200      C     INPUTS TO SUBROUTINE:
29300      C
29400      C     X = COMPLEX NUMBER FOR WHICH MAGNITUDE, ANGLE ARE REQUIRED
29500      C
29600      C     OUTPUTS GENERATED:
29700      C
29800      C     Y = MAGNITUDE OF THE VECTOR
29900      C     Z = ANGLE OF THE VECTOR
30000             COMPLEX X
30100             R1=REAL(X)
30200             X1=AIMAG(X)
30300             Z1=ATAN(X1/R1)
30400             Z=Z1/.017453
30500             IF(R1.LT.0.)Z=Z+180.
30600             Y=CABS(X)
30700             RETURN
30800             END
30900
31000
31100
31200
31300             SUBROUTINE NAROWB(XK,TMS,TML)
31400      C     ------------------------------
31500      C
31600      C     SUBROUTINE TO COMPUTE:
31700      C     (a)SOURCE AND LOAD IMPEDANCES FOR SIMULTANEOUS MATCHING
```

Fig. 7-24. (*Continued*)

```
31800    C      (b)MAXIMUM AVAILABLE POWER GAIN WHEN STABILITY FACTOR,K>1
31900    C
32000    C      NOTE: THE SUBROUTINE IS CALLED ONLY WHEN CONDITIONS FOR
32100    C            SIMULTANEOUS SOURCE AND LOAD MATCHING ARE SATISFIED
32200
32300    C      INPUTS TO SUBROUTINE:
32400    C
32500    C      XK = STABILITY FACTOR,K
32600    C
32700    C      OUTPUTS GENERATED:
32800    C
32900    C      TMS = REQUIRED SOURCE IMPEDANCE
33000    C      TML = REQUIRED LOAD IMPEDANCE
33100    C
33200
33300           COMMON/BLOCK1/S11(20),S12(20),S21(20),S22(20),FREQ(20)
33400           COMPLEX S11,S12,S21,S22,D,XM,XN,TMS,TML,ZS,ZL
33500
33600           WRITE(6,11)
33700    11     FORMAT(//35X,'OPTIMUM NARROW-BAND DESIGN'/35X,
33800         1 '-------------------------')
33900    C
34000    C      COMPUTE OPTIMUM SOURCE AND LOAD IMPEDANCES
34100    C
34200           D=S11(1)*S22(1)-S12(1)*S21(1)
34300           AD=CABS(D)
34400           XM=S11(1)-D*CONJG(S22(1))
34500           AXM=CABS(XM)
34600           XN=S22(1)-D*CONJG(S11(1))
34700           AXN=CABS(XN)
34800           S1=CABS(S11(1))**2
34900           S2=CABS(S22(1))**2
35000           B1=1.+S1-S2-AD*AD
35100           B2=1.+S2-S1-AD*AD
35200           SIGN=-1.
35300           IF(B1.LT.0.)SIGN=1.
35400           SRHS=(B1+SIGN*SQRT(B1*B1-4.*AXM*AXM))
35500           SRHS=SRHS/(2.*AXM*AXM)
35600           TMS=(CONJG(XM))*SRHS
35700           XLRHS=B2+SIGN*SQRT(B2*B2-4.*AXN*AXN)
35800           XLRHS=XLRHS/(2.*AXN*AXN)
35900           TML=(CONJG(XN))*XLRHS
36000    C
36100    C      CONVERT IMPEDANCE TO NORMALIZED (R+jX) FORM
36200    C
36300
36400           CALL STOZ(TMS,ZS)
36500           CALL STOZ(TML,ZL)
36600    C
36700    C      COMPUTE MAXIMUM AVAILABLE POWER GAIN
36800    C
36900           GAMAX=(CABS(S21(1))/CABS(S12(1)))*(XK+SIGN*SQRT(XK*XK-1.))
37000           GAMAX=10.*ALOG10(GAMAX)
37100           CALL MAGTHT(TMS,RS,THS)
37200           CALL MAGTHT(TML,RL,THL)
37300           WRITE(6,21)ZS,ZL,GAMAX
37400    21     FORMAT(1H0,15X,'THE OPTIMUM IMPEDANCES OF THE NARROW-BAND'
37500         1 ,' AMPLIFIER FOR MAXIMUM GAIN'/16X,'-------------------------'
37600         1 ,' ---------------'/1H0,5X,'SOURCE IMPEDANCE(R+JX) ='
37700         1 ,2F8.2//5X,'LOAD IMPEDANCE(R+JX)=',2F8.2//
37800         1 5X,'MAXIMUM GAIN (DB)=',F6.2)
37900    10     RETURN
38000           END
38100
38200
38300
38400
 8500
 8600
 8700           SUBROUTINE STOZ(S,Z)
 8800    C      --------------------
 8900    C
 9000    C      SUBROUTINE TO CONVERT IMPEDANCE (REFLECTION COEFFICIENT FORM) TO
 9100    C      NORMALIZED (R+jX) FORM
```

Fig. 7-24. (*Continued*)

```
39200   C
39300   C   INPUT TO SUBROUTINE:
39400   C
39500   C   S = IMPEDANCE IN REFLECTION COEFFICIENT FORM
39600   C
39700   C   OUTPUTS:
39800   C
39900   C   Z = IMPEDANCE IN NORMALIZED (R+jX) FORM
40000
40100           COMPLEX S, Z
40200           Z=(1.+S)/(1.-S)
40300           RETURN
40400           END
40500
40600
40700
40800
40900           SUBROUTINE GS21(S,GAIN)
41000   C       ----------------------
41100   C
41200   C   SUBROUTINE TO COMPUTE FORWARD TRANSDUCER GAIN OF A DEVICE
41300   C
41400   C
41500   C   INPUTS TO SUBROUTINE:
41600   C
41700   C   S = S-PARAMETER, S21, OF A DEVICE IN REAL, IMAGINARY FORM
41800   C
41900   C   OUTPUTS:
42000   C
42100   C   GAIN = FORWARD TRANSDUCER GAIN OF THE DEVICE IN DB
42200   C
42300
42400           COMPLEX S
42500           GAIN=(CABS(S))**2
42600           GAIN=10.*ALOG10(GAIN)
42700           RETURN
42800           END
42900
43000
43100           SUBROUTINE GMAX(S,GNMAX)
43200   C       -----------------------
43300
43400   C   SUBROUTINE TO COMPUTE THE MAXIMUM INCREASE IN GAIN POSSIBLE WITH THE
                                                                            DEVICE
43500   C   BY MATCHING THE SOURCE OR LOAD
43600   C
43700   C   INPUTS TO SUBROUTINE:
43800   C
43900   C   S = S11 (FOR SOURCE MATCHING) OR S22 (FOR LOAD MATCHING) PARAMETER
44000   C       OF THE DEVICE IN REAL, IMAGINARY FORM
44100   C
44200   C   OUTPUTS:
44300   C
44400   C   GNMAX = MAXIMUM INCREASE IN GAIN POSSIBLE BY USING A MATCHING NETWORK
44500   C
44600           COMPLEX S
44700           XMAX=1./(1.-((CABS(S))**2))
44800           GNMAX=10.*ALOG10(XMAX)
44900           RETURN
45000           END
45100   C
```

Fig. 7-24. (*Continued*)

A typical design procedure uses the following steps (steps 3 through 5, however, must be performed at several frequencies for a broadband amplifier):

1. Characterize the amplifier in terms of its gain, center frequency, bandwidth, and RF power.

2. Select a suitable device. The transistor's S-parameters can be obtained by measurement or from data sheets, but measuring the parameters can improve the accuracy of the design. (If Y-parameters are available, they can be converted to S-parameters)[3].
3. Using subroutine STABLE, calculate the stability factor K (equation 4 of Table 7-4), and determine whether the conditions for absolute stability are satisfied (equation 3).
4. Calculate the centers and radii of the input and output stability circles (equation 1, subroutine STABLE).
5. For broadband designs calculate the mismatch necessary to obtain the required gain at each frequency (equation 6, subroutine GS21). For a narrowband design, a simultaneous matching of the source and load is possible if the stability factor is more than 1 (equation 8, subroutine NAROWB). In this case the maximum available gain is obtained from equation 9 (subroutine NAROWB).
6. Calculate the increase in gain that is possible by matching the source ($G1_{max}$) or the load ($G2_{max}$) using the subroutine GMAX (equation 7).
7. Using the values of $G1_{max}$, $G2_{max}$, and the required match (step 5), decide whether the gain requirement can be met using input matching, output matching, or input-plus-output matching.
8. Plot the stability and gain circles (at each frequency, for a broadband design) on a Smith chart.
9. The matching network can now be designed graphically using standard techniques.[4,5] The unstable regions of the chart should be avoided, since the loads in those areas are likely to cause oscillations.

The design of a narrow-band amplifier providing 15 dB of gain at 200 MHz provides a good illustration of the program's capabilities. A 2N918 was chosen for the example, mainly because it is readily available. The transistor Y-parameters at 200 MHz (V_{ce} = 15 V, I_c = 10 mA) were taken from data sheets and converted to s-parameters.

Feeding these data into the program produced the information needed to use the Smith chart (Figure 7-25). The required gain through matching was 0.67 dB, the maximum gains available through input and output matching were 0.46 and 2.64 dB, respectively. Since input matching could not provide the required gain, an output matching circuit was designed.

The stability circles and the 0.67-dB circle were plotted on the chart, and an LC network was chosen to provide the matching (Figure 7-26). The actual values of the L and C (shown in parentheses on Figure 7-27) deviate from the ideal, mainly because of factors such as parasitic reactances, nonideal components, and deviations in S-parameters in the actual transistor.

The amplifier's measured gain was 14 dB, compared with the design value of 15 dB. The difference was attributed to the insertion loss of the matching

Table 7-4. Equations Used by the Matching Program

Parameter	Subroutine	Equation
Stability circle on input plane (a) center (rs_1) (b) radius (RS_1)	STABLE	$rs_1 = \dfrac{C_1^*}{\|S_{11}\|^2 - \|\Delta\|^2}$ (1a) $RS_1 = \dfrac{\|S_{12}S_{21}\|}{\|S_{11}\|^2 - \|\Delta\|^2}$ (1b)
Stability circle on output plane (a) center (rs_2) (b) radius (RS_2)	STABLE	$rs_2 = \dfrac{C_2^*}{\|S_{22}\|^2 - \|\Delta\|^2}$ (2a) $RS_2 = \dfrac{\|S_{12}S_{21}\|}{\|S_{22}\|^2 - \|\Delta\|^2}$ (2b) where $C_1 = S_{11} - \Delta S_{22}^*$ $C_2 = S_{22} - \Delta S_{11}^*$ $\Delta = S_{11}S_{22} - S_{12}S_{21}$
Conditions for absolute stability	STABLE	$\|S_{11}\| < 1$ (3a) $\|S_{22}\| < 1$ (3b) $\left\| \dfrac{\|S_{12}S_{21}\| - \|M^*\|}{\|S_{11}\|^2 - \|D\|^2} \right\| > 1$ (3c) $\left\| \dfrac{\|S_{12}S_{21}\| - \|N^*\|}{\|S_{22}\|^2 - \|D\|^2} \right\| > 1$ (3d) where $D = S_{11}S_{22} - S_{12}S_{21}$ $M = S_{11} - DS_{22}^*$ $N = S_{22} - DS_{11}^*$
Stability factor (K)	STABLE	$K = \dfrac{1 + \|D\|^2 - \|S_{11}\|^2 - \|S_{22}\|^2}{2\|S_{12}S_{21}\|}$ (4)
Constant-gain circles (a) center (r_i)	CIRCLE	$r_i = \dfrac{g_i\|S_{ii}\|}{1 - \|S_{ii}\|^2(1 - g_i)}$ (5a) where

network, which implies that it is a good practice to design for 1 or 2 dB more gain than required. Overall, however, the results indicate that the program is a very useful tool for simplifying the design of RF amplifiers.

7.7. SIMULATING A SERVO SYSTEM

The analysis of feedback systems is commonplace in the world of electronics. From simple amplifier circuits to industrial robotics, varying degrees of sophis-

Parameter	Subroutine	Equation
(b) radius (ρ_i)		$g_i = G_i(1 - |S_{ii}|^2)$; $i = 1,2$ G_i = required gain circle $\rho_i = \dfrac{\sqrt{1-g_i}\,(1-|S_{ii}|^2)}{1-|S_{ii}|^2\,(1-g_i)}$; $i = 1,2$ (5b)
Forward transducer gain of a device	GS21	$G_o = |S_{21}|^2$ (6a)
Required match to obtain given gain, G		$X_m = G - G_o$ (6b)
Maximum unilateral transducer gain	GMAX	$G_{imax} = \dfrac{1}{1-|S_{ii}|^2}$; $i = 1, 2$ (7)
Source (TMS) and load (TML) for simultaneous match when stability factor > 1	NAROWB	$TMS = M^* \left[\dfrac{B_1 \pm \sqrt{B_1^2 - 4|M|^2}}{2|M|^2}\right]$ (8a) $TML = N^* \left[\dfrac{B_2 \pm \sqrt{B_2^2 - 4|N|^2}}{2|N|^2}\right]$ (8b) where $B_1 = 1 + |S_{11}|^2 - |S_{22}|^2 - |D|^2$ and $B_2 = 1 + |S_{22}|^2 - |S_{11}|^2 - |D|^2$ (Use minus sign when B_1 is positive, plus sign when B_1 is negative.)
Maximum available power gain, GA_{max}, when $K > 1$	NAROWB	$GA_{max} = \left\|\dfrac{|S_{21}|}{|S_{12}|}\left(K \pm \sqrt{(K^2-1)}\right)\right\|$ (9) (Use minus sign when B_1 is positive, plus sign when B_1 is negative.)

Note: Subroutines IMAG, MAGTHT, and STOZ are required for complex math.

tication and accuracy are required from the system. Thus, being able to simulate servo systems on PCs is a most desirable capability.

Don Stauffer has developed a fine servo system simulation program (Figure 7-28), originally published in *BYTE* (February 1985). The appended code listing and its explanation, as well as the description of the theory of servo control, are adapted from that article.

A *servo mechanism* is essentially a small motor that controls a larger motor. A servo-control system consists of the logical instructions needed to guide the

```
                    ***** RF AMPLIFIER DESIGN *****



                         S-PARAMETERS (MAG, ANGLE)
                         -------------------------
   FREQ(MHZ)        S11              S12             S21              S22

     200.0      0.316, -72.6     0.066, 68.1     5.207, 100.1     0.675, -20.8


   AMPLIFIER GAIN = 15.0 DB         CENTRE FREQ = 200.0 MHZ   (NARROW BAND DESIGN)

                            STABILITY PARAMETERS
                            --------------------
   FREQUENCY = 200.0 MHZ
   STABILITY FACTOR (K) =   0.92
   CENTRE OF INPUT STABILITY CIRCLE (MAG, ANGLE)=   3.14, -48.40     RADIUS =   4.03
   CENTRE OF OUTPUT STABILITY CIRCLE (MAG, ANGLE)=  2.22,  31.12     RADIUS =   1.27

                         GAIN REQUIRED THROUGH MATCHING
                         ------------------------------
    S.NO.     FREQ(MHZ)     G1MAX(DB)     G2MAX(DB)    REQD GAIN IMPROVEMENT (DB)

     1         200.0          0.46          2.64              0.67

                                GAIN CIRCLES
                                ------------
         MATCHING REQUIRED FOR LOAD ONLY

              FREQ     CENTRE(MAG, ANGLE)     RADIUS
              ----     ------------------     ------
              200.0       0.514,  20.8        0.395
```

Fig. 7-25. RF amplifier program calculated results.

servo mechanism. Some examples of currently popular servo mechanisms are automobile cruise controls, model airplane radio-controlled systems, stereo turntables and tape decks, kitchen appliances, and home workshop tools. The design of such systems remains one of the most intricate of the electrical engineering sciences. However, the computer's simulation ability has simplified the designer's job; simulation is now a common part of the system designer's tool kit, and it no longer requires a large computer but can be effected on PCs.

An example best explains both the theory of servo-control systems and the code that allows the design of such systems at our desks, or in our studies. To

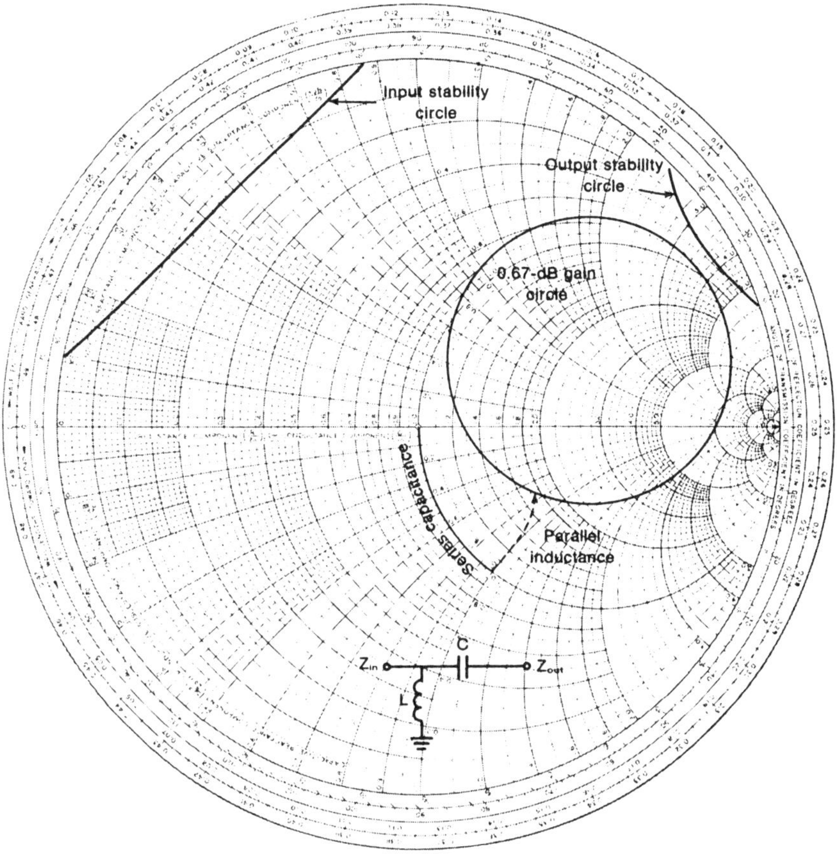

Fig. 7-26. Smith-chart plot of results.

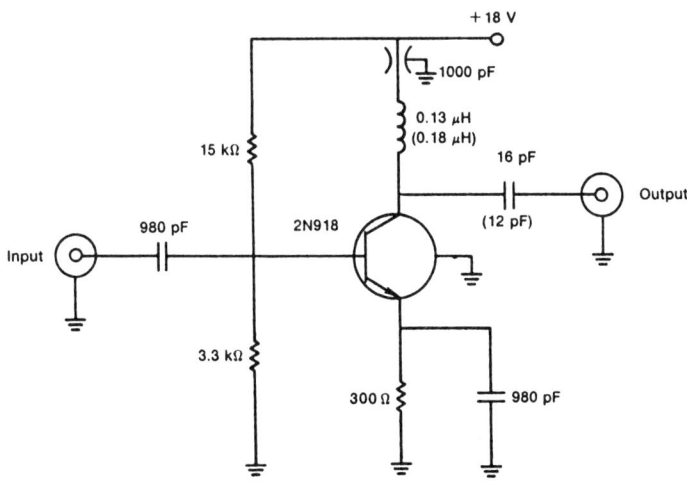

Fig. 7-27. Circuit design of RF amplifier.

240 ANALYSIS AND DESIGN OF ELECTRONIC CIRCUITS USING PCs

```
40 REM FROM BYTE 2/85 BY DON STAUFFER
50 CLEAR 200
60 REM EDIT ASSIGNMENT STATEMENTS TO ALTER CONTROL CONSTANTS
70 REM*** LINES 120-440 SET ANALYSIS CONDITIONS AND VARIABLES***
80 PR=0 : REM PRINT CONTROL
85 PRINT"Do you want the results printed on the screen? (0),"
90 PRINT"a hard copy of the calculations? (1),"
100 PRINT"a hard copy plot of the indicated weight Vs time? (2),"
105 PRINT"(If you want both, first type 1, then run again and type 2)"
110 INPUT"or, end of session? (3). Type 0, 1, 2 or 3.";PR
115 IF PR = 3 THEN GOTO 590
120 REM PR=0 : REM PRINT CONTROL SCALE
130 REM IF PR=0, THEN NO HARD COPY OF SCREEN RESULTS
140 REM IF PR=1, THEN HARD COPY OF LISTING SHOWN ON SCREEN
150 REM IF PR=2, THEN GET A GRAPHIC TRACE OF THE INDICATED WEIGHT
160 REM IF PR=3, THEN END
170 TH=0 : REM SCALE BALANCE BEAM ANGLE
180 TM=0 : REM BEAM ANGLE DURING LAST ITERATION
200 PRINT"The following constants are defaulted in the program"
210 PRINT"Beam moment of inertia = 5, Distance from pivot to"
220 PRINT"weight or solenoid = 5, Scale factor, solenoid current"
230 PRINT"to force = 10, Proportional servo constant = -0.4,"
240 PRINT"Rate servo constant = -8, and, Lag servo constant = -8."
250 PRINT"To change the pan weight type  '340 W = new value'   "
260 PRINT"To change the beam moment type  '350 JS = new value'"
270 PRINT"To change the distance type  '360 D = new value' "
280 PRINT"To change the scale factor type  '370 K = new value'"
290 PRINT"To change the prop servo constant type  '380 K1 = new value'"
300 PRINT"To change the rate servo constant type  '390 K2 = new value'"
310 PRINT"To change the lag servo constant type  '400 K3 = new value'" : PRINT
330 PRINT" TO RUN THE ANALYSIS TYPE < CONTinue >"
335 STOP
340 W=5 : REM INITIAL WEIGHT OF PAN
350 JS=5 : REM BEAM MOMENT OF INERTIA
360 D=5 : REM DISTANCE FROM PIVOT TO WEIGHT OR SOLENOID
370 K=10 : REM SCALE FACTOR, SOLENOID CURRENT TO FORCE
380 K1=-.4 : REM PROPORTIONAL SERVO CONSTANT
390 K2=0
400 K3=0
410 DT=0.2 : REM TIME INCREMENT
420 T=0 : REM INITIAL TIME
430 T7=16
440 ST=0 : REM STOP PARAMETER
450 REM *****  BEGIN SIMEULATION LOOP *****
460 IF PR > 1.5 GOSUB 1080
470 REM *****   CHECK FOR INPUT    *****
480 GOSUB 600
490 REM ***** COMPUTE CONTROL FORCE *****
500 GOSUB 680
510 REM ***** COMPUTE MOTION *****
520 GOSUB 740
530 REM *****DISPLAY AND PRINT INPUT
540 GOSUB 850
550 REM ***** UPDATE TIME *****
560 T=T+DT
570 IF T>T7 GOTO 590
580 IF ST < 0.5 THEN 470
590 STOP
600 REM **** CHECK FOR INPUT ****
610 IF PEEK(14400)= 128 THEN GOTO 610
620 IF PEEK(14340)=8 THEN ST=1
630 IF PEEK(14340)<>128 THEN RETURN
640 PRINT @65,"";
650 INPUT "CHANGE WEIGHT";W
660 IF W<0 THEN W=0
670 RETURN
680 REM **** COMPUTE CONTROL CURRENT ****
690 ER=TH
```

Fig. 7-28. Program for servo system simulation.

```
700 IF ER<-10 THEN ER=-10 ELSE IF ER>10.0 ER=10.0
710 I=K2*(TH-TM)/DT+K1*ER+K3*(ER+EM)
720 EM=ER
730 RETURN
740 TM=TH
750 J=JS+W*D[2
760 F=K*I
770 LC=F*D
780 LW=W*D
790 AA=(LC-LW)/J
800 WD=WD+AA*DT
810 TH=TH+WD*DT
820 IF TH<-10 THEN TH=-10 ELSE IF TH>10 THEN TH=10
830 RETURN
840 REM **** OUTPUT INSTRUCTIONS ****
850 CLS
860 PRINT"ACTUAL WEIGHT=";W
870 PRINT@384,"TIME","ANGLE","CURRENT","WEIGHT"
880 PRINT@448,T,TH,I,F
890 IF PR<0.5 THEN RETURN
900 IF PR>1.5 GOTO 970
910 REM LPRINT "TIME =";T,"WIEGHT =";W
920 IF T>0 THEN GOTO 940
930 LPRINT "TIME","ANGLE","CURRENT","WEIGHT" : LPRINT
940 LPRINT T,TH,I,F
960 RETURN
970 NF=T : NI=INT(NF) : RX=NF-NI
980 IF (RX<DT) THEN LPRINT CHR$(43); ELSE LPRINT CHR$(58);
990 PP=10*F : IF PP<2 THEN PP=2
1000 IF PP>134 THEN PP=134
1010 IFPP=134 THEN CH=74 ELSE CH=42
1020 NS=PP-2 : IF NS<0 THEN NS=0
1030 FOR NZ=1 TO NS : LPRINT CHR$(32); : NEXT NZ
1040 LPRINT SP$;CHR$(CH)
1050 RETURN
1060 REM LINE 1080 PUTS PRINTER IN CONDENSED (17 CHARACTERS PER
1070 REM INCH) MODE AND 112 LINES PER INCH VERTICALLY
1080 LPRINT CHR$(27)+CHR$(81):LPRINT CHR$(27)+CHR$(84);"12"
1090 REM LINE 1100 PUTS PRINTER INTO GRAPHICS MODE
1100 LPRINT CHR$(27)+CHR$(35)
1110 FOR N=1 TO 75
1120 NF=N/10 : NI=INT(NF) : RX=NF-NI
1130 IF (RX<1E-2) THEN LPRINT CHR$(43); ELSE LPRINT CHR$(45);
1140 NEXT N
1150 LPRINT CHR$(10)
1160 RETURN
```

Fig. 7-28. (*Continued*)

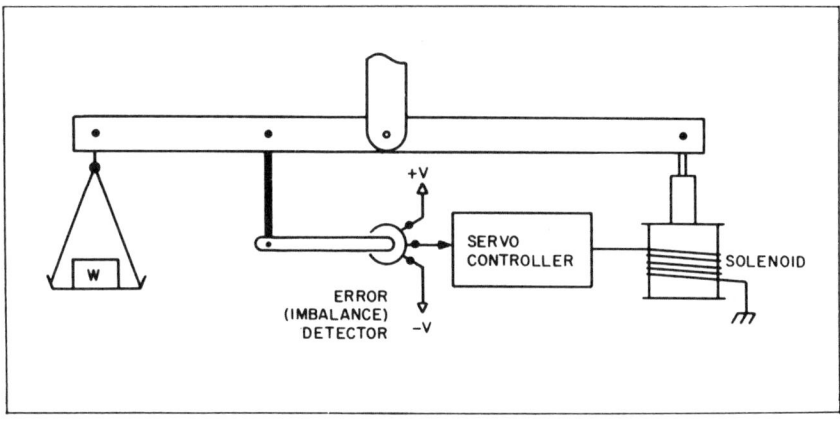

Fig. 7-29. Scale servo system example.

demonstrate, an electronic weighing scale will be designed. Figure 7-29 is an illustration of how such a scale would be arranged. A balance beam forms the main part of the scale along with the pan on the left. On the right side, instead of the normal balance weights, a solenoid is attached. The solenoid is designed so that the pull on the armature is directly proportional to the current in the coil. A sensor, such as a low-friction potentiometer, forms an error detector which give a voltage that is proportional to the angle by which the scale is out of balance. The servo-control system uses this error signal to change the current through the solenoid to eliminate the imbalance. The current in the solenoid is now proportional to an equivalent weight in the pan, and the current meter can be calibrated to read in weight units.

Figure 7-30 is the type of diagram that the designer would draw for this kind of feedback servo-control system. The circle at the left represents a summing junction. The output at the right of the junction is the sum of the inputs to the other two (or three) quadrants. As shown here, the junction indicates the difference between the command or desired quantity, Q_c, and the actual quantity, Q. The servo-control computer operates on this difference and outputs a voltage to the actuator. The actuator is a physical device, usually a force transducer, that drives the quantity to be controlled either up or down so that the actual value equals the desired value. At this point the system is balanced, and the error signal (feedback) will remain at zero unless some perturbing force displaces the system, or a new input value is commanded.

A servo-control engineer is concerned with several aspects of the system's behavior. First and foremost is stability. That is, does the system indeed act to reduce the error and not cause the error to increase? Then how soon will the system reach a new equilibrium? If it takes too long to settle down, the system may not be usable in practice. Is the amount of error that remains after the system reaches equilibrium sufficiently small? Ideally, there should be no error but in practice some will always exist and it must be decided if it is tolerable.

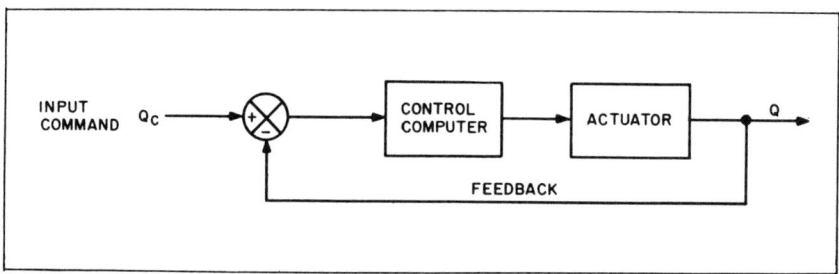

Fig. 7-30. Scale servo system block diagram.

Without simulation, complicated differential equations must be solved to try to predict a mechanism's behavior. Computer-based simulation does the math. In addition, simulation allows the design of more complex servos whose behavior could not be predicted easily by normal differential equation methods. Figure 7-31 charts a typical simulation. After setting the initial conditions, the program enters the iterative loop (input, model, output, update). It scans user or process input to see if conditions are to be changed. If the simulation is supposed to be continuous, such as the physical simulation in the example being examined, input is best done with a keyboard monitoring routine to keep the program running between inputs.

The heart of the simulation is the next step—the math model. In this block, the computer performs its mathematical operations on the equation that describes the system being simulated. Almost any system or situation that can be mathe-

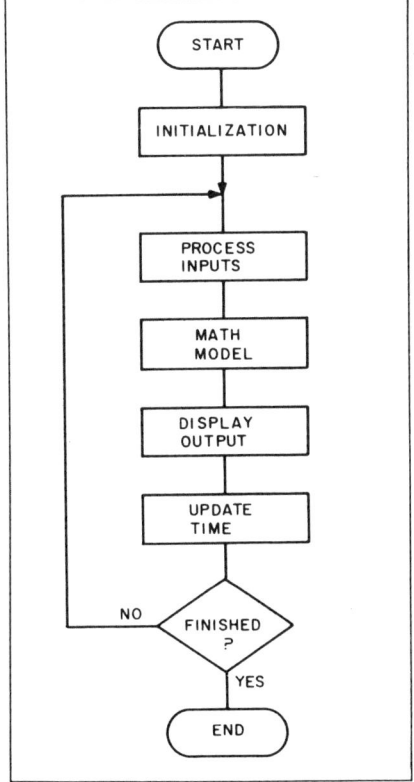

Fig. 7-31. Typical simulation program flowchart.

matically described in a cause-and-effect relationship can be simulated by a computer. Next, the program displays or prints an output. The time variable is incremented and, if the program is not terminated by some condition that reaches or exceeds a limit, the program repeats.

The code listing that is provided (Figure 7-28) follows this flow chart closely. The program is written in TRS-80 Level II BASIC, but can be adapted to many of the BASIC dialects; Subsection 7.7.2 provides more information on these adaptations. Lines 120-440 set the physical constant's values and give initial values to variables. The stop variable ST (in line 440) is used to terminate the program on command. The program must be edited to change the value of any of the constants. This is done interactively by instructions provided in the program. PR is a variable output control. When PR = 0, the calculated results will be displayed on the computer screen. When PR = 1, a hard copy of the calculated results will be provided; the results will also be displayed on the screen. When PR = 2, a hard copy of a graphic trace of the results will be printed. When PR = 3, the session will be terminated. Line 1080, referenced if PR > 1.5 (line 460), is used to set up the scale of the printer and print the axes when a plot is desired.

At line 470, the main loop begins. Line 600 looks for a user input. The function of the space bar and the S key will be described later. The subroutine starting at line 600 is the math model of the control computer block in Figure 7-30. This block will be better understand when the servo simulation exercise is started. The subroutine that begins at line 740 is also part of the math model and represents the physics of the scale in the example. It represents Newton's second law of motion as applied to rotating systems. Details on the physics math model are presented at the end of this description under that title. The force applied to the solenoid equals the current after it is multiplied by a scale factor (line 760). Torque is equal to the product of a force (F) multiplied by a distance (D), line 770. Assuming that the distances from the pivot to the weight and the solenoid are equal, line 780 calculates the torque due to the weight. Then line 790 determines the angular acceleration by finding the net difference between the torque due to the weight and the torque due to the solenoid current. That net difference is divided by the moment of inertia. Lines 790 and 800 integrate the acceleration to angular velocity and angle.

The subroutine starting at line 850 displays the output to the screen. The program displays elasped time, the deflection angle, the solenoid current, and the indicated weight. For reference, the actual weight is also displayed at the upper left corner. If a hard copy of the calculations is desired, (PR=1), the print routine is initiated per lines 100 and 140. Lines 910-960 output the tabulated data. The printed output slows execution considerably, so it should be used only when a hard copy of the results is required. The graphic output, (PR=2), per lines 100 and 150 is begun by the command at line 970. The values listed are those for a C. Itoh ProWriter.

7.7.1. Servo Theory

Figure 7-32 shows the simplest form of a servo controller; it is known as a *proportional control system*. The controller merely takes the error signal ($Q_c - Q$) and multiplies it by a constant called the "gain constant." In the example it is desired that the angle of the scale be zero. Thus the commanded value of $Q(Q_c)$ will always be zero, and the error will always be $-Q$, where Q is the scale's actual angle. The output signal to the actuator and, as mentioned previously, the restoring force on the scale, are proportional to the error.

Consider how it is desired for the scale to behave. Obviously, the weight readout should be close to the actual weight in the pan. Also, the balance should not take too long to settle out nor should it continue to oscillate about its final value for long times. Electronic scales can oscillate in a similar fashion; therefore it is desirable to have the scale settle down quickly. Additionally, if the scale comes to rest with the beam not level, there will be an inaccuracy in the results. With these three criteria in mind, the program should be run with the initial values in the code and the scale performance should be observed.

At time equal to zero, the scale is in balance and everything stays at rest with the scale at zero angle. Some very small time later, the 5-ounce weight puts the scale out of balance, and the beam swings to a negative angle. The control system senses this angular error and increases the solenoid current. This attracts the beam and slows it down. Now the current-generated force exceeds the weight, and the beam's angle moves back towards zero. When this happens, the solenoid current stops and the cycle repeats. The scale, with the physical values defined by the parameters shown, represents a good oscillator. The simulation would provide the same varying results forever. Figure 7-33 is a plot of one cycle of this condition. This, of course, does not represent the real world since pivot friction would eventually reduce the oscillations. However, it would take a very long time; also, this friction would have very small effect on a properly designed system. Therefore, pivot friction has been omitted from consideration in the simulation model. Varying with the value of K1 will affect the period of the oscillation but will not eliminate it.

External oscillation is eliminated by adding "rate damping" to the system. Figure 7-34 shows a proportional-plus-rate system. The symbol $\dot{Q}$ (pronounced

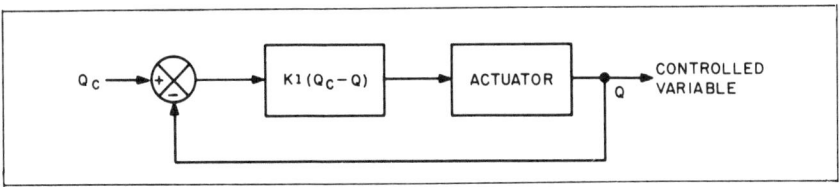

Fig. 7-32. Proportional control servo system.

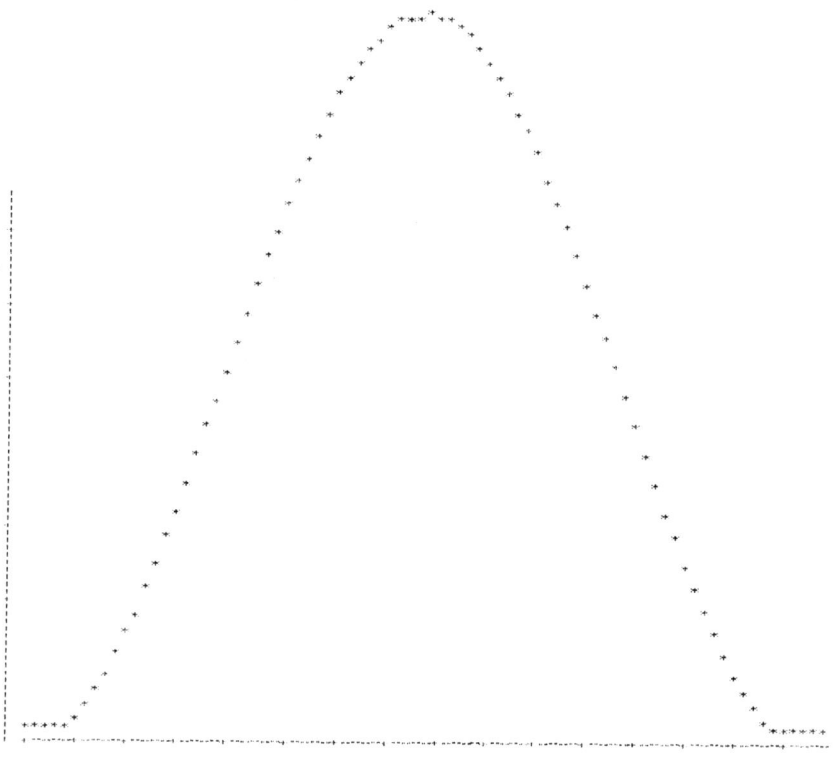

Fig. 7-33. Oscillation in Servo System.

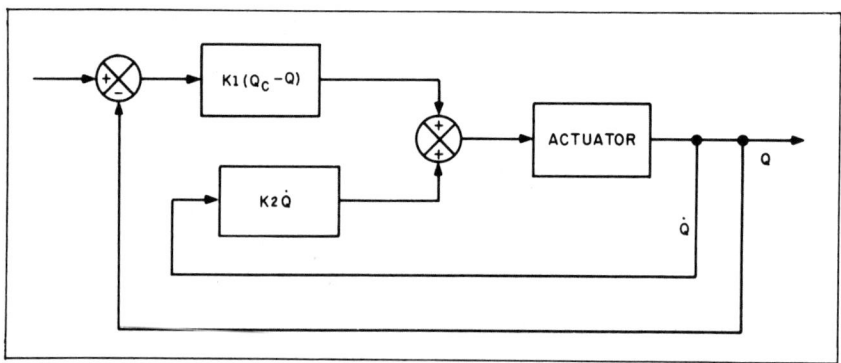

Fig. 7-34. Proportional-plus-rate servo system.

"Q dot") represents Q's rate of change over time. In calculus, this is the time derivative. Remember, Q is the controlled variable, the angle of the beam. Rate damping is added to the system by setting K2 to some nonzero value (try a value of -4). Figure 7-35 shows a typical result. Most of the oscillation has been removed, although a small amount of excess motion remains. This will eventually stop, but will take a long time. This scale could be considered practical although it still has undesirable features. However, in addition to sluggish response and excess motion, there is an additional problem. As the system approaches equilibrium, an angle error of about 1 degree exists. Although this is not excessive, in most cases it would be unacceptable.

To correct for this error, another block must be added to the system to create a proportional-plus-rate-plus-integral, or, proportional-plus-rate-plus-lag system (Figure 7-36). This proportional/rate/lag servo is created by changing K3 to a nonzero value (set K3 to -3 in line 400). Run the example again. This speeds up the response and increases the excess motion. But when the system damps out, a significant reduction in residual angle is observed. If K2 is again changed, to a value of -8, some excess motion occurs, but equilibrium is quickly obtained, and the angle error is less than $\frac{1}{10}$ degree. The default values shown

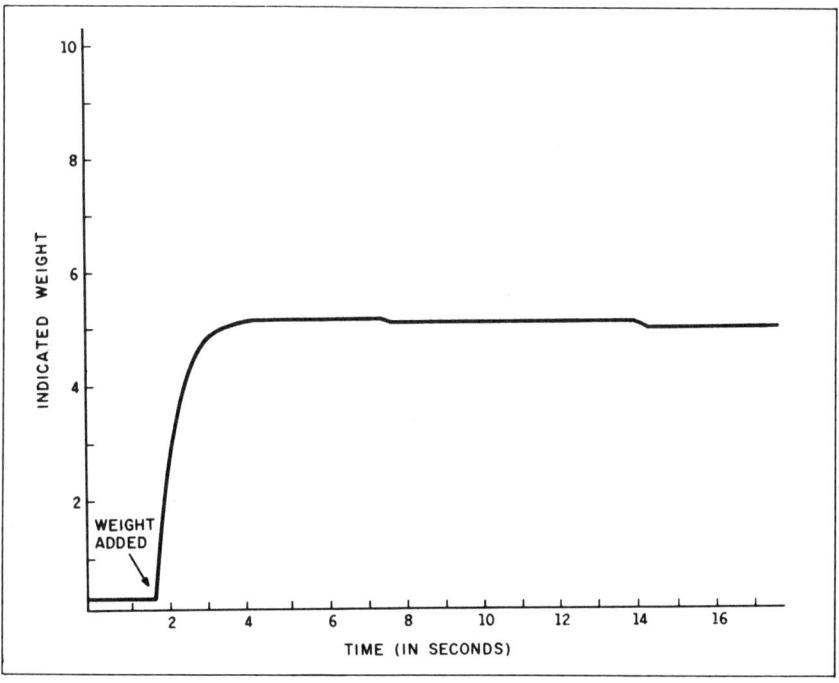

Fig. 7-35. Rate-of-change feedback effect.

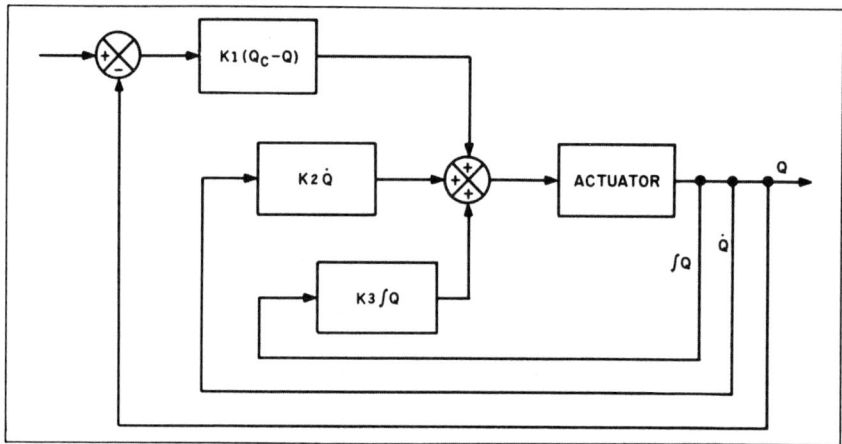

Fig. 7-36. Proportional-plus-rate-plus-log servo system.

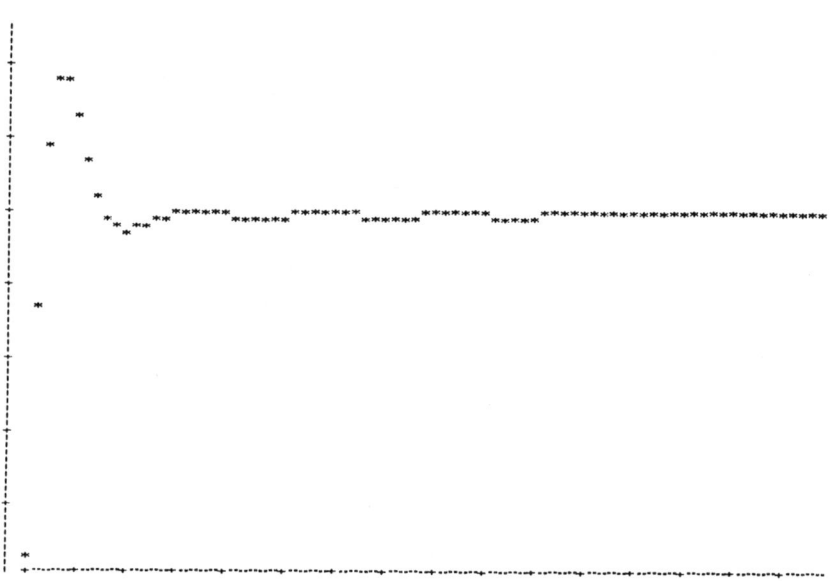

Fig. 7-37. Program analysis results—plot.

on the code listing produce the results shown in Figure 7-37. A tabulation of Figures 7-37 values are shown in Figure 7-38.

7.7.2. Program Changes

The BASIC used in this program, Radio Shack Level II BASIC, can easily be converted for use on other computers. The CLEAR command in line 50 (Figure 7-28) clears for string space and is needed only for the graphic print option.

TIME	ANGLE	CURRENT	WEIGHT
0	-7.69231E-03	0	0
.2	-.0230296	3.07692E-03	.0307692
.4	-.0459174	9.21183E-03	.0921183
.6	-.076215	.018367	.18367
.8	-.113736	.030486	.30486
1	-.158249	.0454944	.454944
1.2	-.209481	.0632997	.632997
1.4	-.267116	.0837924	.837924
1.6	-.330799	.106846	1.06846
1.8	-.400139	.13232	1.3232
2	-.474709	.160056	1.60056
2.2	-.55405	.189884	1.89884
2.4	-.637674	.22162	2.2162
2.6	-.725066	.25507	2.5507
2.8	-.815688	.290026	2.90026
3	-.908983	.326275	3.26275
3.2	-1.00438	.363593	3.63593
3.4	-1.10128	.401751	4.01751
3.6	-1.1991	.440513	4.40513
3.8	-1.29724	.479641	4.79641
4	-1.39508	.518894	5.18894
4.2	-1.49203	.558032	5.58032
4.4	-1.58749	.596812	5.96812
4.6	-1.68087	.634996	6.34996
4.8	-1.77161	.672349	6.72349
5	-1.85913	.708642	7.08642
5.2	-1.9429	.743651	7.43651
5.4	-2.02241	.77716	7.7716
5.6	-2.09717	.808964	8.08964
5.8	-2.16671	.838867	8.38867
6	-2.23061	.866684	8.66684
6.2	-2.28848	.892245	8.92245
6.4	-2.33996	.915392	9.15392
6.6	-2.38473	.935983	9.35983
6.8	-2.42251	.95389	9.5389
7	-2.45308	.969005	9.69005
7.2	-2.47625	.981233	9.81233
7.4	-2.49187	.9905	9.905
7.6	-2.49985	.996749	9.96749
7.8	-2.50014	.99994	9.9994
8	-2.49273	1.00006	10.0006
8.2	-2.47768	.997093	9.97093
8.4	-2.45507	.991072	9.91072
8.6	-2.42505	.982028	9.82028
8.8	-2.38779	.970019	9.70019

Fig. 7-38. Listing of plotted results

9	-2.34354	.955117	9.55117
9.2	-2.29255	.937414	9.37414
9.4	-2.23515	.91702	9.1702
9.6	-2.17168	.894059	8.94059
9.8	-2.10255	.868673	8.68673
10	-2.02816	.841018	8.41018
10.2	-1.94899	.811265	8.11265
10.4	-1.86552	.779597	7.79597
10.6	-1.77826	.746208	7.46208
10.8	-1.68775	.711303	7.11303
11	-1.59454	.675099	6.75099
11.2	-1.49922	.637816	6.37816
11.4	-1.40236	.599686	5.99686
11.6	-1.30456	.560943	5.60943
11.8	-1.20643	.521824	5.21824
12	-1.10856	.482571	4.82571
12.2	-1.01157	.443425	4.43425
12.4	-.916042	.404627	4.04627
12.6	-.82257	.366417	3.66417
12.8	-.731729	.329028	3.29028
13	-.644077	.292692	2.92692
13.2	-.560154	.257631	2.57631
13.4	-.480476	.224061	2.24061
13.6	-.405533	.19219	1.9219
13.8	-.335787	.162213	1.62213
14	-.271668	.134315	1.34315
14.2	-.213568	.108667	1.08667
14.4	-.161847	.0854273	.854273
14.6	-.116822	.0647388	.647388
14.8	-.0787704	.0467288	.467288
15	-.0479264	.0315082	.315082
15.2	-.0244798	.0191706	.191706
15.4	-8.57478E-03	9.7919E-03	.097919
15.6	-3.09356E-04	3.42991E-03	.0342991
15.8	2.65669E-04	1.23743E-04	1.23743E-03
16	-6.85325E-03	-1.06268E-04	-1.06268E-03

Fig. 7-38. (*Continued*)

The keyboard scanning routine in lines 610-630 checks the keyboard for depressed keys. The PEEKs look at the memory area of the memory-mapped TRS-80 keyboard. The Apple II uses the same technique, although the memory locations are different. Line 610 looks for the space bar and freezes the program as long as the key is depressed. Line 620 looks for the S key; line 630 for the W key. For the Commodore 64, use the GET command.

If a printer is used with any other computer, the TRS-80 LPRINT commands must be changed. For the Apple II, change all LPRINTs to PRINTs, precede each with a PR#1, and follow it with a PR#0. For a Commodore 64, use the OPEN command before each output to the printer, followed by an OPEN 1, 3 to return the output to the screen.

The other area of the program that requires modification for use with other printers, as well as to accommodate different computer systems, is that which contains the graphic commands to the ProWriter.

7.7.3. Physics Math Model

The code in lines 740-830 is the mathematical model of the physics of the scale. The scale operates according to Newton's second law of motion, but is expressed in a form for angular motion, which may make it seem slightly unfamiliar. The law is normally expressed as $F = MA$. For angular, or rotary, motion it is expressed as $AA = L/J$, where AA is the angular acceleration (degrees per second squared); L is the net torque (difference between the torques in opposite directions); and J is the moment of inertia, which is the resistance to a change in rotation and is the rotary equivalent of mass. The moment of inertia is a function of the beam's structure and the weight added to the pan (line 750). Torque equals force times distance. For the scale, it is assumed that the distance between the weight and the pivot is the same as the distance between the pivot and the point at which the solenoid applies its force. Thus, line 770 represents the torque generated by the solenoid, while line 780 represents the torque from the applied weight. Line 790 calculates the angular acceleration. Line 800 integrates the acceleration to find the angular velocity; line 810 integrates once more to find the angle. Line 820 represents mechanical stops that prevent the beam from rotating more than 10 degrees in either direction.

REFERENCES

1. S-parameter technique for faster, more accurate network design.'' *Hewlett-Packard Journal*, 18(6) (Feb. 1967).
2. Froehner, W. H., "Quick amplifier design with scattering parameters.'' *Electronics*, Oct. 16, 1967.
3. S-parameters... circuit analysis and design.'' Hewlett-Packard Application, Note 95.
4. Smith, P. H., *Electronic Applications of the Smith Chart in Waveguide, Circuits, and Component Analysis*. New York: McGraw Hill, 1969.
5. Carson, R. S., *High-frequency Amplifiers*. New York: Wiley-Interscience, 1975.

Appendix A
Glossary of Terms

A-Plus A logic description language created by Altera for their devices.

ABEL (Advanced Boolean Expression Language) A logic compiler for PLDs from I/O Data Corporation.

Access time The time required to obtain data from the main memory or a storage device such as a diskette or a hard disk.

Acoustic coupler A device for transmitting or receiving data over telephone lines by changing electrical signals to acoustic signals, or vice versa.

Active-Hi/Active-Lo Refers to specific signal polarity assertion levels at which the meaning carried by the signal name will come true when that signal goes high (5 volts) or low (0 volts).

ADC Analog-to-digital converter.

AMAZE (Automatic Map And Zap Equations) A symbolic assembler created by Signetics for their IFL logic devices.

Analysis See *Simulation*.

Application A specific task or program, such as evaluating or sorting lists of engineering data, to which a computer solution can be applied.

Application program A computer program designed to meet specific user needs, such as a program that simulates the performance of a circuit or monitors a manufacturing process.

Architecture The design or organization of computer central processor unit (CPU).

ASCII (The American Standard Code for Information Interchange) A code that has assigned a binary number for each alphanumeric character as well as several nonprinting characters, that are used to control communication devices and printers. The binary code (the assigned number) for each alphanumeric character is called the ASCII code.

ASCII Keyboard A keyboard that sends an ASCII character to the computer when the appropriate key is pressed.

Assembler Transforms low-level instructions or logic descriptions into a functional description in the native format of a machine. Instructions or logic descriptions written in an assembler language are usually not portable from one machine to another.

Asynchronous A communication method in which data is sent as soon as it is ready as opposed to being sent at fixed intervals.

Backup copying of files onto a storage medium for safekeeping in the event the original file becomes lost or damaged.

BASIC (Beginners' All-purpose Symbolic Instruction Set) A widely used interactive programming language that is especially suited for use on personal computers.

Batch processing The process of executing a set of computer programs without human interaction during the execution.

Baud A unit of speed for transmitting/receiving data approximately equal to a bit per second. Common baud rates used with personal computers are 110, 300, 1200, 2400, 4800, and 9600.

BCD (Binary Coded Decimal) A binary coding system in which each digit of a number is stored as its four bit binary equivalent.

Bidirectional (Computers) The ability to transmit data in two directions on a data bus. (Printers) The ability of the print head to print while moving either left to right or right to left.

Binary A number system with only two digits, 1 or 0. Each digit in a group of digits represents a decimal power of two.

Bit (Binary digIT) Unit of data that can have only two possible values, 1 or 0. It is the smallest unit of data recognized by a computer. All alpha-numeric data used by the computer is expressed in some combination of 1s and 0s.

Bit-map graphics A technique that allows for the control of individual pixels on a display screen to produce high-resolution graphics. This technique permits accurate representation of circles, arcs, sine waves, and so forth, which can then be reproduced with appropriate printers.

Bipolar A semiconductor technology comprised of bipolar transistors as opposed to MOS devices.

Board (also known as PWB—(Printed Wire Board) or circuit board A plastic or glass material in combination with resins, on which electronic components are mounted and interconnected.

Boolean The algebra of logic. Originally created by George Boole, a nineteeth-century British mathematician.

Buffer A temporary storage area for data. Buffers are normally used to hold data passed to other devices, such as printers, which operate at slower speeds.

Bus A group of parallel connections that carry signals between computer functions, or external devices.

Byte The number of bits used to represent a character. For personal computers, 8, 16, or 32 bits are usually used.

CAD (Computer Aided Design) See *CAE*.

CAE (Computer Aided Engineering) Usually referring to analog and logical design and analysis.

Cathode-ray tube A vacuum tube that generates and guides electrons onto a fluorescent screen to produce graphic displays of symbols, figures, or text; the basic element of most video displays.

Central processing unit (CPU) Some of the electronic components in a computer that as a group control the transfer of data in the computer as well as perform logic and mathematical calculations.

Character A single printable alphanumeric character (A–Z or 0–9, or symbol (%, $, *)), used to represent data or instructions.

Character code A code that assigns numerical values to characters, i.e., the ASCII code.

Character printer A printer that prints one character at a time in the same manner as a typewriter, i.e., a line printer.

Checksum The sum of all binary values of data in a record, used to verify error-free data transmission from one computer to another.

Chip A piece of semiconductor material containing integrated circuits.

Circuit A collection of semiconductors and passive electronic components interconnected to perform an electronic function.

Circuit optimization Obtaining solutions by iterative techniques.

Circuit synthesis Obtaining the desired results by creating and expanding close-form expressions.

COBOL (COmmon Business Oriented Language) A programming language that was developed particularly for business applications requiring complex data handling as well as the generation of large amounts of printed data.

Combinatorial A logic circuit consisting of connected logic gates, with no digital storage elements, i.e., flip-flops.

Command A user instruction to the computer, generally input from a Keyboard, that causes the computer to perform the defined operation.

Comments Textual descriptions added to logic descriptions to help the reader understand the intent of the designer.

Compatibility The ability of a set of instructions, or a program to be run on different computers. Also, the ability of different types of computers to work together.

Compiler A program that transforms high-level instructions or descriptions into the native function for some machine. These descriptions are usually portable from one machine to another.

Computer network An interconnection of computers and/or terminals tied together by a communication facility.

Configuration The interconnection of different components of a computer system, i.e., disk drives, terminals, printers, etc., to create a particular functioning system.

Constant gain circle Solution of terminations in high-frequency circuits that results in a equal amount of gain.

Core An older type of nonvolatile memory made of ferrite rings used to represent binary data as determined by the direction of magnetization of the ferrite.

CP/M (Control Program for Microcomputers) An operating system used by many personal computers. Also the earliest disk operating system for mircoprocessors.

CPU See *Central processing unit*.

Cross junction Joining four transmission lines, causing discontinuities.

CRT See *Cathode-ray tube*.

CUPL (Compiler, Universal, Programmable Logic) A high-level logic compiler for PLDs from P-CAD.

Cursor A movable, blinking marker on the video screen that determines the next point for character entry or change.

DAC Digital-to-analog converter.

Daisy-wheel A print head that prints full characters rather than characters formed of dots. The process is similar to that of a typewriter.

Data Numbers, letters, and symbols stored in a computer. These data can be manipulated by the computer to perform desired operations.

Database A collection of organized information required to perform a task.

Data communication The movement of coded data from a sender to a receiver by means of electrically transmitted signals, i.e., via a modem.

Data diskette See *Diskette*.

Data processing Application in which the computer works with numerical data as opposed to text.

Device A piece of hardware that performs some specific function. A keyboard is an input device; a printer is an output device.

Diagnostic A part of a program that checks the operation both of specific hardware and of computer programs for errors and reports its findings as an error message.

Direct memory access A method for transferring information to or from a computer's memory without CPU intervention.

Discontinuities Disturbances and reflections caused by unequal transmission line lengths or impedances. This leads to circuit losses.

Disk A rigid flat plate, normally circular, with a magnetic coating, on which to store data that in turn can be randomly accessed.

Disk drive A unit used to write data to a disk/diskette, or read data from the disk/diskette.

Diskette A flexible, flat circular plate with a magnetic coating on which data are stored.

Dispersion A frequency-dependent effect that changes transmission line impedance and effective dielectric constant.

Display screen The surface of a CRT that provides the visual representation of information.

Distributed circuits Components that cannot be modeled directly by lumped elements principally at rf/microwave frequencies.

Dot matrix printer A printer that forms its characters from a two-dimensional array of dots.

Double density A special method of recording information on a diskette that permits twice as much data to be stored as does a single density recording.

GLOSSARY OF TERMS 257

Downtime The period of time a computer, or another associated device, is not working.

Drive A peripheral device that holds a disk or diskette so that the computer can read or write information.

Effective dielectric constant The true physical average of a nonhomogeneous dielectric field.

Emulator A program that allows a computer to imitate a different system. This permits different systems to use the same programs and data.

Ergonomics The science of human engineering that combines the study of human body mechanics with industrial psychology.

Error function Description of the difference between desired and actual response during circuit optimization.

Error message Text displayed by a computer when it discovers an error in either input or processing of a program.

File A collection of logically related records or data that are treated as a single item. Files are usually stored as entities on disks/diskettes for ready recall when needed.

Filename A sequence of alphanumeric characters assigned by the user to enable the identification of a file by both the computer and the user.

Flexible disk See *Diskette*.

Flip-flop A device that can store a logic value from one point in time to another.

Floppy disk See *Diskette*.

Font A set of symbols, letters, and numbers of the same typestyle. It is used for either video display or printing.

Formfeed A symbol that causes the printer to advance the paper automatically to the top of the next sheet.

FORTRAN (FORmula TRANslation) A widely used programming language especially suited for problems that can be expressed in algebraic or other mathematical forms. It is generally used in scientific or engineering applications.

FPLA (Field Programmable Logic Array) Typically refers to an AND/OR structure with programming arrays at the inputs of AND gates and also between AND and OR gates.

Function Key A single Keyboard Key that causes a computer to perform a function such as initiating a run, clearing the screen, or displaying letters in all caps.

Fuseable links Typically made of some type of metal that are actual fuses that are blown (separated) by a rapid infusion of current during the programming process.

GaAs FET (Gallium Arsenide Field Effect Transistor) Frequently used in active microwave circuits.

Gate array A semi-custom integrated circuit initially fabricated to the point at which a matrix of separate gates exists, the final function being programmed by way of metal mask interconnects between gates thereby creating logic elements and functions.

Graphics The presentation of lines and figures on the display screen, as opposed to text or other characters.

Hard copy Output in a permanent form, i.e., printed, as opposed to the temporary CRT display.

Hard disk A disk that is not flexible; for example, a Winchester disk. Hard disks are capable of storing considerably more information than are floppy disks.

Hardware The physical equipment that constitutes a computer system.

Hard wired A physically permanent as opposed to an electronically controlled connection between two points on a circuit.

Head A component that reads, writes, and/or erases information on a hard or soft disk.

Help A series of messages stored in the computer that provides assistance for some of the more common problems encountered either in specific programs or with the operating system.

HELP A logic assembler created by Harris Semiconductor for their PLDs.

Hexidecimal A representation in number base 16.

High-level Descriptions closer in form to the original thoughts of the designer.

I/O Input/Output.

Impedance matching Transformation of the source termination to the complex conjugate of the load, for maximum power transfer.

Indexed variable A variable name that contains a decimal number on the end indicating its position or value.

Inpact printer A printer that prints on the paper by striking an inked ribbon with a preformed character, as a typewriter, daisywheel printer, or dot matrix printer.

IC See *Integrated circuit*.

Instruction A command that tells the computer what operation to perform next.

Integrated circuit (IC) A complete circuit integrated on a single chip.

Interface An electronic component, normally a connector, that joins an external device such as a modem or a printer, to the computer.

Intermediate variable A variable which, once defined, is used in expressions of other variables. For PLDs these variable names appear neither in input nor output pin lists.

Job A task for the computer to perform. A job may be running a program or printing the results.

K In data processing the symbol used to represent 2 to the 10th power, or 1024. "K" is generally interpreted to mean 1000, as does "k," for "kilo," in most metric measures.

Karnaugh map A form of logic truth table whereby a process of graphically combining tabular elements provides a methodology that allows logic reduction, originally created by Karnaugh.

Kbyte Kilobyte, or 1024 bytes.

Keyboard The set of keys on a terminal that allows alphanumeric characters or symbols to be transmitted to the computer when a key is depressed.

Kuroda transformation Changing to a new load configuration through lossless transmission line manipulations.

Large-scale integration (LSI). The combining of more than 1000 circuits on a single integrated circuit chip.

Letter-quality printer A printer that produces hard copy of computer output of a quality comparable to that of a typewriter.

Line printer A high-speed printer that prints a whole line at one time as opposed to a PC printer that prints a character at a time.

Logic analyzer An electronic test instrument capable of storing the values of multiple logic signals at the same time.

Logic reduction A process whereby the number of gates in a proposed logic circuit are reduced without affecting the logic function itself. In a PLD this usually refers to fulfilling the same function with fewer number of product-terms.

Logic minimization See *Logic reduction.*

LSI See *Large-scale integration.*

Magnetic tape A magnetic device for storing very large quantities of data. It is a serially written and read device. In general, it is not used with PCs, except for cassette tapes.

Mainframe A physyically very large computer capable of running programs that are too large to run on PCs, or would require too much time for practical usage on the PC.

Main memory See *Memory.*

Mass storage A device like a hard disk or a magnetic tape that can store large amounts of data and be readily accessible to the CPU.

Mbyte Megabyte, or 1,048,576 bytes of memory. Normally interpreted as 1 million bytes.

Memory A device on which data can be stored and from which it can easily be retrieved. It is also the main area in which files, data, and temporary results associated with programs are stored.

Menu A displayed list of options from which the user selects an action to be performed by typing a single letter or number as defined by the program being run.

Microcomputer A computer that is physically very small, based on LSI technology. A PC is a microcomputer.

Microprocessor A central processing unit (CPU) that is fabricated on a single LSI chip.

Microstrip line An approximate transmission line realization using an nonhomogeneous dielectric around the center conductor.

Microwave Generally refers to the frequency range above 1 gigahertz.

Minicomputer A computer that is physically smaller than a mainframe, but much larger than a microcomputer.

Minimum insertion loss (MIL) Used in matching network analysis to specify loss at frequencies where reflection zeros occur.

Mnemonic A short, easy-to-remember name or abbreviation. FORTRAN and BASIC are examples.

Modem (MOdulator/DEModulator) A device that converts computer signals into phone transmittable signals, and vice versa.

Monitor A TV-like device that is used as an output display device.

MOS (Metal Oxide Semiconductor) A semiconductor technology device to perform electronic functions in the same manner as bipolar transistors.

Monte Carlo analysis Statistical random pertubation of components to simulate tolerance effects.

Multimodal optimization Random-search technique to find best solution when the error function has several minimums.

Multiplexer A logic element capable of steering one of a number of data signals to a single output line.

Network A group of computers, or terminals, connected to each other to share information and resources.

Network analyzer Test equipment to measure swept frequency (scattering) s-parameters.

Next state A particular combination of logic values contained at the inputs of state-bit flip-flops.

Nodal analysis Evaluation of circuit response by examination of node voltages and branch currents.

Nonvolatile memory Memory that does not lose its information when power is interupted, lost, or removed.

Norton transformation Changing to a new load impedance through lossless circuit element manipulation.

On-line Under the control of the computer.

Operating system A collection of computer programs that control the performance of the computer. The programs perform such functions as allocating memory, controlling input and output, and so forth.

Operators The symbols/characters reserved as logical and arithmetic functions/operations for a particular language.

Oscilloscope An electronic test instrument capable of displaying voltages on one axis versus time on another axis. This display is typically performed on a CRT.

PAL (Programmable Array Logic) A PLD having a programmable AND array and a fixed OR array.

PALASM The PAL assembler created by Monolithic Memories for their PAL logic devices.

Parallel transmission Sending more than one bit at a time. In PCs, eight bits are normally sent at one time.

Parasitic elements Reactive portions of terminations or components, generally negligible at low frequencies.

Parity A one-extra-bit code added to a transmission to verify that the correct information has been transmitted.

Peripheral A device that is external to the computer but connected to it to perform such functions as printing or display.

Pin assignment The process of assigning particular numbered pins of a chip to specific variable means as part of their logic description.

Pin declaration Same as *Pin assignment*.

Pixel (picture elements) Defineable locations on a display screen (CRT) that are used to form images. For graphics, screens with more pixels can display at a higher resolution.

PLAN A PAL asembler created by National Semiconductor for their PAL devices.

PLD (Programmable Logic Device) Usually refers to integrated circuit with an AND/OR structure.

Port (Computers) An area on the computer for connecting printers, modems, and so on. (Microwave) A pair of nodes within a circuit. See *Port analysis*.

Port analysis Examination of circuit response between pairs of nodes. Generally one of the nodes of each port is grounded.

Potential instability Possibility of generating negative resistance that may lead to oscillation of the circuit.

Power supply A circuit that converts ac power to dc power to energize the electronic circuits in a computer.

Present state A particular combination of logic values contained at the output of state-bit flip-flops.

Primary memory See *Main memory*.

Printer A device that produces a paper copy—or hard copy—of an analysis result, a graph, or a letter.

Printhead The element on a printer that forms the printed character.

Printout An informal expression used to represent anything printed by a computer peripheral device.

Printwheel See *Daisywheel*.

Processor The functional part of a computer that reads, interprets, and executes instructions; also known as the central process unit, or CPU.

Product term A group of variables ANDed together. In the AND/OR structure of a typical PLD, this typically refers to the number of AND gates required for a particular function.

Program A complete sequence of instructions needed to solve a problem or execute other computer directions.

Program disk A disk containing the instructions of a program.

Programming array A matrix of cells or fuses on a PLD chip that when programmed,

Significance (bit) Indicates the value of a bit position according to a binary weight with the more significant/more heavily weighted bit on the left and the least significant bit on the right.

Simulation (logic) A process whereby stimulus patterns for a circuit are compared with

RAM (Random Access Memory) Memory that can be both read and written to during normal computer operation. RAM is the type of memory normally used in most PCs to store instructions and data of currently used programs. All information stored in RAM is lost when power is removed from the system.

Realtime Taking place at the time of the event. Realtime refers to computer programs/systems that perform a computational function during the time an event being monitored occurs. An example is monitoring and recording an industrial process.

Record A collection of related data items.

Reflection coefficient Ratio of the reflected and incident waves at a given point in a circuit. A perfect match has zero reflections.

Reflection zero The condition in which the reflection coefficient is at a minimum.

Remote Connected by means of temporary (removable) lines or wires. Examples are modems, printers, etc., that are located some distance from the computer.

Reverse video A feature on a display that produces the opposite combination of characters and background from that which is normally displayed. For example, black characters on a white background, if white characters on a black background are normal.

ROM (Read Only Memory) Computer memory that can be read only and not written to, containing a permanent storage of data and/or instructions.

RS-232 A standard describing electrical characteristics and pin assignments for serial communications.

Scattering parameters *N*-port circuit definition using ratios of traveling waves instead of voltages and currents.

Scope See *Oscilloscope*.

Screen The display surface of a video monitor.

Self-test A procedure whereby a program or peripheral checks its own operation.

Sequential access Refers to storage devices, such as disks or tapes, from which data or instructions can be retrieved only by a passing search through all storage locations between the one currently being accessed and the desired one.

Serial access See *Sequential access*.

Serial transmission Sending one bit at a time.

Sequential logic A logic circuit including flip-flop/register storage elements whose outputs feed back into the logic. This causes the next value of a flip-flop output to depend on the previous value, creating a sequence-dependent function.

Sequencer A sequential logic circuit.

Signal polarity assertion Refers to the voltage level to which a logic signal will go when the meaning carried in the name of the signal becomes true.

GLOSSARY OF TERMS 263

allows selective interconnects at the different points in the matrix, thereby configuring the logic function.

Programming language Words, symbols and mnemonics with associated rules used in constructing computer programs. Examples are BASIC, FORTRAN, COBOL.

response patterns according to the logic function for the circuit to determine proper operation.

Single density Descriptive of one of the normal recording densities of PC diskettes. The other is double density.

Single-sided Describes the recording on only one side of a diskette.

Smith chart Polar plot mapped from rectangular impedance or admittance plane.

Soft copy Alphanumeric and/or graphic data presented in nonpermanent form, as on a CRT.

Software The temporary or permanent programs that cause a computer to perform a function.

Sort Rearrangement of the records on a file so that the order is convenient to the user, i.e., alphabetic, numeric, etc.

State A particular combination of state bits of a state machine.

State assignment Refers to the process of assigning specific numerical values to the state bits that represent symbolic state names.

State bits The individual bits contained in a flip-flop that comprise the state of a state machine.

State machine A sequential circuit design done in a regular way such that a group of bits known as state bits feed back into the logic array to determine their value.

Storage unit A place where documents can be saved for later use, such as a drawer in a cabinet, a tape, or a disk.

Stubs Transmission lines with one end open or short-circuit terminated. May be used in series or parallel configuration.

Symbolic translator A program that translates a logic expression in sum-of-products notation to a fuse-map programming pattern such that there is a one-to-one correspondence between the variables in the expression and the programmed connections in the array.

Syntax The basic format and rules for a design language.

System A combination of hardware and software that perform processing operations.

T-junction A joining of three transmission lines, causing discontinuities.

Tape A recording medium for computer data or programs. For PCs, this usually implies casette tapes.

Target device The particular type of device that is chosen for a specific logic description implementation.

Task A program in execution.

Terminal An input/output device used either to enter data into a computer, or to obtain data from the computer. Input terminals are normally keyboard-like devices. Output terminals can be CRTs or hardcopy printers.

Terminal emulation A communication method in which a terminal or computer acts as though it were of a different design.

Timesharing Provides computer services simultanously to more than one user.

Track The portion of a disk, diskette, or tape that is accessible to a given read/write head position.

Transfer function The process of equating a logic expression to an output variable where the output variable represents a storage element and is updated at each clock pulse. For analog applications, it indicates the gain between input and output nodes.

Tri-state (three-state) Typically refers to a semiconductor device component whose output is in the high-impedance state rather than driving high or low.

TTL (Transistor Transistor Logic) Basically, the 7400 series of logic elements, popular for the past 2 years.

Turnkey system A stand-alone system that is ready for use without the addition of software or hardware; it is considered a complete system.

Typeface See *Font*.

Unconditional stability Absence of negative resistance at all ports for all possible passive source and load terminations.

Unimodal optimization Deriving the error function that has a unique minimum value.

Universal Supporting devices from all manufacturers (typically the characteristic of a compiler).

Unix The first portable operating system, originally developed by Bell Laboratories and written in the C programming language.

Upgrade To expand or improve the capability of a PC by the addition of new or improved features.

Vector A set of logic values representing more than one bit.

Video monitor See *Cathode-ray tube*.

Video terminal See *Cathode-ray tube*.

Volatile memory Memory that loses its contents when power is removed from the system.

VSWR (Voltage Standing Wave Ratio) A measure of impedance match between two terminations ranging from one (perfect match) to infinity (total mismatch).

VT (Video Terminal) Another name for a CRT.

Wait state A time period whereby a microprocessor is held mid-cycle until a slow peripheral is ready to transfer data.

Weighting factor Used in circuit optimization to increase or decrease the effects of various error terms.

Winchester disk A hard disk permanently sealed in a drive unit to protect it from dust or external physical damage. The disk is capable of storing very large amounts of data.

Word The greatest number of bits that a computer is capable of handling in one operation. Each word is usually divided into bytes.

Word processing system A system that processes text, peforming such functions as paging, paragraphing, left and right justification, etc., of the printed text.

Word wrapping The automatic shifting of words from a single line that is too long to the next line.

Appendix B
BIAS-D Reference Manual*

Brian Biehl

E-1. INTRODUCTION

BIAS-D is a computer-aided circuit-analysis program written in FORTRAN IV for minicomputers with a minimum of 32 kwords of internal memory. It can perform ac, dc, and transient analysis of a 30-node circuit that contains up to 150 elements—resistors, capacitors, inductors, voltage sources, current sources, and transistors. For transistor circuits, BIAS-D converges to a solution by linearizing the built-in Ebers-Moll transistor model about an operating point in much the same manner as done in larger circuit-analysis programs such as BIAS-3, SLIC, and SPICE.

Circuit data are typed into the keyboard in a semifree input format. Error messages are given for recoverable data errors enabling immediate corrections. Transistor parameters, temperature coefficients, and transient sources are entered by specifying one or more of five available model types.

BIAS-D executes in a semi-interactive mode in which elements or models are altered, temperature varied, and elements inserted between existing nonsource nodes. The program is structured so that the circuit size and element capacity can be easily modified in accordance with the available memory size. Execution time for a dc solution of a 10-node, 5-transistor circuit is approximately 0.6 s on a PRIME 400 minicomputer.

E-2. INPUT DATA

The input data are divided into two categories: circuit data and control statement data. The circuit element data (e.g., resistors, transistors, etc) are input by specifying the element symbol (R, Q, M, etc) followed by the required data for that element. The control statement data are characterized by a dot (.) followed by the desired operation (.TRAN, .ALTER, etc). Control statements do not affect the results of the analyses—they only enable the user to direct the analysis procedure.

---

*See reference 8 for full citation of source, from which this reference manual is adapted.

E-2.1. Circuit Data

Certain general instructions must be followed to input circuit data.

  a. Each circuit element must begin in column 1.
  b. Spaces are used as delimiters between data fields.
  c. Scientific notation may be used (i.e., 1000 = 1E3).
  d. Decimal points are not required (i.e., 2 = 2.0).
  e. The ground node must be node 0 (zero).
  f. Compact node numbering is not required (i.e., node numbers may be skipped).
  g. The maximum allowable node number is 99.
  h. Element values are to be in basic units (i.e., ohms, farads, volts, amperes, hertz, seconds).
  i. Abbreviated notation may be used as follows:

$P = 10^{-12}$ $K = 10^{3}$
$N = 10^{-9}$ $ME = 10^{6}$
$U = 10^{-6}$ $G = 10^{12}$
$M = 10^{-3}$
(e.g., $10U = 1.0E-5$)

E-2.1.1. Resistors, Capacitors, Inductors.

General form:

$$\text{RXX N1 N2 VALUE MYY}$$
$$\text{CXX N1 N2 VALUE MYY}$$

where XX is any two-character name, N1 and N2 are the node numbers (order not important), and VALUE is the resistor, capacitor, or inductor value in ohms, farads, or henries. YY denotes the model name. VALUE cannot be zero.

E-2.1.2. Independent Sources—Voltage, Current.

General form:

$$\text{VXX N+ N- VALUE MYY}$$
$$\text{IXX N+ N- VALUE MYY}$$

where XX is any two-character name, N+ and N- are the positive and negative source nodes, respectively, and VALUE is the source value in volts or amperes. The letter M followed by an integer from YY denotes the model name. For voltage sources, either N+ or N- must be grounded (node 0). For example,

$$\text{V+ 3 0 5 M1}$$

and

$$V+\quad 0\quad 3\quad -5\quad M1$$

are equivalent.

For current sources, current flows from the positive node through the source to the negative node. The letter M followed by the model name, YY may be omitted. However, a default number of zero is assigned.

E-2.1.3. Bipolar Transistors

General form:

$$QXX\ NC\ NB\ NE\ MYY$$

where XX is any two-character name, and NC, NB, and NE are the collector, base, and emitter node numbers, respectively. The letter M followed by an integer from 1 to 9 denotes the model name (see sect. E-2.1.4). The letter M followed by the model name [YY] may be omitted. However, a default number of zero (0) is assigned.

E-2.1.4. Models.

General form:

$$MYY\ MMM\ F1\ F2\ F3\ F4\ F5\ F6$$

where YY is the model name model [name] designated on the source or element. MMM is a three-letter name designating one of five available model types as follows.

1. NPN   npn transistor parameters
2. PNP   pnp transistor parameters
3. PUL   pulse source specifications
4. SIN   sinusoidal source specifications
5. EXT   external source model
6. TEM   element temperature coefficients

F1, F2, $\cdots$, F6 are the data fields for specifying the above model parameters. These fields are defined below.

1. NPN—transistor parameters

Field	Parameter	Default value
F1	Forward dc beta (B_F)	100
F2	Reverse dc beta (B_R)	1

Field	Parameter	Default value
F3	Saturation current (I_S)	$1E-15$
F4	Early voltage (V_A)	$1E+12$
F5	Recombination current parameter (collector current at which beta = $B_F/2$)	0
F6	Not used	—

2. PNP—transistor parameters (same as NPN)
3. PUL—pulse source specifications

Field	Parameter	Default value
F1	Initial source value at t = 0	0
F2	Pulsed value	0
F3	Pulse delay time	T_{step}
F4	Pulse rise time	0
F5	Pulse duration (width)	0
F6	Pulse fall time	0

4. SIN—sinusoidal source specification

Field	Parameter	Default value
F1	dc source value (offset)	0
F2	Source amplitude (0-P)	0
F3	Source frequency (Hz)	0
F4	Time delay	T_{step}
F5	Phase shift (deg)	0
F6	Not used	—

The value of the sinusoidal source is determined by the equation

$$F(t) = F1 + F2 \cdot \sin\left[2\pi F3(t - F4) + F5\right].$$

5. EXT—External source parameters are to be defined by the user in a subroutine.
6. TEM—element temperature coefficients

Field	Temperature coefficient	Default value
F1	Resistor (T_{C1})	0
F2	Resistor (T_{C2})	0
F3	Capacitor (T_{C1})	0
F4	Capacitor (T_{C2})	0
F5	Transistor beta (T_{C1})	0
F6	Transistor beta (T_{C2})	0

The element value at temperature T is determined by the equation

$$E(T) = E(T_0)[1 + (T - T_0)T_{C1} + (T - T_0)^2 T_{C1}],$$

where $T_0 = 300$ K. T_{C1} and T_{C2} are the element's first- and second-order temperature coefficients, respectively. The dimensions of T_{C1} and T_{C2} are in decimal percentages per degree Celsius (a decimal percentage of $0.002 = 2000$ ppm/C).

E-2.1.5. Comment Statement

General form:

*any comment

A comment may be inserted at any line in the input circuit by using an asterisk (*) in column 1 followed by any message up to 80 characters long.

E-2.1.6. END statement. END terminates the inputting of circuit data. If a default transistor model is used, it may be necessary to use END twice in succession.

E-2.2. Control Commands

After completion of each type of analysis, program control is returned to the operator. This is indicated by "INPUT DATA" appearing on the display. At this time it is possible to initiate a new analysis. This is done by using one of the control commands described in the following sections; all control commands are prefixed by a dot (.).

E-2.2.1. .AC. .AC command initiates the small-signal frequency response. This analysis can be obtained as follows.

```
.AC
"VIN FSTRT FSTOP PTS/DEC TYPE"
    (enter "V"—input node, starting
    frequency, final frequency, frequency
    points per decade, and type of output; may
    also be current input—IIN)

"VXX PRT/PLO XMIN XMAX VMIN VMAX"
    (enter "V"—output node, PRT—print, or
    PLT—plot)
```

For Print, no other parameters are necessary, and both the magnitude gain (TYPE = 0) or decibel gain (TYPE = 1) and phase of node XX are printed. For Plot, X and Y scale parameters are necessary (defaults are used if none are given). The plot type is determined by the value of TYPE.

$$\text{TYPE} = \begin{matrix}0\text{—magnitude gain}\\1\text{—decibel gain}\\2\text{—phase}\end{matrix}$$

E-2.2.2. .ALTER. The .ALTER command enables element values, models, and model parameters to be altered. This is done as follows.

```
          .ALTER
          RXX VALUE
          VXX VALUE
              ⋮
          END
```

where XX is a valid element name (i.e., has been previously defined) and VALUE is the new element value. One or more element values may be altered using a single .ALTER command. An END statement terminates the alter operation. Models and model parameters may be altered in the same manner as the elements. Model types may be changed by entering a different model designation (see sect. E-2.1.4). For example, a pulse source PUL can be changed to a sinusoidal source, SIN, etc. All model parameters must be entered or they will be set to their default values. Both models and elements can be altered at the same time.

An additional .ALTER command permits sweeping element values over a specified range of values. This can be done as follows.

```
          .ALTER
          VXX EI EF DEL
```

where EI is the initial element value, EF the final element value, and DEL the increment value (DEL can be negative). This must be the last statement in a .ALTER command. It is then necessary to define an output node, a PRT/PLT specification, and so on (see sect. E-2.2.1). At the end of this analysis the altered value is returned to its original value.

E-2.2.3. .END. The .END command permits entering a new circuit without terminating the program. At this time all previous circuit values, names, and nodes are erased from memory.

E-2.2.4. .INSERT. The .INSERT command permits elements, models, or additional nodes to be inserted into an existing circuit. Any element or model may be inserted with this command. The .INSERT command is used as follows.

.INSERT
RXX N1 N2 VALUE
QXX NC NB NE MYY
MYY [MMM] F1 F2 F3 F4 F5 F6

⋮

END

The format for the elements and models is the same as described at the beginning of this section.

E-2.2.5. .LOAD. The .LOAD command permits loading a circuit directly from a disc file. This is done as follows.

.LOAD
"ENTER FILENAME"
(enter file name)

Several circuits may be merged or models entered by successively using the .LOAD command. This command is terminated by an END statement (either in a file or via keyboard). Note that when several circuits are merged, unique node numbering must be maintained.

E-2.2.6. .PRINT. The element names and values can be displayed at any time by using the .PRINT command. Note that the node numbers displayed are a correct set of node numbers but are not necessarily the original set of numbers. If the original set of node numbers is necessary, the following sequence of commands can be used.

.INSERT
.PRINT

E-2.2.7. .SAVE. The .SAVE command is similar to the .LOAD command except that the circuit is written to a disc file. The contents of this file will be identical to that printed by a .PRINT command.

E-2.2.8. .TEMP. The analysis of the circuit at a temperature other than 27 C is obtained as follows.

.TEMP
"T(DEG C)"
(enter temperature)

This procedure is repeated for each new temperature. If a TEM model has not been defined, "TEMP. MODEL NOT SPECIFIED" will be displayed. This model can be

274 APPENDIX B

inserted with the .INSERT command. Note that any subsequent analysis will be performed at the last temperature specified.

E-2.2.9. .TRAN. A transient analysis can be obtained using the .TRAN command as follows.

$$.\text{TRAN}$$
$$\text{``TR'' TO TF TSTEP}$$
$$(\text{enter ``}T_R\text{'' to } T_{\text{step}})$$

where T0 is the initial transient time, TF the final transient time, and TSTEP the output time increment. In order for the transient analysis to be meaningful, one or more source models (SIN, PUL, EXT) must have been specified. Voltage or current sources as well as models can be inserted once the initial circuit has been entered (see sect. E-2.2.4).

Note: Any control command (except .LOAD and .SAVE) will override a previously initiated control command. If a reply is expected, the command should be entered twice; the first time will cause an error message which can be ignored.

E-3. MISCELLANEOUS

Sometimes convergence to the desired accuracy is not attained. If this happens a "MEAN ERROR" message will appear. These results may or may not be correct. If, during a dc analysis, a more accurate solution is desired, the following procedure can be used.

$$.\text{ALTER}$$
$$\text{END}$$

This does not change the circuit but does allow at least four more iterations to occur.

In the general version of BIAS-D, several system-dependent subroutines have been commented out. These routines are OPNFL, CLSFL, GRAPH, IPACK, and SECND. Although BIAS-D will run without these routines, their implementation is desirable. A summary of these subroutine functions is as follows.

OPNFL
CLSFL   Permits storage and retrieval of disc files.

GRAPH   Permits graphical output on any refresh graphics or storage tube graphic terminal.

IPACK   Permits use of two-character element names.

SECND   Gives elapsed execution times.

Index

8088, 142
A-PLUS, 141
ABEL, 141, 154, 156, 157
Active HI/active LO, 145
ADC, 178
Address bus, 10
Admittance matrix, 42, 88
ALTERA, 141
AMAZE, 141
Analysis, 139, 140, 141
APPLE™, 12
ASCII, 8
Assembler, 139
AUTOART™, 86, 123
Available power gain, 130

Balance/scale servo system design example, 242
Bandpass filter design, 104, 222
Bandpass filter synthesis with FILSYN, 93
BASIC, TRS-80 level II, 161, 244, 249
BIAS-D, 29
BIAS-D Reference Manual, 267
Binary coded decimal, 9
Bipolar, 145
Bipolar junction transistor, 66
Bit, 7
Boolean, 139
Bryant and Weiss, 110, 111
Bus, 10, 149, 151

CAD, 22
CAE, 138, 140, 141, 157
Capacitor model, 57
Circuit analysis, 87
Circuit convergence, 45
Circuit layout, 124
Circuit layout with MICAD, 124
Circuit modeling, 119

Circuit optimization, 113
Circuit optimization with TOUCHSTONE, 117
Circuit Q, 96, 105
Circuit setup, 39
Circuit simulator architecture, 37
Circuit synthesis, 92, 109
CLD™, 86, 104
Combinatorial, 145
Comments, 147, 149, 153, 154, 157
Commercial circuit analysis programs, 88
Communications, 23
COMPAC 386™, 12
Computer interface, 159
Computer-aided circuit layout and mask generation, 122
Constant-gain circles, 131
Constant-noise circles, 131
Coplaner waveguide, 111
CP/M, 138
CPU, 11
CRT, 15, 142, 143
CUPL, 141, 149, 152, 154
Current source model, 55
Curve fitting, 192

DAC, 178
Data bus, 10
Data I/O, 141
DC operating point, 46
Delay equilization, 98
Dependent current source model, 56
Device modeling, 119
Dielectric loss, 111
Diode model, 62
Discontinuities, 101
Disk systems, 14
Dispersion, 111
Distributed circuits, 111

275

INDEX

Ebers-Moll equations, 67
ECM™, 86, 104
Edge-coupled filter, 104
Effective dielectric constant, 111, 113
Equalizer sections, 98
Error functions, 114, 122
ESYN™, 86

Feedback amplifiers, 98
Field Effect Transistor, 70
Fifth order lowpass filter synthesis, 98
Filter design, 92, 93
Filter synthesis combined with optimization, 104
FILSTN™, 92
Flip-flop, 139, 145
Floating point coding, 9
FORTRAN, 161
Forward substitution, 44
FPLA, 144
Fuseable link, 145

GaAs FET, 120, 121, 134
Gate array, 140
Gauss elimination, 42
General order polynomial fitting code, 192
Graphics, 21

Help, 141
Hexidecimal, 8, 157
High level, 141
Historical review of microwave CAD, 85

I/O bus, 10
IBM PC/AT/XT, 12, 138
IDD™, 86, 104
Illustrative examples of microwave CAD, 126
Impedance mapping, 88
Impedance matching, 108, 210
Impedance matching to a complex load, 126
Independent current source model, 56
Indexec variable, 149
Inductor model, 59
INTEL™, 11
Interface, 13
Interface, serial, 171
Intermediate variable, 49

JFET Model, 70

Karnaugh map, 147, 149
Kuroda transformations, 93, 100

L-Network design, 210
Lange coupler design, 91
LC Network combinations—design, 218
LINECALC™, 86, 101, 111
Logic analyzer, 141
Logic minimization, 140, 147
Logic reduction, 140
Low noise amplifier design, 130
Low pass filter design, 98
Low-noise amplifiers, 130
LU factorization, 42

Macromodels, 78
Markowitz reordering, 44
Mass storage, 13
Matching network synthesis, 108
Matrix decomposition, 44
McIntosh™, 12
Memory ROM/RAM, 12
MICAD™, 86, 123, 124, 135
MICROCAP-11, 44
Microstrip analysis with MSTRIP+, 111
Microstrip line, 112
Microwave CAD, 85
Microwave capabilities, 87
Microwave Spice, 86, 124
MIL, 86, 110
Minimum insertion loss, 86, 108
Mocroprocessor, 11
Model, capacitor, 57
Model, current source, 55
Model, dependent current source, 56
Model, diode, 62
Model, independent current source, 56
Model, inductor, 59
Model, JFET, 70
Model, MOSFET, 73
Model, resistor, 54
Model, Schichman-Hodges, 70
Model, voltage source, 61
Modeling, device, 119
Modeling, transistor equivalent, 121
Modem, 17
Modified nodal analysis, 62
Monolithic memories, 139, 141, 144
Monte Carlo analysis, 88, 117, 130
MOSFET Model, 73
MOTOROLA, 11
Mouse, 16
MSTRIP+, 86, 111
Multimodal optimization, 115
Multiplexer, 142, 143

INDEX 277

National Semiconductor™, 141
Network analyzer, 85
Networking, 23
Newton-Raphson linearization, 62
Next state, 151, 152
Nodal analysis, 87
Noise figure, 130
Non-linear circuit design, 124
Norton transformation, 93, 94, 96
Number systems, 7
Nyquist theorm, 163

Operating power gain, 130
Operating systems, 19
Operators, 151
Optical disk, 15

PAL, 139, 145, 147
PAL16L8, 139
PAL16R4, 139
PAL16R6, 139, 143, 147, 154
PAL16R8, 139
PALASM, 141, 144, 145, 147, 153
Parasitic absorption, 108
Parasitic elements, 108, 128
PCAD, 141
Peal investigation, 177
PI network, 94
PI Pad design, 196, 201
Pin assignment, 145, 154
Pin declaration, 149
PLAN, 141
PLD, 138–145, 147, 249, 157
PMS, 121
Port Model Synthesis, 121
Predistortion, 101
Present state, 151, 152
Printer, 16
Product-term, 147
Programming array, 145
Programming, real time, 160
Proportional control servo system, 245
Proportional-plus rate-plus-lag servo system, 247
Proportional-plus-rate servo system, 246
PSPICE™, 32

Q effects, 96, 97, 105

Random trials, 117
Reflection coefficient, 114
Reference Manual BIAS-D, 267

Reflection zero, 109
Register, 11
Relative dielectric constant, 108
Resistor model, 54
RF Amplifier design, 226
Ripple, 108
Robot, CS, 181

S-parameter generation with Microwave Spice, 124
S-parameter useage, 227
Saturation current, 63
Scale/balance system design example, 242
Scattering parameters, 88
Schichman-Hodges model, 70
Scope, 141
Sequencer, 141
Sequential logic, 139
Servo system simulation, 236
Servo system theory, 245
Signal polarity assertion, 149
Signetics™, 141
Significance of a bit, 151, 154
Simulation, 139, 141, 145, 147, 154, 157
Small signal ac analysis, 48
Smith chart, 91, 235, 239
Software, 17
Speed comparisons, 36
SPICE, 124
Stability analysis, 90, 91
Stability analysis with SUPERCOMPACT PC, 91
Stability circles, 91
State, 149, 151, 154, 156
State assignment, 151
State bits, 149, 151, 154
State machine, 149, 151, 152, 154, 156, 157
Statistical analysis, 117
Stripline, 111
Stubs, 98, 100, 132
Substrate, 108
SUPERCOMPACT™, 86, 91
SUPERCOMPACT PC™, 91
Suspended substrate, 111
Symbolic translator, 145
SYNMAT™, 126, 136
Syntax, 149, 154, 156

T-Pad design, 196, 201
Target device, 147
TEE network, 94
Terminal emulation, 23

The FILSYN program, 92
The future of microwave CAD, 137
The LINECALC program, 111
The Microwave Spice program, 111
The MSTRIP+ program, 111
The SYNMAT program, 126
Third order lowpass filter synthesis, 98
Three-state (Tri-state), 142, 143, 145
Timesharing, 85
Tolerance analysis, 117
TOUCHSTONE™, 86, 89, 108, 121
Transducer power gain, 120
Transfer function, 145, 156
Transient analysis, 46
Transistor equivalent circuit modeling, 121
Transistor modeling, 121
Transmission line analysis and synthesis, 110
Transmission line synthesis with LINECALC, 111
Transmission lines, 111
Transmission zeros, 94
Tri-state (Three-state), 142, 143, 145
TRS-80 level II BASIC, 244, 249
TTL, 139, 141

Tuning, 88, 90
Tuning "with Touchstone", 88
Two's complement, 9

Unconditional stability, 92
Unimodal optimization, 115
Unit element, 98, 101
Universal, 141
Unix, 139
Unloaded Q, 105

Vector, 147, 153, 154
Voltage source model, 61
USWR, 106

Wait state, 143, 144
Weighting factor, 114
Word, 7
Worst case analysis—band pass filter, 222

XON/XOFF, 175

Yield analysis, 117
Yield analysis with SUPERCOMPACT, 117